C0-ARS-773
01 13 $34.00

LIGHT

Volume 1

LIGHT

Volume 1
WAVES, PHOTONS, ATOMS

H. HAKEN

Institut für Theoretische Physik, Stuttgart

NORTH-HOLLAND PUBLISHING COMPANY
AMSTERDAM · NEW YORK · OXFORD

ISBN: 0 444 86020 7

Publishers:
North-Holland Publishing Company
Amsterdam · New York · Oxford

Sole distributors for the U.S.A. and Canada:
Elsevier North-Holland, Inc.
52 Vanderbilt Avenue
New York, N.Y. 10017

Library of Congress Cataloging in Publication Data

Haken, H
Light.

Bibliography: p.
Includes index.
CONTENTS: v. 1. Waves, photons, atoms.
1. Light. 2. Lasers. 3. Nonlinear optics.
4. Quantum optics. I. Title.
QC355.2.H33 535 80-22397
ISBN 0-444-86020-7 (v. 1)

Printed in The Netherlands

Preface

In the 20th century, the classical discipline of optics has been the subject of two incisive revolutions, namely the discovery of the quantum nature of light, and the invention of the laser. The concept of energy quanta introduced by Planck at the turn of this century has deeply influenced our fundamental understanding of light and matter. The laser made it possible to generate light with entirely new properties. These in turn led to the discovery of completely new types of optical processes such as frequency transformations in matter and many other "non-linear" phenomena. In this way, whole new branches of physics called quantum optics, laser physics, and non-linear optics rapidly developed. There is hardly any other field in physics in which a profound understanding of the fundamental physical processes is so intimately interwoven with technical and physical applications of great importance, as in modern optics. This connection makes this branch of physics particularly attractive for scientific study. At the same time, the need for a coherent text arises which, starting from fundamental principles of the physical nature of light, presents the physics of laser action, and finally gives a transparent account of non-linear optics. The present text, which will be subdivided into three volumes, is meant to meet this need.

At the same time, this text offers a new pedagogical approach to quantum optics by giving a self-contained and straightforward access to the quantum theory of light. The present Volume 1 begins at a rather elementary undergraduate level and requires no prior knowledge of quantum mechanics. It thus (and in other ways) differs from usual texts which develop quantum *mechanics* with its applications to the physics of atoms in great detail. There the quantization of the light field is often presented more or less in the form of an appendix. A detailed treatment of quantized light fields is mostly left to texts on relativistic quantum field theory whose study requires a good deal of mathematical knowledge. By contrast, the present text develops the quantum mechanics of *matter* only insofar as the results are directly relevant to the interaction between light and matter, but leaves aside all the superfluous material of atomic or relativistic physics. From the very beginning, this text focusses its attention on the physical nature of *light*. In particular, the present volume deals with the coherence properties of light, its seemingly conflicting wave and particle aspects and its interaction with *individual atoms*. This interaction gives rise to absorption and spontaneous and stimulated emission of light (the latter process being fundamental to laser action), and to numerous other effects. This book will also be of interest to graduate students and research workers. It includes, among others, most recent results on quantum beats and the dynamic Stark effect and it clearly mirrors a shift of emphasis which has been taking place in quantum physics in recent years. While originally quantum theory emphasized *stationary states*, its interest is becoming more and more concentrated on *processes*. In addition, we recognize that *isolated* quantum systems often represent too great an idealization. Rather, quantum systems interact all the time with their surrounding. This leads to a number of quantum statistical effects. Because of their fundamental importance to laser physics and nonlinear optics, we give a detailed presentation of methods to cope with these phenomena.

Volume 2 will deal with the laser. Here we will get to know the properties of laser light and how it is produced within the laser.

Volume 3 will then be dedicated to the action of intense coherent light on matter, where we will find a whole new world of non-linear phenomena.

I wish to thank my coworker, Dipl. Phys. H. Ohno, for his continuous and valuable assistance in the preparation of the manuscript. In particular, he carefully checked the formulas and exercises, contributed some in addition, and drew the figures. Dr. Chaturvedi and Prof. Gardiner critically read the manuscript. I am indebted to Prof. Gardiner for numerous highly valuable suggestions on how to improve the text. My particular thanks go to my secretary, Mrs. U. Funke, who in spite of her heavy administrative work always managed to type the various versions of this manuscript both rapidly and perfectly. Her indefatigable zeal constantly spurred me on to bring it to a finish.

The writing of this book (and of others) was made possible by a program of the Deutsche Forschungsgemeinschaft. This program was initiated by Prof. Dr. Maier-Leibnitz. The Bundesministerium für Bildung und Wissenschaft provided the funds and the Baden-Württembergische Ministerium für Wissenschaft und Kunst and the University of Stuttgart awarded me a sabbatical year. I wish to thank all of them for their unbureaucratic and efficient support of this endeavor.

H. Haken

Contents

Preface v
Contents vii
List of symbols xi

1. What is light? 1

A brief excursion into history and a preview on the content of this book 1

1.1. The wave–particle controversy 1
1.2. Geometrical optics 1
1.3. Waves 3
1.4. The oscillator model of matter 4
1.5. The early quantum theory of matter and light 8
1.6. Quantum mechanics 12
1.7. An important intermediate step: The semiclassical approach 14
1.8. Quantization of the electromagnetic field: Quantum electrodynamics (QED) 16
1.9. The wave–particle dualism in quantum mechanics 18
1.10. The wave–particle dualism in quantum optics 21
1.11. Coherence in classical optics and in quantum optics 24
1.12. Spontaneous emission and quantum noise 28
1.13. Damping and fluctuations of quantum systems 29
1.14. Photon numbers and phases. Coherent states. 31
1.15. The crisis of quantum electrodynamics and how it was solved 32
1.16. How this book is organized 34
1.17. Laser and nonlinear optics 34

2. The nature of light: Waves or particles? 37

2.1. Waves 37
2.2. Classical coherence functions 44
2.3. Planck's radiation law 50
2.4. Particles of light: Photons 56
2.5 Einstein's derivation of Planck's law 58

3. The nature of matter. Particles or waves? 63

3.1. A wave equation for matter: The Schrödinger equation 63
3.2. Measurements in quantum mechanics and expectation values 71
3.3. The harmonic oscillator 80
3.4. The hydrogen atom 93
3.5. Some other quantum systems 102
3.6. Electrons in crystalline solids 104
3.7. Nuclei 110
3.8. Quantum theory of electron and proton spin 112

4. Response of quantum systems to classical electromagnetic oscillations 119

4.1. An example. A two-level atom exposed to an oscillating electric field 119
4.2. Interaction of a two-level system with incoherent light. The Einstein coefficients 122
4.3. Higher-order perturbation theory 126
4.4. Multi-quantum transitions. Two-photon absorption 130
4.5. Non-resonant perturbations. Forced oscillations of the atomic dipole moment. Frequency mixing 133
4.6. Interaction of a two-level system with resonant coherent light 137
4.7. The response of a spin to crossed constant and time dependent magnetic fields 141
4.8. The analogy between a two-level atom and a spin $\frac{1}{2}$ 145
4.9. Coherent and incoherent processes 155

5. Quantization of the light field 157

5.1. Example: A single mode. Maxwell's equations 157
5.2. Schrödinger equation for a single mode 164
5.3. Some useful relations between creation and annihilation operators 165
5.4. Solution of the time dependent Schrödinger equation for a single field mode. Wave packets 169
5.5. Coherent states 170
5.6. Time-dependent operators. The Heisenberg picture 174

5.7. The forced harmonic oscillator in the Heisenberg picture 178
5.8. Quantization of light field: The general multimode case 179
5.9. Uncertainty relations and limits of measurability 189

6. Quantization of electron wave field 195

6.1. Motivation 195
6.2. Quantization procedure 196

7. The interaction between light field and matter 201

7.1. Introduction: Different levels of description 201
7.2. Interaction field–matter: Classical Hamiltonian, Hamiltonian operator, Schrödinger equation 204
7.3. Interaction light field–electron wave field 208
7.4. The interaction representation 212
7.5. The dipole approximation 218
7.6. Spontaneous and stimulated emission and absorption 222
7.7. Perturbation theory and Feynman graphs 230
7.8. Lamb shift 242
7.9. Once again spontaneous emission: Damping and line-width 251
7.10. How to return to the semiclassical approach. Example: A single mode, absorption and emission 254
7.11. The dynamic Stark effect 256

8. Quantum theory of coherence 265

8.1. Quantum mechanical coherence functions 265
8.2. Examples of the evaluation of quantum mechanical coherence functions 271
8.3. Coherence properties of spontaneously emitted light 275
8.4. Quantum beats 277

9. Dissipation and fluctuations in quantum optics 285

9.1. Damping and fluctuations of classical quantities: Langevin equation and Fokker–Planck equation 285
9.2. Damping and fluctuations of quantum mechanical variables: Field modes 295
9.3. Quantum mechanical Langevin equations. The origin of quantum mechanical fluctuating forces 297
9.4. Langevin equations for atoms and general quantum systems 304
9.5. The density matrix 316
Mathematical Appendix 335
References and further readings 339
Subject index 349

List of symbols

A	Einstein coefficient of spontaneous emission
$\boldsymbol{A}(\boldsymbol{x}, t)$	vector potential
$\boldsymbol{A}^{(+)}(\boldsymbol{x}, t), \boldsymbol{A}^{(-)}(\boldsymbol{x}, t)$	positive and negative frequency parts of vector potential
$A(t)$	expansion coefficient
a_j, a_j^+	fermion annihilation and creation operators
a_0	Bohr-radius
a	lattice constant
$a_{k,e}^j$	Einstein coefficient for spontaneous emission
B_{jk}	Einstein coefficient for stimulated transition $j \to k$
$\boldsymbol{B}(\boldsymbol{x}, t)$	magnetic induction
B_z	z-component of magnetic induction
$\boldsymbol{B}_0$	spatially and temporally constant magnetic induction
$\boldsymbol{B}^p$	oscillating magnetic induction
B_ω, B_ω^+	heatbath annihilation and creation operators
$b_{k,e}^j$	Einstein coefficient for stimulated transitions
b, b^+	annihilation and creation operators of harmonic oscillator
b_λ, b_λ^+	photon annihilation and creation operators of mode λ
C	integration constant
c	velocity of light in vacuum
$c_j(t)$	expansion coefficient
$c_n^{(k)}(t)$	kth iterate in perturbation expansion
$\boldsymbol{D}(\boldsymbol{x}, t)$	dielectric displacement
$\boldsymbol{D}$	expectation value of dipole moment
$D(\nu)\mathrm{d}\nu$	number of modes per unit cavity volume

$D_j(t); D_{j,\lambda}(t)$	expansion coefficients
$d_{j,s}; d_{j,\lambda,s}$	Laplace transforms of expansion coefficients
$d_j(t)$	expansion coefficients in the interaction picture
$\mathrm{d}^3\boldsymbol{x}, \mathrm{d}V$	volume elements
$\boldsymbol{E}(\boldsymbol{x}, t)$	electric field strength
$\boldsymbol{E}_j(\boldsymbol{x}, t)$	partial electric wave
$\boldsymbol{E}^{(+)}(\boldsymbol{x}, t)$; $\boldsymbol{E}^{(-)}(\boldsymbol{x}, t)$	positive and negative frequency part of electric field strength
e	elementary charge
$\boldsymbol{e}_\lambda$	unit vector of polarization of mode λ
$F(t)$	driving force, fluctuating force
$f(v, t)$	probability distribution function
$f_{\mathrm{st}}(v)$	stationary solution of Fokker–Planck equation
$f(t)$	external force
f	Hooke's constant
$G(1,2)$	mutual coherence function
$G_{jk,lm}$	correlation coefficients of fluctuating forces
$g; g_\lambda; g_{jk}; g_{\lambda,jk}$	coupling constants
H	Hamiltonian
H_0	unperturbed Hamiltonian
H^{p}	perturbation Hamiltonian
H^{p}_{mn}	matrix elements of perturbation Hamiltonian
$\hat{H}$	Hamiltonian in the interaction picture
H_λ	Hamiltonian for single cavity mode λ
$\hat{H}_I(\tau)$	interaction Hamiltonian in 2nd quantization
H_{B}	Hamiltonian of heatbath
$H_n(\xi)$	Hermitian polynomial
$\boldsymbol{H}(\boldsymbol{x}, t)$	magnetic field strength
$h, \hbar = \frac{h}{2\pi}$	Planck's constant
$I; I(\nu), I(\omega)$	intensity
i	index, integer
i	imaginary unit, $i^2 = -1$
j	index, integer
$\boldsymbol{j}(\boldsymbol{x}, t)$	current density
$K(t, \tau)$	kernel of integral

$k(t)$	external force acting on a particle
k	index, integer
k	Boltzmann's constant
$\boldsymbol{k}$	wave vector
L	length of cavity
L_j	periodicity interval in j-direction
$\mathfrak{L}_s[x]$	Laplace transform
$\mathfrak{L}, \mathfrak{L}_j$	operators of angular momentum
l	index, integer
m	index, integer
m	mass of particle (electron)
m_0	electron rest mass
m^*	effective mass of electron
N	number of atoms
$N_j(t)$	occupation number of state j
$\mathfrak{N}, \mathfrak{N}_\lambda$	normalization factor (of mode λ)
n	index, integer; index of refraction
n_j	number of photons in mode j
$\bar{n}_\omega(T)$	average number of photons with frequency ω at temperature T
$\{\boldsymbol{n}\} = \{n_1, n_2 \cdots\}$	set of photon numbers
$\bar{n}$	average photon number
$n(\nu), n(\omega)$	photon numbers
$P(t)$	probability
$\boldsymbol{P}(\boldsymbol{x}, t)$	electric polarization density
P_n	occupation probability for state n
P_{jk}	projection operator
p_n	occupation probability for state n
$\boldsymbol{p}$	momentum of particle
p_λ	probability of finding a photon in mode λ
$\boldsymbol{p} = \frac{\hbar}{i}\nabla$	momentum operator
$\boldsymbol{p}_{jk}$	matrix element of momentum operator
$p_\lambda(t)$	coefficient in cavity mode expansion (electric field)
Q	diffusion constant in Fokker–Planck equation
q	displacement of particle

$q_\lambda(t)$	coefficient in cavity mode expansion (magnetic induction)
r	spatial distance
r, ϑ, φ	spherical polar coordinates
s	independent variable in Laplace transforms
$\tilde{s}$	pseudo spin of two-level atom
$\tilde{s}_1, \tilde{s}_2, \tilde{s}_3$	components of pseudo spin
$\boldsymbol{s}$	spin vector
s_x, s_y, s_z	components of spin vector, spin matrices
s_+, s_-	spin-flip operators
T	period of oscillation
T	temperature
T_1, T_2	longitudinal and transverse relaxation times
Tr	trace
Tr_B	trace over heatbath variables
t	time
$U = \exp[-iH_0t/\hbar]$	unitary operator
$U(\boldsymbol{x})$	energy density of radiation field
$\overline{U}$	energy of radiation field in volume V
$u(\nu)$	spectral energy density of radiation
$u_{\boldsymbol{k},n}(\boldsymbol{x})$	Bloch wave function
$\boldsymbol{u}_\lambda(\boldsymbol{x})$	cavity mode function
V	volume
$V(\boldsymbol{x})$	potential energy
V	visibility of fringe pattern
v	velocity of particle
$\boldsymbol{v}_\lambda(\boldsymbol{x})$	cavity mode function
W	energy
W_n	energy eigenvalue of state n
W_{jk}	transition rates $k \to j$
$\overline{W}$	energy expectation value, averaged energy
x, y, z	Cartesian coordinates
$\boldsymbol{x}_{jk}$	matrix element of vector $\boldsymbol{x}$
y	coordinate
Z	partition function
z	Cartesian coordinate
Z	number of elementary charges per nucleus

z_k	zero of polynomial
α	angle, complex variable
β	amplitude of driving mode
$\beta = \frac{1}{kT}$	normalized inverse temperature
Γ	damping constant
$\Gamma(t, \tau)$	correlation function
$\Gamma_{jk}(t); \Gamma_{jk}^{+}(t)$	fluctuating forces for atomic variables
γ	atomic linewidth, phase factor
γ_0	decay rate, damping constant
$\gamma(1,2)$	complex degree of coherence
γ_{jk}	damping constant for transition $j \leftarrow k$
Δ	Laplacian
$\Delta p, \Delta x$	uncertainty in momentum, position
$\Delta\varepsilon$	energy shift
$\boldsymbol{\nabla}$	gradient
δ	complex constant
δ_{jk}	Kronecker's symbol
$\delta(\boldsymbol{x} - \boldsymbol{x}')$	Dirac's delta function
ε	dielectric constant of matter
ε_0	dielectric constant of vacuum
$\varepsilon_n = W_n/\hbar$	frequency corresponding to energy W_n
θ	angular variable
ϑ	angular variable
$\boldsymbol{\vartheta}$	dipole moment
$\boldsymbol{\vartheta}_{jk}$	matrix element of dipole moment
κ	damping constant
κ_λ	damping constant of mode λ
λ	wavelength
λ, λ_j	mode index
$\mu = \frac{e\hbar}{2m}$	Bohr's magneton
μ_0	permeability of vacuum
$\nu = \frac{1}{T}$	frequency of oscillation
ξ	complex constant, dimensionless coordinate of harmonic oscillator
π	$3.14159\cdots$

π	dimensionless momentum of harmonic oscillator
$\rho(\boldsymbol{x}, t)$	charge density
ρ	density matrix
$\rho(\nu)$	energy density of radiation
ρ_{B}	density matrix of heatbath
ρ_{nm}	matrix element of density matrix
$\tilde{\rho}$	reduced density matrix
σ	electric conductivity
τ	time variable, lifetime of excited state
τ_j	integration variable
φ	phase
$\varphi_n; \vert\varphi_n\rangle$	eigenfunction of unperturbed Hamiltonian
$\varphi_\uparrow; \varphi_\downarrow$	spin functions
$\Phi(\boldsymbol{x}, t); \vert\Phi(t)\rangle$	general solution of Schrödinger equation
Φ_j	random phase lag
$\phi_n; \vert\phi_n\rangle$	eigenfunction of unperturbed Hamiltonian
$\phi_0; \vert\phi_0\rangle$	vacuum state
$\phi_\alpha; \vert\phi_\alpha\rangle$	coherent state
$\psi(\boldsymbol{x}, t); \vert\psi(t)\rangle$	general solution of Schrödinger equation
$\psi(\boldsymbol{x}), \psi^+(\boldsymbol{x})$	annihilation and creation operators for fermion fields
$\psi_n; \vert\psi_n\rangle$	eigenfunction of unperturbed Hamiltonian
$\psi_{n,l,m}$	hydrogen wave function
$\vert\chi\rangle$	state vector
Ω	frequency
Ω_p	plasma frequency
$\Omega_j; \Omega^{(j)}$	general operator
$\Omega_{nm}^{(j)}; \Omega_{nm}$	matrix element of general operator
Ω^+	adjoint operator
$\omega = 2\pi\nu$	circular frequency
$\omega_j; \omega^{(j)}$	eigenvalues of general operators
$\overline{\omega}; \omega_0$	center frequency
ω_λ	circular frequency of mode λ
$\omega_{jk} = (W_j - W_k)/\hbar$	transition frequency $j \to k$

1. What is light?

(A brief excursion into history and a preview on the content of this book)

1.1. The wave–particle controversy

Most of the information we receive from our surroundings passes through our eyes. This information is carried by light. Thus it is not astonishing that the physical nature of light has been a subject of scientific study for many centuries.* Two important and contrasting concepts are due to Newton (1643–1727) and to Huygens (1629–1695). According to Newton, light consists of individual particles which are emitted by light sources and which propagate through space in straight lines. According to Huygens on the other hand, light is described by waves quite similar to water waves. Since these two concepts seemed so different there were serious scientific disputes concerning them. Quite surprisingly, however, we know nowadays that both concepts are correct in a well-defined way, and indeed it is one of the objectives of our book to elucidate this point.

1.2. Geometrical optics

Our daily experience is rather adequately described by geometrical optics. When we put an opaque obstacle into the pathway of light stemming from a point-like light source, we obtain a clear cut shadow (fig. 1.1). This experiment and similar ones lead us to the idea that light propagates in straight lines in the same way as freely moving particles do. When a light

*It is not the purpose of my presentation to give a complete outline of the historical developments which eventually led to the modern quantum theory of light and matter, nor do I always present the development of ideas in exact historical order. I rather wish to illuminate the way to our present understanding of the nature of light.

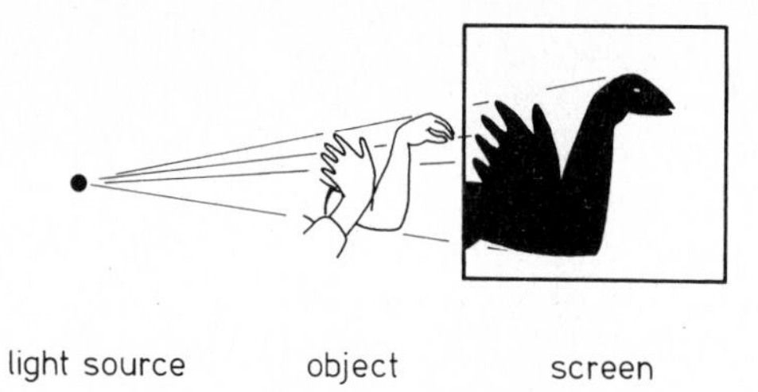

Fig. 1.1. Propagation of light along straight lines; an example from every day life.

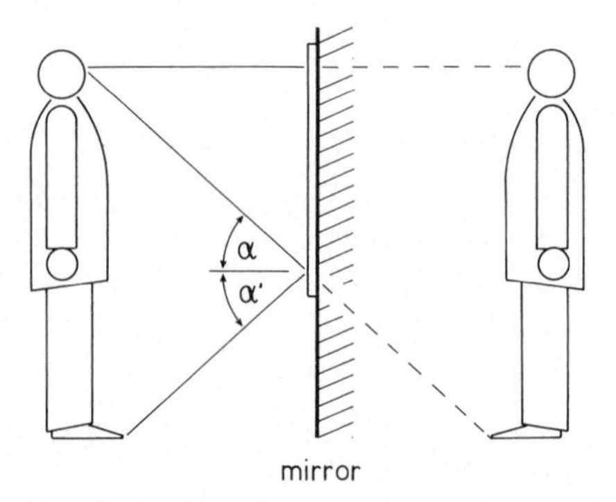

Fig. 1.2. The law of reflection allows us to construct the mirror image.

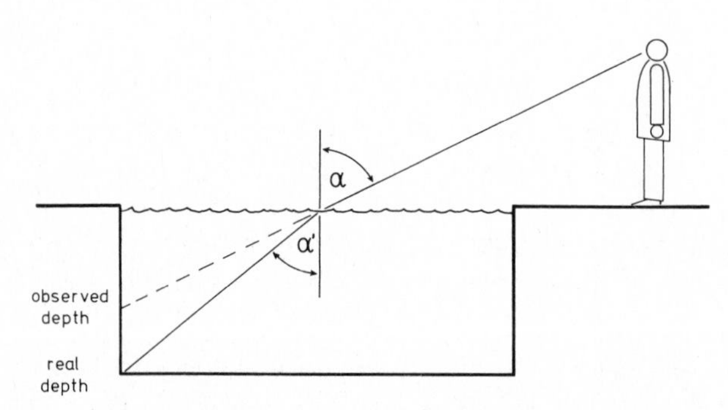

Fig. 1.3. The law of refraction. A swimming pool appears shallower.

beam is reflected by a mirror, we may easily check the law of reflection: incident beam, reflected beam, and the normal to the mirror lie in the same plane, and the angle between the reflected beam and the normal equals that between the incident beam and the normal (fig. 1.2). Standing close to a swimming pool we immediately realize the validity of the law of refraction (fig. 1.3). According to this well-known law, the incident beam, the normal to the surface, and the refracted beam lie in the same plane, and the ratio of the sine-functions of the angles of the incident and the refracted beam equals that of the indices of refraction of the two media.

As we know, geometrical optics can only be considered as a limiting case of the wave picture insofar as we can neglect diffraction and interference. Roughly speaking this approximation holds if the dimensions of obstacles or mirrors etc. are large compared to the wavelength of light.

1.3. Waves

The decisive step towards a decision in favour of the wave picture was taken by Young (1801). He let light pass through two slits in an opaque screen and observed the distribution of light intensity on a second screen (fig. 1.4). His observation showed clearly that light added to light may result in darkness, or in other words, he observed interference, a phenomenon well known in the case of water waves. (We will discuss his experiment in detail in sections 2.1 and 2.2.)

Fresnel (1819) applied Huygen's principle (fig. 1.5) in an improved form to phenomena of diffraction and interference. But what was the nature of these light waves? When Maxwell (1831–1879) established his fundamental equations of electromagnetism it soon turned out that his equations allowed for solutions describing the propagation of electromagnetic waves. In particular, the propagation velocity of these waves could be calculated from purely electromagnetic quantities and this velocity was exactly that of

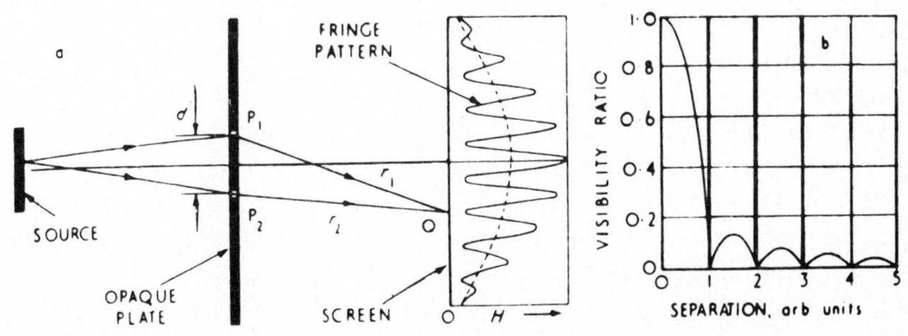

Fig. 1.4. Young's double slit experiment. Figure from his original paper. [T. Young, 1802, Phil. Trans. Roy. Soc. **12**, 387.]

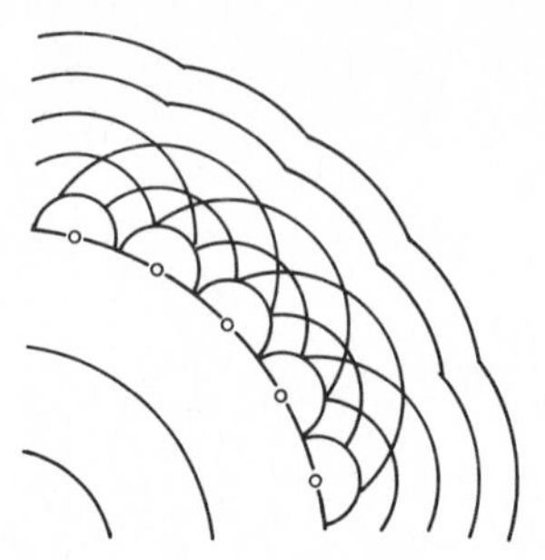

Fig. 1.5. Visualization of Huygen's principle. Each space point hit by a wave becomes a starting point of a new spherical wave. By interference of these spherical waves a new total wave is constructed. For visualization we have chosen only a few discrete points as starting points.

light. A detailed study of these equations also revealed how such electromagnetic waves could be generated, namely by accelerated (or decelerated) charges or in particular by oscillating charges and currents (fig. 1.6). Following up these ideas, Hertz (1888) soon made his fundamental discovery that electromagnetic waves were produced by an electric oscillator circuit. The idea that light is composed of electromagnetic waves had already been adopted by the end of last century. Light represents just a narrow band of a frequency range, which extends from the highest frequencies of γ-rays down to comparatively low frequencies of radio waves.

1.4. The oscillator model of matter

While it had been known for a long time that the laws of reflection and refraction could be derived from wave theory by means of Huygen's principle, the index of refraction was still to be derived theoretically. Here, the model of Drude and Lorentz proved most useful: A piece of transparent matter was assumed to be composed of atoms whose electrons were elastically bound to their nuclei (fig. 1.7). Thus each electron was described as an oscillator which was forced to oscillate under the action of the incident electromagnetic wave. Depending on the ratio between the frequency ν of the incident wave and the natural frequencies of the oscillators, the latter could follow the driving force to a greater or lesser degree. Or, in other words, the oscillator amplitudes became ν-dependent. As can be seen from fig. 1.8, each oscillator represents an oscillating dipole moment. According to Maxwell's equations, each of these oscillating dipoles emits electromagnetic waves with the frequency ν with which the oscillator is driven by the incident field. All these newly emitted waves must be superimposed on the incident wave. In this way a new wave results which moves more slowly with a velocity $v = c/n$, where n is the

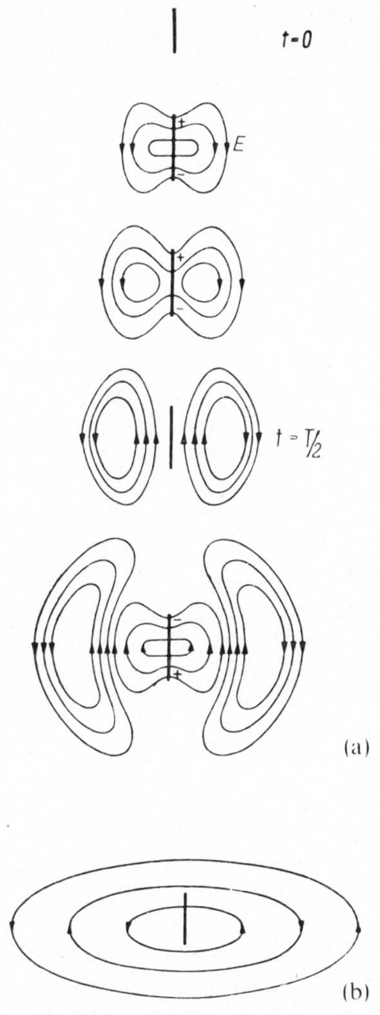

Fig. 1.6. Radiation field of the Hertzian dipole. The dipole is indicated by the vertical bar. (a) The individual pictures show from top to bottom the evolution of the field lines of the electric field strength. (b) Field lines of magnetic induction in horizontal plane through center of dipole.

index of refraction and c the velocity of light in vacuum. As the theory shows, the ν-dependence of the oscillator amplitudes causes the ν-dependence of n: $n(\nu)$ (fig. 1.9). This oscillator model can be easily extended to anisotropic crystals: One has only to assume that there are different electronic oscillators with elastic forces depending on the direction of displacement of the electrons. Such a microscopic model, together

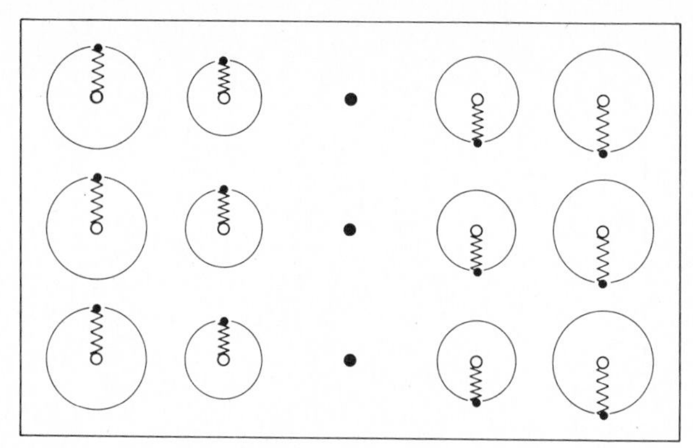

Fig. 1.7. Oscillator model of matter. Each atom of the crystal is represented by an individual oscillator.

with Maxwell's equations, allowed for an adequate theoretical treatment of light propagation in anisotropic transparent media. The emission and absorption of light by atoms was also described within the framework of the oscillator model. To treat emission, one calculates the field generated by an oscillating dipole moment. To treat absorption, one considers the energy transfer of the field to an oscillatory dipole. The term "oscillator strength", still used in spectroscopy, stems from such a model. But why is the same oscillator model capable of simultaneously describing dispersion, i.e. the response of a transparent medium, to the incident field, and absorption, i.e. the response of an opaque medium to the field? The answer is easily given: In the case of dispersion, the frequency of the incident wave is different from the natural frequencies of the oscillators, in the other case, the incident wave is in resonance with the oscillators. As

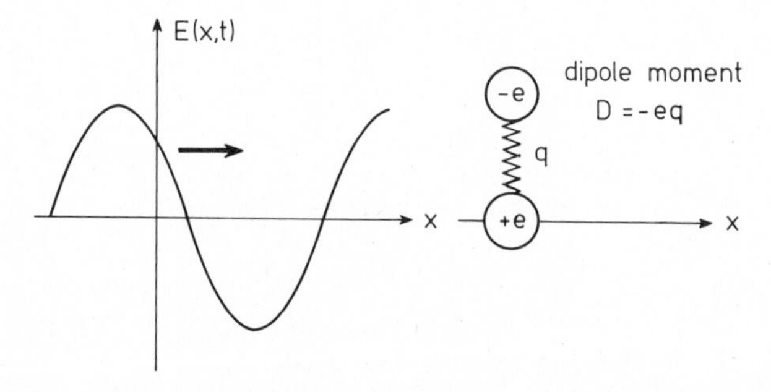

Fig. 1.8. Model of individual atom. Left part – indicates the electric field strength of a light wave impinging on the atom; Right part – displacement of the electron from the positive center is denoted by q.

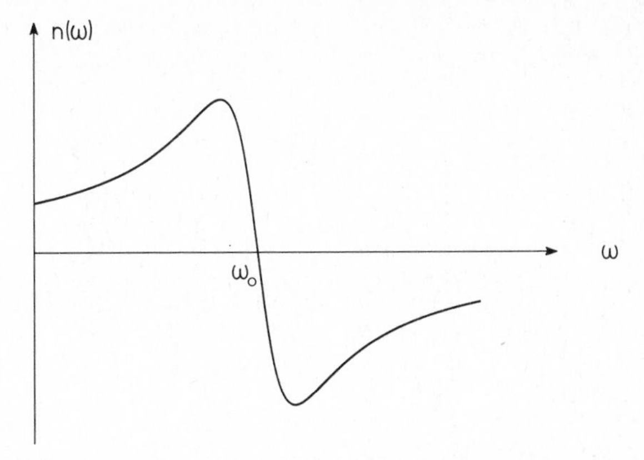

Fig. 1.9. The index of refraction plotted as a function of the frequency of the incident field. ω_0 is the natural frequency of the oscillator. In this picture it is assumed that the oscillator is damped. ($\omega = 2\pi\nu$)

spectroscopy shows the frequencies of light emitted from atoms, (and, of course, also from other light sources) are not sharp, but show a certain spread, or, in other words, a finite linewidth is observed (fig. 1.10). This width can often be accounted for by a damping of the individual oscillators, whose equations of motion thus read:

$$m\ddot{q} + \Gamma\dot{q} + fq = -eE \tag{1.1}$$

(q: displacement of an electron, m: mass of electron, $-e$: its charge, Γ: damping constant, f: Hooke's constant, E: field strength).

After such a damped oscillator is excited, it generates a light wave whose amplitude decreases exponentially. When such a wave is decomposed into

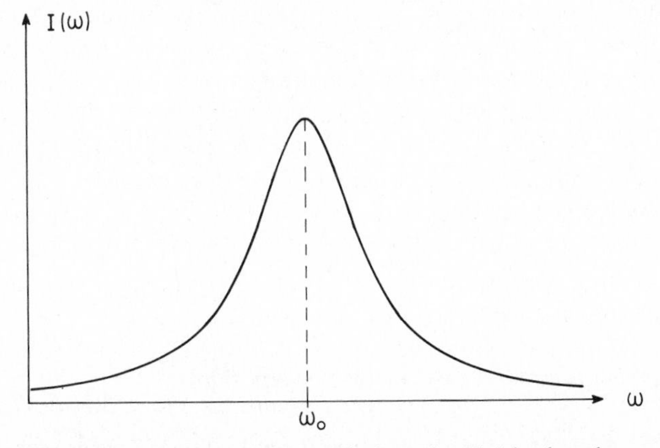

Fig. 1.10. Example of a spectral line. The intensity $I(\omega)$ is plotted versus ω.

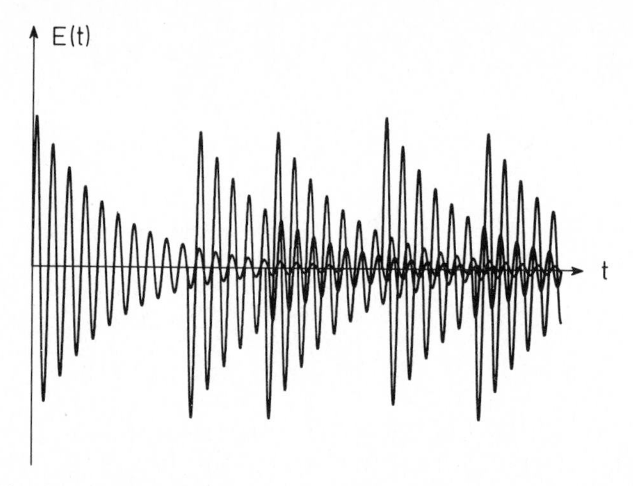

Fig. 1.11. The electric field strength of light from thermal sources. The field is composed of damped wave trains with random phases.

waves with sharp frequencies ν, e.g. by a spectrograph, the distribution of the intensity $I(\nu)$ over ν is measured, and the width $\Delta\nu$ is proportional to the damping constant Γ. Since any light source consists of many atoms, (or in our model oscillators), we have to superimpose on each other the light trains emitted by the individual oscillators. In the conventional light sources the individual oscillators are excited in an uncorrelated fashion, so their wave trains are also uncorrelated, i.e. the field amplitudes possess random phases (cf. fig. 1.11). The light we observe resembles a plate of spaghetti (or a box of chinese noodles), it is by no means just a pure, infinitely extended sine-wave. Of course, it is important to find a quantitative measure of the degree of "interruptedness" or, in physical terminology of "incoherence". Expressed more positively we wish to find a measure for the "degree of coherence". We will come back to this problem extensively later in this book (cf. sections 1.11, 2.2 and ch. 8).

In spite of the success of the oscillator model, serious difficulties soon arose. Just to mention two of them: (i) Planck (1900) found that the conventional oscillator model led to results in conflict with the observed thermal properties of light; (ii) Rutherford (1911) discovered a physical structure of atoms which differs basically from that of oscillators. Let us discuss these problems more closely.

1.5. The early quantum theory of matter and light

Planck tried to unify the theory of light with that of another great branch of physics, namely thermodynamics. He wanted to derive a formula

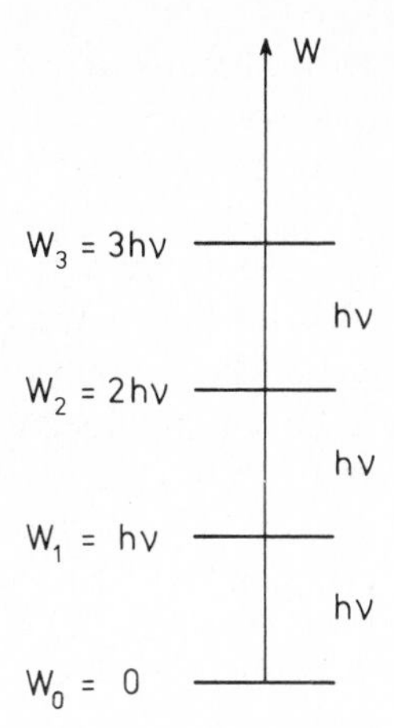

Fig. 1.12. The quantized energy levels of a harmonic oscillator according to Planck's hypothesis.

describing the energy density of the light field which is enclosed in a box and which is in thermal equilibrium with the walls of that box. To this end he represented the light field as a superposition of standing waves fitting into the box. He made a model in which he coupled each of these waves to an individual oscillator kept at temperature T and derived the mean thermal energy of the light wave by means of thermodynamic arguments. But to obtain a formula consistent with experimental data he had to make an assumption which was quite revolutionary at his time. He had to quantize the energy of the oscillators, i.e. he had to assume that the energy W of an oscillator can have only discrete values, $W_0, W_1, W_2, \ldots$. More specifically, he had to assume that these energies are given by $W_n = nh\nu$, $n = 0, 1, 2, \ldots$, where h is Planck's constant and ν the frequency of the oscillator (fig. 1.12). (We will present Planck's theory in section 2.3.) Thus the classical oscillator model had to be abandoned.

Let us now talk about the second failure of the oscillator model. Rutherford was led by his experiments on α-particle scattering by atoms to the following conclusion. An atom contains a positively charged nucleus which is nearly pointlike and Coulomb's law is valid till very small distances inside the atoms. Therefore he assumed that the electrons of atoms are attracted by the nucleus by Coulomb's force. As a consequence, electrons circle around nuclei as planets circle around the sun. But assuming this to be so, further difficulties arose: Since the electrons are charged, and are accelerated all the time due to their circular motion they should emit radiation with a frequency equal to the frequency of their circular motion. But since, according to classical mechanics, any frequency is possible, the atoms should emit a continuum of frequencies. We all know that the contrary is true. Atoms emit quite characteristic discrete spectral lines. Furthermore, as they give up their energy to the radiation field, the electrons should come indefinitely close to the nucleus and eventually collide with it (fig. 1.13). This is in contrast to the experimental finding that atoms have a definite size. To overcome these difficulties, Bohr (1913) extended Planck's ideas on quantized energy levels of oscillators to quantized energy levels of atoms. He assumed like Rutherford that an electron in an atom orbits its nucleus according to classical mechanics, i.e. like a planet around the sun, but attracted by Coulomb's force, $-e^2/(4\pi\varepsilon_0 r)$, r = electron–nucleus distance. However, in addition he made his revolutionary postulates. The electron can move only along certain well-defined orbits with discrete energies W_n, and does so without radiating light. It radiates light when it jumps from one orbit n to another one, n', emitting light with frequency

$$\nu = \frac{W_n - W_{n'}}{h} \tag{1.2}$$

where h is again Planck's constant (fig. 1.14). To bridge the gap between quantum physics and classical physics he formulated his correspondence principle: At high enough quantum numbers, n, the effects predicted by quantum theory must approach those of classical physics. This, together with his postulates, allowed him to calculate the energy levels W_n of the hydrogen atom explicitly and his model was consistent with the observed spectral series of hydrogen.

So far in this section we have been concerned with atoms. But simultaneously, at the beginning of this century, new ideas on the nature of light evolved. An important step was the explanation of Einstein (1905) of the

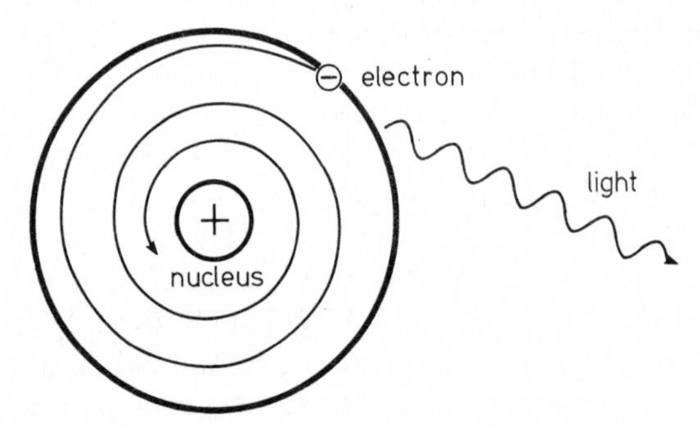

Fig. 1.13. An electron orbiting the nucleus performs an accelerated motion and should all the time emit light according to classical electrodynamics. Because it must give up its energy it comes indefinitely close to the nucleus.

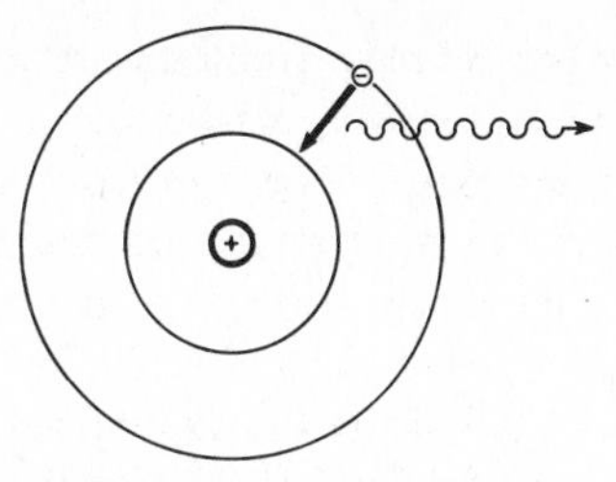

Fig. 1.14. An illustration of Bohr's postulate. An electron moving around its nucleus on a stationary orbit with energy W_2 may suddenly jump to another orbit with energy W_1 whereby a photon is emitted.

photoelectric effect (fig. 1.15 and section 2.4). He assumed that light consisted of particles, photons. According to these ideas, one associates photons of energy $h\nu$ and momentum $h\nu/c$ with a light wave with frequency ν. But what does the word "associate" mean in a precise physical interpretation? We will come back to this highly crucial point later in this introduction and again and again in this book.

Later, Einstein (1917) gave his second fundamental contribution to "quantum optics" when he rederived Planck's law. (We shall present this derivation in section 2.5.) Einstein considered a set of atoms each with one electron. Each electron can move in either of two orbits 1, or 2, with the corresponding energies W_1 and W_2. When jumping from 1 to 2 or 2 to 1, the electron is assumed to absorb or to emit a photon with energy $h\nu = W_2 - W_1$. To derive Planck's law properly, Einstein had to introduce three different processes:
(a) Initially the electron is in its upper level and no photon present: It can emit a photon spontaneously. (b) Initially the electron is in its upper level and n photons of the "same kind" are present: The electron can emit an additional photon by stimulated emission. (As we will see later in Volume 2 stimulated emission is fundamental for the laser process.) (c) Initially the

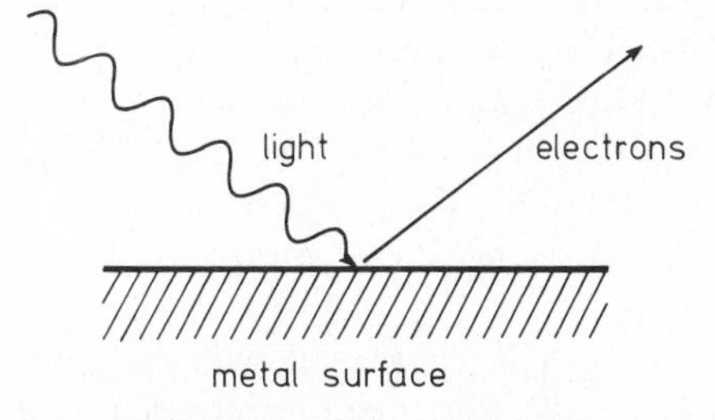

Fig. 1.15. Scheme of photo-electric effect. Light impinging on a metal surface can free electrons. The kinetic energy of the electrons does not depend on the light intensity but on the frequency of the light.

electron is in its lower state and n photons are present: The electron can absorb a photon and jump to its upper level.

To describe the number of jumps the electrons make per second between levels 1 and 2 in connection with the different processes (a), (b), (c), Einstein introduced certain rate constants, now called the Einstein coefficients, phenomenologically.

Thus, within less than two decades, the theoretical description of atoms and the light field changed dramatically. However, theory was still not in a satisfactory state. On the one hand, many of the laws of "classical" physics were still considered valid down to the atomic level, for instance when the electron orbits were calculated. But the new quantization rules had to be applied in addition. It turned out that more atomic quantum numbers than "n" were needed, and Sommerfeld developed additional quantization rules. Similarly, Einstein's postulates were added to the laws of classical physics. Eventually it even became evident that entirely new types of quantum numbers (half-integers) had to be introduced. Thus physicists became more and more aware of the fact that this kind of procedure could only be an intermediate step and that the "true" quantum theory had still to be discovered. (In a way the present situation in elementary particle physics is reminiscent of this picture. Here quantum theory (see below) is applied, but has again to be supplemented by rules which carry a similar "ad hoc" character.)

1.6. Quantum mechanics

The breakthrough was achieved by Heisenberg (1925) and Schrödinger (1926) working independently of each other. Heisenberg invented "matrix mechanics", while Schrödinger formulated quantum mechanics by means of the by now fundamental Schrödinger equation. Shortly afterwards Schrödinger showed that his and Heisenberg's formulations were mathematically equivalent. We shall base our approach mainly on the Schrödinger equation, which we will derive and explain in detail (cf. ch. 3). As we will see, the operator techniques introduced by Heisenberg also play an important role in quantum optics. The Schrödinger equation is non-relativistic but its extension to the relativistic case was given by Dirac (1928). In order not to overload our presentation we will leave aside his approach in our book.

Schrödinger's discovery is connected with an interesting anecdote. In order to elaborate, I must first mention another important intermediate step: as we have seen above, Einstein attributed photons with energy

$$E = h\nu \tag{1.3}$$

and momentum

$$p = h/\lambda \tag{1.4}$$

to a light wave with frequency ν and wavelength λ. De Broglie (1924) had the ingenious idea to apply these relations to particles, for instance to electrons. He interpreted these relations so to speak, in the "opposite" way. A wave with frequency ν and wavelength λ is now attributed to the motion of a free electron with energy E and momentum p by the relations

$$\nu = E/h, \qquad \lambda = h/p. \tag{1.5,6}$$

We will come back to de Broglie's idea and its experimental verification in a later section (cf. section 3.1). But here is the anecdote: When Schrödinger gave a seminar talk on de Broglie's theory at the ETH in Zürich, Debye said: "When there is a wave, there must be a wave equation!" Sure enough, Schrödinger derived this equation a short time later. Soon after the introduction of the Schrödinger equation, quantum theory was applied to the explanation of many fundamental properties of matter, such as the structure of atoms and molecules, and the electronic and atomic states in crystalline solids. Later it turned out that similar approaches could also be applied to nuclei and amorphous solids. The emphasis was primarily on stationary states which are the quantum mechanical analogues of the orbits of electrons in the sense of Bohr. It turned out that the concept of an orbit had to be abandoned. It was replaced by a new mathematical quantity – the wave function ψ which, loosely speaking, can be visualized as describing an electronic "cloud" (fig. 1.16).

We will get to know the precise meaning of these wave functions ψ later in this book (ch. 3). In the case of a single particle, for instance an electron, ψ depends on the coordinate x of that particle and on time t, so that $\psi = \psi(x, t)$. The solutions of the Schrödinger equation are just these wave functions. Simultaneously, the Schrödinger equation fixes the quantized energy levels W_n.

In many cases it is important to know the details of such energies and wave functions. On the other hand, those details are not of central importance for an understanding of the basic methods and results of quantum optics to which this book is devoted. Indeed we shall see that only a few general properties of energies and wave functions are needed for this purpose, with one exception, namely the quantum mechanical harmonic oscillator which we will treat in detail. Therefore, our text will differ from usual texts on quantum mechanics by leaving aside such material. Readers who wish to learn about this or that detail of quantum mechanical wave functions and energies, for instance those of the hydrogen atom, can find them in any standard textbook on quantum mechanics.

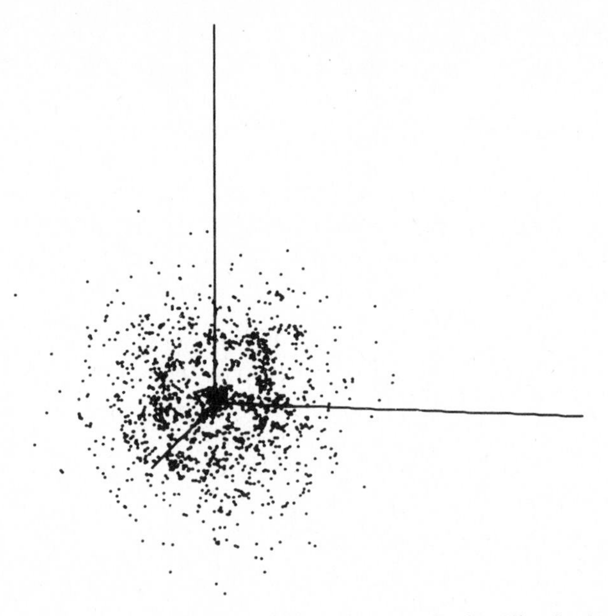

Fig. 1.16. How to visualize an electron cloud in quantum mechanics. The density of dots measures the density of the electron. For a more correct interpretation see below when we discuss Born's probability interpretation.

1.7. An important intermediate step: The semiclassical approach

As I mentioned above, the old classical oscillator model of atoms had been quite successful in treating the absorption or emission of light by atoms as well as dispersion, though classical mechanics eventually could not provide any sound justification of that model. So it was one of the first tasks of the newly developed quantum mechanics to treat this problem again. To this end, one studied the response of electrons, described by the Schrödinger equation, to a prescribed classical electromagnetic field, in particular, to a wave. As we will demonstrate in chapter 4, the answer can be described as follows. In the case of dispersion, i.e. when the frequency ν of the incident wave differs from the frequency $\nu' = (W_2 - W_1)/h$ of the electronic transition, the following happens: Under the influence of the field, the electron leaves its stationary state 1, which is represented by a certain wave function φ_1. The motion of the electron is now described by a superposition of wave functions, e.g. by

$$\psi(x,t) = c_1(t)\varphi_1(x) + c_2(t)\varphi_2(x), \tag{1.7}$$

i.e. by a "wave packet", where φ_2 is the wave function of the excited state of the electron, and c_1 and c_2 are time dependent factors. As we will see, the center of gravity of the charge density described by this wave packet oscillates back and forth as if the charge were that of an oscillator (fig. 1.17). Thus the original oscillator model finds its justification in quite a surprising way, whose details we will learn about in the course of this book.

The new quantum mechanics were also able to treat absorption and stimulated emission of light by atoms. Under the influence of incoherent ("spaghetti-like") light, the mean number of electrons in the initial state 1 or 2 of an ensemble of atoms decreases, while the mean number of the electrons in the other state 2 or 1, increases proportionally to time. [We will give a precise definition of these "mean numbers" later in this book (chs. 3 and 4).] In this way it became possible to calculate the corresponding Einstein coefficients from first principles. Quite new effects occur when the incident light wave is completely coherent and in resonance with the atomic transitions, i.e. $\nu_{\text{field}} = (W_2 - W_1)/h$. According to the classical oscillator model, the oscillator amplitude should increase more and more in the course of time. Quite on the contrary, in quantum mechanics again wave packets eq. (1.7) are formed describing an oscillatory motion of the electron.

Since laser light has a high degree of coherence, it has become possible to observe a variety of these highly interesting new phenomena, especially the "photon echo" which we will treat in section 4.8. The approach described in this section is often called the semiclassical approach. Indeed, it treats the electromagnetic field in the framework of classical physics, while matter, in particular electrons, is treated by means of the Schrödinger equation, i.e. quantum mechanically. In this way we can adequately

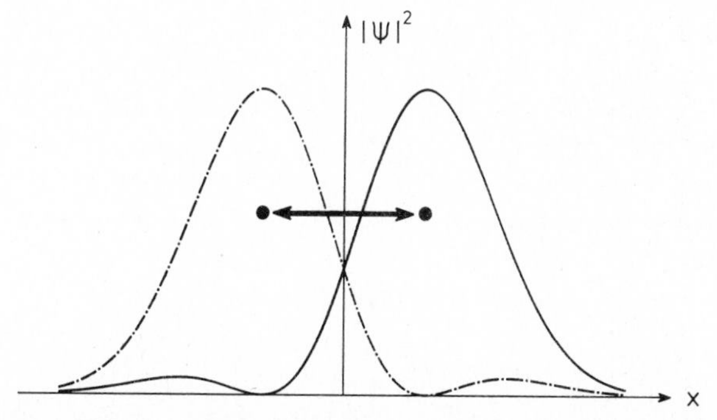

Fig. 1.17. Example of the oscillation of a wave packet $\psi = c_1\varphi_1 + c_2\varphi_2$. Because ψ is generally a complex function we have plotted $|\psi|^2$. The solid curve applies to the initial time, the dot-dashed curve refers to some later time. The displacement of the center of gravity of $|\psi|^2$ is clearly seen. The position of this center oscillates as a function of time.

describe absorption and stimulated emission. However, this kind of approach does not allow us to theoretically describe any spontaneous emission of atoms. This strongly indicates that a major ingredient of the theory is still missing. We will find this missing ingredient when we quantize the electromagnetic field.

1.8. Quantization of the electromagnetic field: Quantum electrodynamics (QED)

We have mentioned above that some experiments, such as the photo-electric effect, can only be explained by the assumption that light consists of particles, photons, which carry quanta of energy. Quantum mechanics made it possible to calculate quantized energy levels for particles, e.g. the electrons. We may therefore ask whether these theoretical methods allow us also to quantize the light field. The basic idea for the solution of this problem is the following.

As usual we describe the light field by its electric and magnetic field vectors. Let us consider the electric field vector in a given direction, say the z-direction, as an example, i.e. $E_z(\boldsymbol{x}, t)$ which is attached to each space point $\boldsymbol{x}$ and time t. We may decompose E_z into a superposition of waves, for instance of waves of the form of standing waves (fig. 1.18)

$$E(\boldsymbol{x}, t) = q(t) \sin \boldsymbol{k}\boldsymbol{x} \tag{1.8}$$

in a cavity, i.e. a metal box. We know that for free fields $q(t)$ represents a harmonic oscillation

$$q(t) = A \sin(\omega t) \tag{1.9}$$

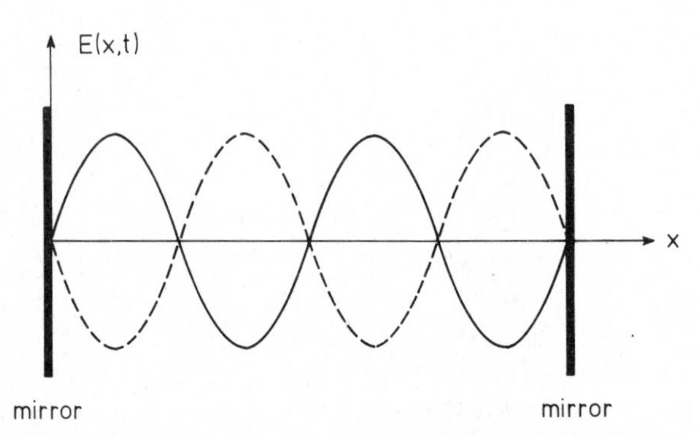

Fig. 1.18. Example of a standing wave of the electric field strength between two mirrors. The solid line and the dashed line represent the field strength at different times.

or in other words, $q(t)$ obeys the equation of the harmonic oscillator. However, quantum mechanics has taught us how to quantize the harmonic oscillator by means of the Schrödinger equation. The energy levels are

$$W_n = nh\nu + \text{const} \tag{1.10}$$

where n is an integer, $n = 0, 1, 2, \ldots$. n can be interpreted as the number of photons with which the wave $\sin \boldsymbol{k}x$ is occupied. The "constant" in (1.10) represents the zero-point fluctuations of the field: even if no photons are present, the field fluctuates all the time. If we quantize all waves in three dimensions, and include the magnetic field as well, we shall succeed in quantizing the light field. (We will present the quantization procedure in ch. 5.) Now we have reached a rather satisfactory level of approach in which both the motions of electrons and the oscillations of the light field are quantized. In particular, the new formalism describes the creation and annihilation of photons. By this new approach we can reproduce not only the former results of absorption and stimulated emission but in addition we find an adequate treatment of spontaneous emission. In quantum mechanics the light field never vanishes entirely, rather it constantly experiences zero-point fluctuations. Because of these, electrons always "feel" the light field, which may then cause an electron in an atom to make a transition to its ground state, emitting a photon spontaneously at the same time. This theory allows one to calculate all Einstein coefficients of emission and absorption from first principles. From the point of view of quantized fields, these coefficients are rate constants describing the production or annihilation rate of photons. In this way light is considered to be composed of particles and no space for any coherence properties of waves seems to be left.

The picture of light as photons dominated the quantum theory of light until the advent of the laser. It appears nowadays that, in fact, physicists long overstressed the notion of individual photons in the treatment of such processes. The problem of reconciling the photon point of view with the aspects of wave optics and coherence was largely ignored. In fact we know that even spontaneously emitted light has certain coherence properties which are manifestations of its wave properties. Furthermore, the laser has made it possible to produce coherent light. Thus we are again confronted with the fundamental question of how to reconcile the wave concept with the particle concept. The solution of this problem is indeed one of the great triumphs of quantum theory which we will now discuss (and in much more detail in our book taking into account the more recent developments, too).

1.9. The wave–particle dualism in quantum mechanics

Since the answer to this problem was first given in quantum *mechanics*, let us return to the seemingly quite simple problem of the motion of a free particle. When we describe it according to classical physics we visualize it, of course, as an extensionless point mass, i.e. the particle can be localized up to any degree of accuracy. At the same time we can measure its velocity (or momentum) precisely. This assumption is indeed the basis of Newtonian or Hamiltonian mechanics. On the other hand, describing a particle quantum mechanically by means of a wave function means quite a different thing. According to de Broglie's fundamental assumption, (cf. section 1.6) in quantum mechanics we attribute an infinitely extended wave

$$\psi(x) = A \exp(2\pi i x/\lambda - 2\pi i \nu t) \tag{1.11}$$

to a freely moving particle with momentum p so that $\lambda = h/p$. The wave must be of infinite extent, otherwise it would not have a definite wavelength, but an infinitely extended wave does not determine any definite space points (fig. 1.19). The localization of the particle has been lost entirely. As is known from wave theory, by a superposition of plane waves we may form "wave packets" which are concentrated around certain space points, thus describing the localization of a particle. But when building up a wave packet we must use a variety of wave lengths or wave vectors $k = 2\pi/\lambda$ and thus, according to $\lambda = h/p$, a variety of momenta p.

As Heisenberg has shown there is a fundamental limit connecting the spread Δx, called the uncertainty of the space coordinate x and Δp, the uncertainty of the momentum p. Heisenberg's uncertainty relation reads

$$\Delta x \Delta p \geqslant \hbar/2. \tag{1.12}$$

This tells us that any measurement which determines the particle coordinate x within a certain accuracy Δx excludes a measurement of the momentum p to a better accuracy Δp than given by eq. (1.12) (cf. fig. 1.20).

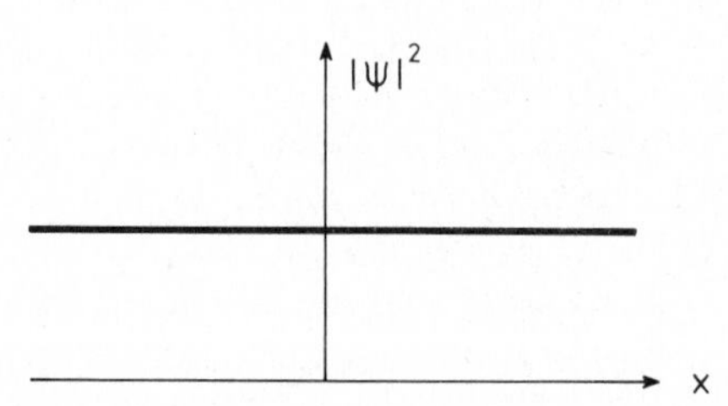

Fig. 1.19. The absolute square of the plane wave function plotted versus x. No space point x is distinguished from any other.

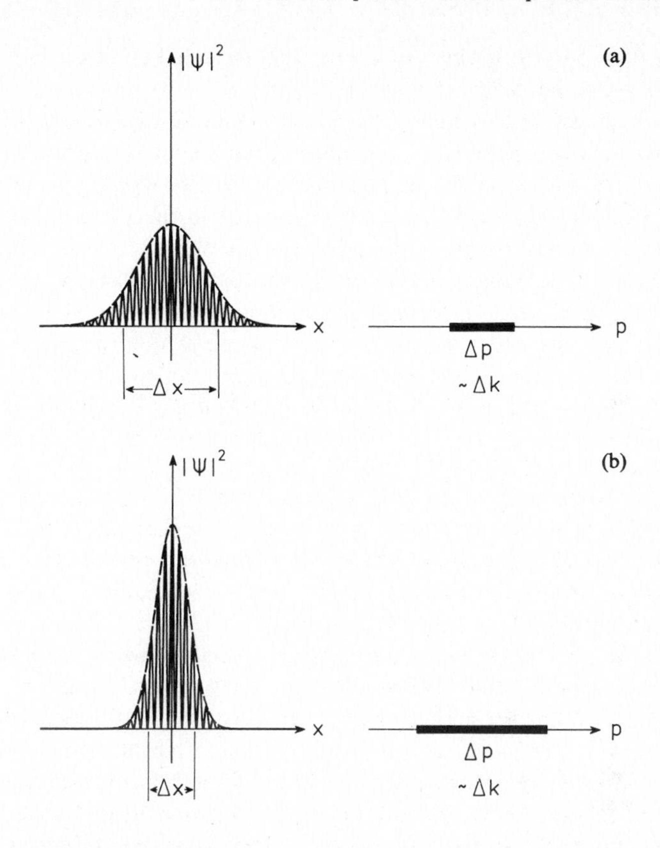

Fig. 1.20. How to visualize Heisenberg's uncertainty relation (1.12). The left upper and lower parts show the absolute square, $|\psi|^2$, of the wave function ψ of the wave packet. The right upper and lower parts show the spread of k-values which is needed to build up the wave packets. (a) Upper part: large uncertainty Δx and small uncertainty Δp. (b) Lower part: small uncertainty Δx and large uncertainty Δp.

The uncertainty relation has far reaching consequences which we will also come across again and again in the quantum theory of light. Namely it implies that the observer perturbs the system by observing it. For instance, by a *precise* observation of the localization point x of the particle, we induce a complete uncertainty in its momentum. The resulting uncertainty of the particle's momentum is by no means just a lack of our knowledge. After the measurement of the particle's position its future motion cannot in principle be predicted. We can also put the results of this discussion in other words. Namely by choosing an appropriate experiment we can determine, for instance, the position of the particle within the desired degree of accuracy, i.e. at our will. At the same time, however, we interfere with the further temporal development of the quantum system and thus cannot make exact predictions about the momentum which is a variable complementary (or "conjugate") to the space variable. Is it really true that we cannot make any statement on that complementary (conjugate) variable? Born found the correct answer to this problem. According to him we can make only probability statements. Let me explain this in more detail – the simplest access to this probability statement is by means of the wave function $\psi(x, t)$ itself. According to Born,

$$|\psi(x,t)|^2 \mathrm{d}x \tag{1.13}$$

(one dimensional example) gives us the probability of finding the particle in the region $x \ldots x + \mathrm{d}x$ (fig. 1.21). In a similar way we can give a precise rule for the probability of finding that the particle has a certain momentum. We shall derive and apply this probability interpretation in detail later in this book (ch. 3, especially section 3.2, and many other sections). According to this interpretation, making quantum mechanical measurements is quite similar to throwing a die, and we will exploit this analogy in our later chapters.

The philosophical consequences of Born's probabilistic interpretation of quantum mechanics and of Heisenberg's uncertainty relation are probably more incisive than even Einstein's theory of relativity. Owing to this, quantum theory versus a "classical interpretation" is still being checked by more and more sophisticated experiments. A set of such experiments has been devised to check "Bell's inequality". Its discussion is beyond the scope of our book, but the interested reader may find references to the corresponding articles on page 333. So far all known experiments are in

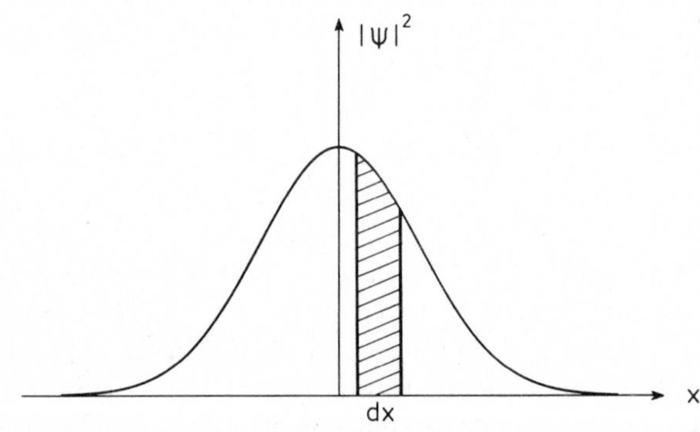

Fig. 1.21. The probability interpretation of quantum mechanics. The shaded area $|\psi|^2 \mathrm{d}x$ gives us the probability of finding the particle in the range $\mathrm{d}x$ around that specific space point.

accordance with quantum theory and in this author's opinion there is no reason to doubt its correctness at the atomic level. Of course, new developments will certainly take place at the level of elementary particle physics, but this is not our concern here.

1.10. The wave-particle dualism in quantum optics

Let us return to our original problem, namely to the question, what light is composed of "in reality". Is it composed of particles or of waves? To this end, let us consider again Young's double-slit experiment (fig. 1.22). Since we observe an interference pattern on the screen, we conclude that light is composed of waves. But let us study the interference pattern more carefully. We replace the screen by a set of photon-counters (photomultipliers). Furthermore, we lower the light intensity so much, that a maximum of only one photon could be present at a time. Then the following happens: At a certain moment, one of the counters indicates that it was hit by a photon. This event takes place in a very localized region, and indicates the energy transfer of a single quantum. Repeating this experiment, each time a photon arrives, the counter is triggered. However, the position at which the photon arrives is statistically distributed, i.e. it cannot be predicted with certainty. When we repeat this experiment many times, and plot the number of counts versus space, we eventually obtain an intensity distribution which is the same as that of the classical interference fringes. Thus we

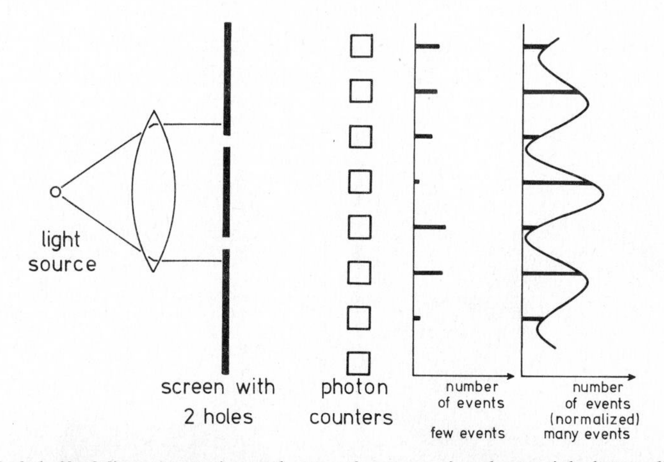

Fig. 1.22. Left half of figure: experimental set-up for measuring the spatial photon distribution behind the screen. Right half of figure: The number of events indicated by the individual photon-counters in case of a small total number of events and in the case of a very big total number of events (schematic).

are inclined to say: clearly, the particle aspect is more fundamental! But if light is composed of particles, we must be able to follow up the path of each particle.

So far we have had two holes in the screen (double slit experiment). To track the path of the particles, i.e. the photons, we just close one of the two holes and repeat the experiment. Again and again one of the photon counters clicks, but when we eventually plot the intensity distribution, quite a different picture results, namely no interference pattern occurs (fig. 1.23). Thus by our different experimental arrangement (closing one hole) we have entirely changed the outcome of the experiment! On the other hand, to obtain the interference pattern with two holes we must not close either of the two holes, i.e. in that case we cannot in principle track the path of the photon. Here the wave character of light is decisive. These results demonstrate that we cannot track the path of a photon, i.e. of a particle, and simultaneously perform the double-slit experiment to prove the wave character of light. It can show the one or the other aspect depending on the kind of experiment we perform. Note that in this context the notion of an experiment must be taken in a wide sense. For instance, as we have seen, even closing a hole in the screen before we start sending light on it, must be considered as part of the experiment. By building up any experimental setup, for instance by installing a screen or photocounters, we predetermine the outcome of the experiment! Note, however, that the outcome cannot be predicted with certainty, at least not in general, but only in the probabilistic sense.

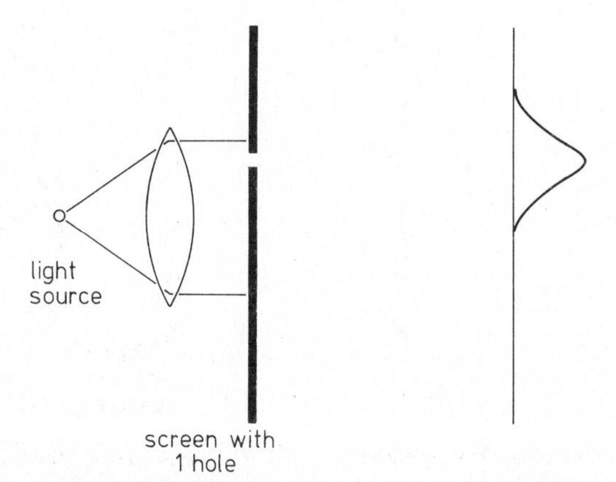

Fig. 1.23. Experimental set up to track the path of a single photon. In this case the former interference pattern has disappeared.

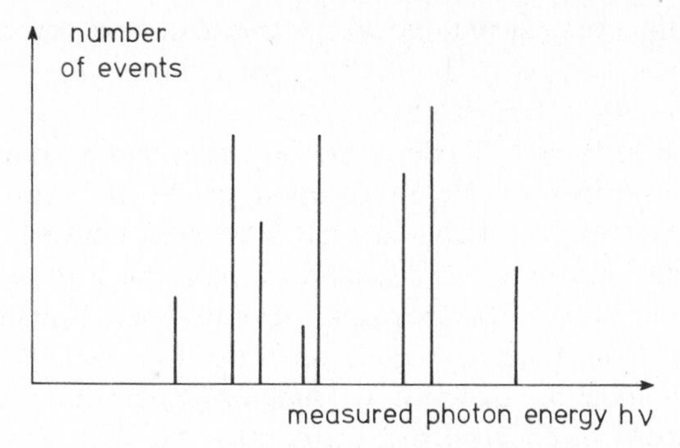

Fig. 1.24. The photon energy of spontaneously emitted light. Measurement of only a few events.

In view of these results we may also reconsider the physical properties of spontaneously emitted light. In an experiment corresponding to Einstein's theory, by which he rederived Planck's formula, we want to measure the emitted photons. More specifically, we want to measure the energy of the emitted photon for each emission act. This energy may differ from experiment to experiment (fig. 1.24). When we repeat this experiment many times and plot the numbers of the events when a definite quantum energy is measured, versus energy, we find the classical frequency distribution of spontaneously emitted light (fig. 1.25). Thus to make contact between the quantum mechanical picture and the classical theory we must either repeat the experiment many times or, use an ensemble of identical atoms. In this case we assume that the atoms do not influence each other (see also below). The ensemble then gives the classical intensity distribution. We

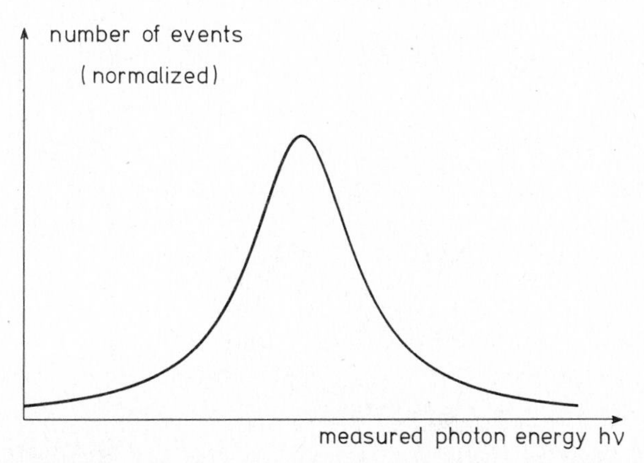

Fig. 1.25. The same as in fig. 1.24 but in the case of very many events. The ordinate was correspondingly reduced.

shall present the detailed theoretical treatment of this problem in sections 7.9 and 8.3.

Let us return for a moment to the classical oscillator model describing the emission of radiation. To describe the observed linewidth (the natural linewidth of the emission line), a damping of the classical oscillator was postulated. As a result, a damped light wave was emitted. When such a wave is decomposed spectroscopically into its frequency spectrum, a finite width results. From the spectroscopic point of view, such a wave has a certain degree of coherence.

Can we still speak of damping in quantum theory and in what sense? Can we still speak of coherence in quantum optics? To answer the second question first, let us briefly discuss how we treat coherence in classical physics.

1.11. Coherence in classical optics and in quantum optics

The concept of coherence is closely related to the phenomenon of interference. Let us therefore start this discussion by studying the way in which interference patterns arise. We consider an experimental arrangement by which we superimpose two waves on each other at a certain point $\boldsymbol{x}$ in space (compare fig. 1.26). Let the waves 1 and 2 have the amplitudes E_1 and E_2, respectively. The individual intensities are $I_1 = E_1^2$ and $I_2 = E_2^2$, respectively. Interference patterns result from the fact that we should not add the individual intensities of the two waves but rather their amplitudes, i.e. we have to first form $E = E_1 + E_2$. The intensity of the total wave, which is measured by a photographic plate or a photomultiplier, is given by

$$I = E^2 = E_1^2 + E_2^2 + 2E_1E_2. \tag{1.14}$$

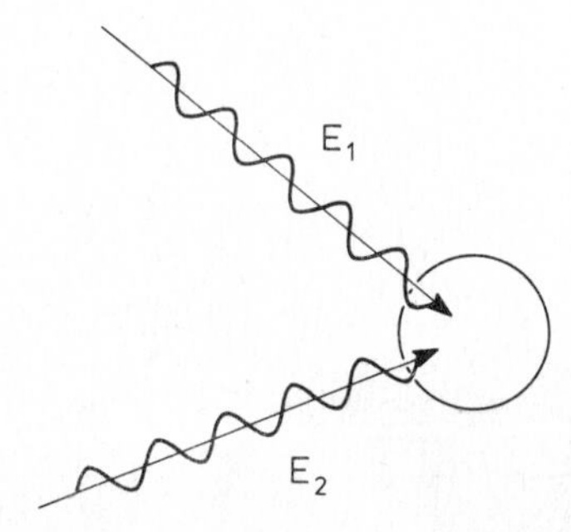

Fig. 1.26. Two waves E_1 and E_2 being superimposed on each other within a region indicated by the circle.

In the case that $E_2 = -E_1$, we evidently find $I = 0$, i.e. light added to light may result in darkness (fig. 1.27). At those points on the screen in Young's double slit experiment we find dark fringes. Consider now such an experiment for a somewhat extended period of time and assume that in the course of time the waves E_1 and E_2 change their amplitudes E_1 and E_2 independently of each other within a certain range. Then clearly the relation $E_2 = -E_1$ cannot be maintained all the time and the total intensity I no longer vanishes (figs. 1.28a, b). The independent temporal changes or fluctuations of the amplitudes just mean that the two waves are not exactly coherent. This incoherence is connected with the decrease of the quantity $2E_1E_2$ as compared to $E_1^2 + E_2^2$.

This leads us to the idea that the quantity

$$E_1E_2 \tag{1.15}$$

can be considered as an appropriate quantity for measuring the degree of coherence. We shall see in section 2.1 of this book that this interpretation is correct except for a slight modification which we will discuss at the end of this section. For our purpose it suffices, however, to consider the expression (1.15) as the relevant physical quantity. How can we measure E_1E_2 experimentally? To this end we have to use a beam splitter so that we can measure the intensities I_1, I_2 and I separately (compare fig. 1.29). Forming then $\frac{1}{2}(I - I_1 - I_2)$ we obtain E_1E_2. Because any measurement is done over a certain time, for instance by the photographic plate, we actually measure in classical physics a time average over E_1E_2, which we denote by $\langle E_1E_2 \rangle_t$.

What is important in our present context is the fact that we can measure the degree of coherence simply by measuring intensities. However, we can

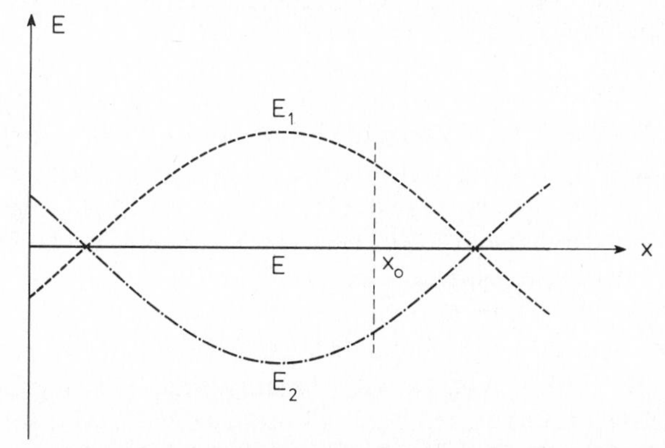

Fig. 1.27. An example of negative interference. The wave amplitudes of E_1 and E_2 result in E which vanishes everywhere.

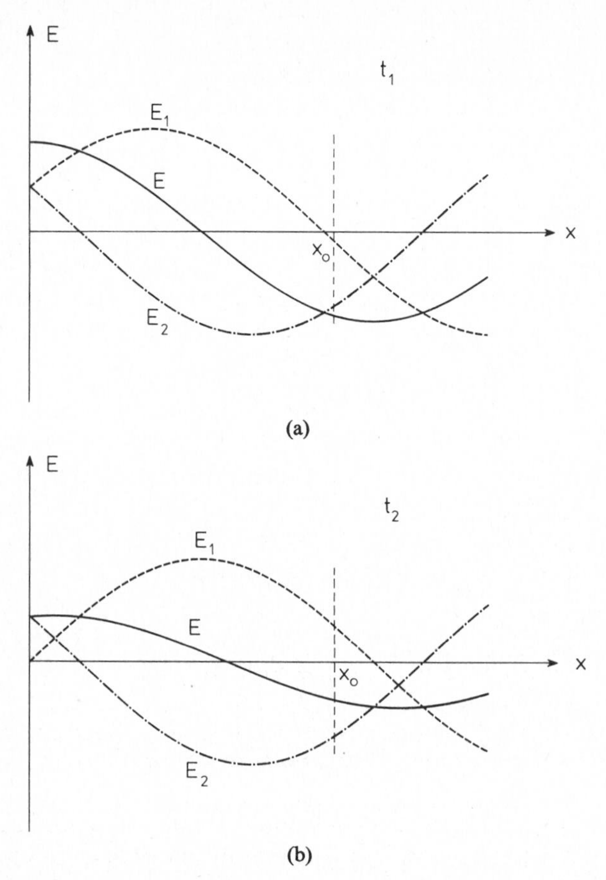

Fig. 1.28a, b. If in the course of time E_1 and E_2 are not completely correlated, a nonvanishing sum E results.

measure intensities at the "quantum level" also, namely by measuring photon numbers and their energies. This allows us to theoretically define and experimentally measure the degree of coherence also in quantum physics. The former time average of classical physics $\langle E_1 E_2 \rangle_t$ is then to be replaced by an "ensemble average" which we shall come across later in this book in more detail (see especially ch. 8).

Our considerations can also be used to measure or to calculate the coherence of a single field amplitude $E(t)$. To this end we correlate the field amplitude E at a time, t_1, with the amplitude E at a later time, t_2. Experimentally, this can be achieved, for instance, by a beam splitter and a delay line (cf. fig. 1.30). Identifying $E(t_1)$ with E_1 and $E(t_2)$ with E_2, where E_1 and E_2 are the two amplitudes considered above, we can repeat all steps

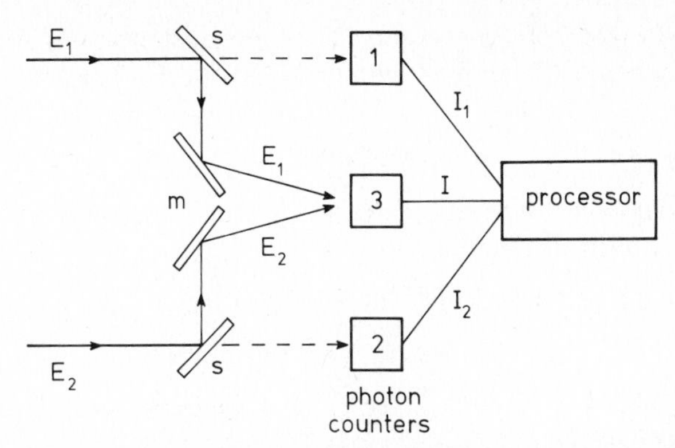

Fig. 1.29. Scheme of a possible experimental arrangement to measure E_1E_2. s denotes beam splitters, m mirrors. Note, that the optical path must be arranged so that no time lag between E_1, E_2 and E occurs.

and introduce $E(t_1)E(t_2)$ or, more precisely speaking, its average, as coherence function. We shall compute this coherence function for spontaneously emitted light fully quantum theoretically (cf. section 8.3). We will find that the result agrees with that of the classical calculation based on the damped harmonic oscillator model! Therefore, as long as we deal with "ensemble averages" we recover the classical results.

Is this agreement purely accidental or can we speak of damping in quantum physics also? We shall discuss this important question in section 1.13.

As we have indicated above, the expressions for the coherence functions are only preliminary. As a later detailed discussion in our book (sections 2.1, 2.2) will show, we have to modify the definition somewhat. To this end we decompose the electric field strength E into a positive frequency part

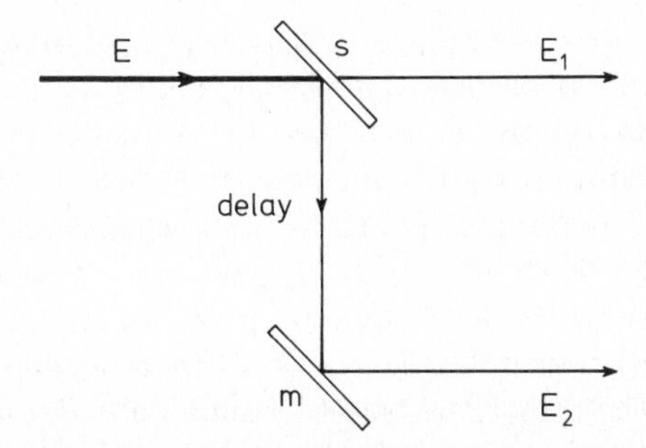

Fig. 1.30. Possible arrangement to measure the coherence of a single wave E at two different times t_1 and t_2. The original wave is split by a beam splitter s, a delay line and a mirror m into two partial waves E_1 and E_2 which now represent E at different times.

$E^{(+)}$ and a negative frequency part $E^{(-)}$. For instance, in the case that E oscillates harmonically in time with frequency ω, $E^{(+)}$ and $E^{(-)}$ are of the form

$$E^{(+)} = B\,\mathrm{e}^{-\mathrm{i}\omega t}, \qquad E^{(-)} = B^*\,\mathrm{e}^{\mathrm{i}\omega t}. \tag{1.16, 17}$$

Now we can give the precise definition of the coherence function which replaces the former definition:

$$\langle E_1(\boldsymbol{x},t)E_2(\boldsymbol{x},t)\rangle \rightarrow \langle E_1^{(-)}(\boldsymbol{x},t)E_2^{(+)}(\boldsymbol{x},t)\rangle + \langle E_2^{(-)}(\boldsymbol{x},t)E_1^{(+)}(\boldsymbol{x},t)\rangle. \tag{1.18}$$

Instead of the sum we introduce as a more fundamental quantity the following coherence function:

$$\langle E_1^{(-)}(\boldsymbol{x},t)E_2^{(+)}(\boldsymbol{x},t)\rangle. \tag{1.19}$$

1.12. Spontaneous emission and quantum noise

The spontaneous emission of photons by atoms is a beautiful example of quantum noise. After we have excited an atom and then want to measure the photon spontaneously emitted by it, for instance by means of a quantum counter, we cannot precisely predict the arrival time of the photon. We can only make probability statements of the following kind: The photon will be counted (or will be present) at a given time t with probability $P(t)$ (which we will calculate later in our book). An ensemble of initially excited atoms which are assumed not to interact with each other (not even via the light field) so that they cannot emit light in a correlated fashion will therefore emit a "rain" of photons whose arrival times are random (fig. 1.31).* Such a process is a typical example of noise. Because it is connected with the emission of quanta (i.e. the photons) we call it quantum noise.

Two features of this process are of particular importance. This noise is unavoidable and its stochastic nature is inherent in quantum physics. This point must be particularly stressed because in some other branches of physics or mathematics it is sometimes assumed that noise is not basically stochastic but only seemingly so, due to complicated underlying processes which in reality are deterministic. Though spontaneous emission is a causal process, it is by no means deterministic.

Furthermore, not only students, but even quite a number of scientists are inclined to ignore the phenomenon of noise entirely or to assume that

*There are a number of very important cooperative effects when photons are emitted. We shall discuss them in detail in the second volume.

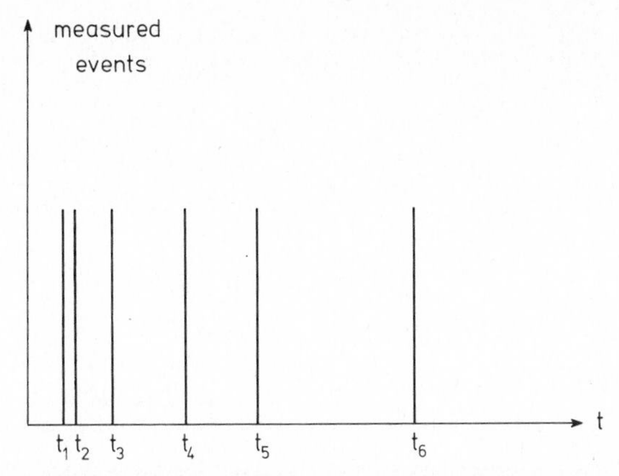

Fig. 1.31. An ensemble of atoms all initially excited at the same time spontaneously emits light. Shown are the arrival times of photons measured by a photon counter.

it plays only a minor role. Spontaneous emission noise tells us that quite the contrary is true. If that noise were not present we would sit in the dark. All light sources emit this kind of noisy light, with one exception – the laser. Its light has features quite different from those of "ordinary" light. This is one of the reasons that makes the laser so interesting from the physical point of view and we will treat these properties in the second volume.

1.13. Damping and fluctuations of quantum systems

Let us come back to the question which we asked at the end of section 1.11. Can we speak of damping in quantum physics also? To this end let us consider again the spontaneous emission process of an atom whose electron has been excited to its upper level. We have seen above that we cannot predict precisely at what time the photon will be emitted. Similarly we cannot predict precisely when the electron will jump from the upper to the lower level, but again we can make probability statements (cf. fig. 1.32). It is indeed possible (and we will perform such calculations later in section 7.9) to calculate the probability with which the occupation number of the upper level decreases. This means that considering an ensemble of many atoms allows us to study the decrease of the mean value of the occupation number of the upper level. It turns out that this decrease obeys the usual exponential law (fig. 1.33)

$$N_2(t) = N_2(0)\,\mathrm{e}^{-2\gamma t}. \tag{1.20}$$

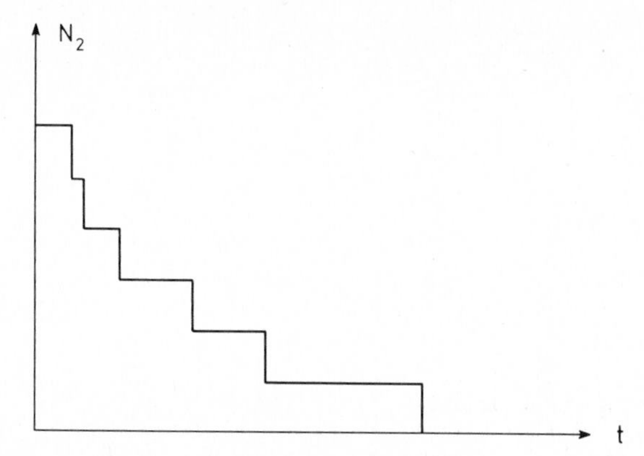

Fig. 1.32. Each time a photon is measured we conclude that the occupation number N_2, i.e. the number of occupied states of the atoms has decreased.

This "atomic" damping is transferred to the emitted field. In this way the quantum mechanical average values ("expectation values") of light emitted by atoms give the same results as we obtain them for classical fields emitted by damped oscillators. What causes this damping? In the case of spontaneous emission the answer is easily found. Because by each spontaneous emission act the photon carries away the energy from an atom, the mean occupation number of the ensemble of atoms decreases. Thus the damping of the atomic system is caused by its coupling to the quantized field. In the terminology of thermodynamics the field acts like a heat bath or reservoir on the atoms. In the special case of spontaneous emission, where initially no photons are present, the lightfield corresponds to a

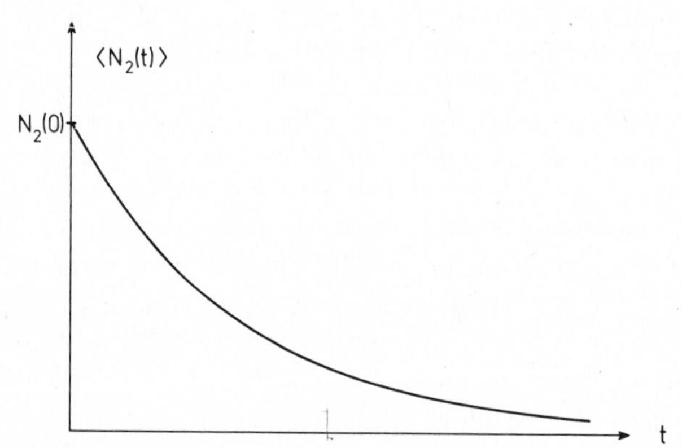

Fig. 1.33. Exponential decrease of occupation number N_2 (corresponding to the case of very many emission acts).

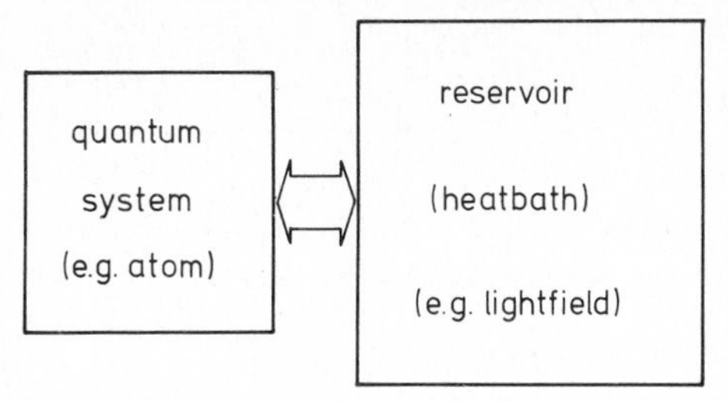

Fig. 1.34. Scheme of coupling of a quantum system such as an atom to a reservoir, for instance a light field.

heatbath at zero temperature. While the mean occupation number decreases exponentially, i.e. the system is damped, the individual photons are emitted at random. These random processes are also called fluctuations. In our present case the occupation number fluctuates while its mean decreases exponentially.

The phenomena of damping and fluctuations are by no means restricted to the process of spontaneous emission. Indeed we will see in chapter 9 that these phenomena are quite general and of fundamental importance for a proper treatment of many processes in quantum optics, especially in laser physics and non-linear optics. The reason for this lies in the fact that hardly any system is entirely isolated from its surroundings but rather interacts with its environment which acts as a heatbath or a set of heatbaths (fig. 1.34). In this way for most realistic quantum systems we have to take into account damping and fluctuations. The correct mathematical treatment of these effects has substantially contributed to the growth of a new discipline, namely the quantum statistics of systems far from thermal equilibrium.

1.14. Photon numbers and phases. Coherent states

The problem of the relation between the particle- and wave-picture is so fundamental to quantum optics that we will discuss it still further. Let us consider a specific example in which the classical field strength $E(x, t)$ has the form of a standing wave

$$E(x, t) = q(t) \sin kx. \tag{1.21}$$

In the case of a freely oscillating field, $q(t)$ can be written as

$$q(t) = A \sin(\omega t + \varphi) \tag{1.22}$$

(A and φ real).

In classical physics, we can simultaneously measure the phase φ and the intensity $I \propto A^2$ with absolute precision, at least in principle. As we have

photon number	known	unknown
phase	unknown	known

Fig. 1.35. Scheme showing possible results of simultaneous measurements of photon number and phase.

mentioned above, in quantum mechanics the photon number n appears as a quantity analogous to the intensity $I \propto A^2$. While the concept of "phase φ" is typical for wave-phenomena, i.e. for the wave-picture of light, the concept of "photon number n" is typical for the particle-picture of light. In the course of this book we will show that φ and n play in quantum optics roles strongly reminiscent of the roles particle coordinate x and momentum p play in quantum mechanics. It is impossible to simultaneously measure both with absolute precision (cf. fig. 1.35).

In quantum optics we shall find "wave functions" describing a quantum state of the electromagnetic field in which a definite number of photons is present. In such a case, the phase is entirely undetermined. (Note that these "wave functions" have a meaning entirely different from that of the classical wave (1.21)!) On the other hand, we can construct wave-packets of wave functions containing different numbers of photons. As we will demonstrate (cf. ch. 5), these wave-packets can be constructed in such a way that they describe an electromagnetic wave with a well defined prescribed phase. Yet at the same time, the photon number is not fixed but shows a spread, Δn. These wave-packets are called coherent states. They were originally introduced by Schrödinger into quantum mechanics and, in the sixties, by Glauber into quantum optics.

1.15. The crisis of quantum electrodynamics and how it was solved

In the preceding sections we have seen that quantum mechanics together with the quantization of the light field has brought about a very satisfactory theory. It solved all the difficulties classical theory had been confronted with. This new theory could adequately treat dispersion, emission and absorption and the question of coherence and above all it gave a solution to the century-old puzzle of the question whether light consists of particles or waves. Of course, there was a price to be paid for the reconciliation of the wave and particle picture, namely the probabilistic interpretation of quantum theory. As the reader will learn later in this book the new theory has a beautiful symmetry in itself and he will learn very quickly how to use these new concepts.

However, in the forties this beautiful theory was marred by very great difficulties. These difficulties stem from the following fact. When one treats the interaction between the quantized light field and an electron in an atom, quite unexpected processes happen. Let us consider for simplicity an atom with two levels only. So far we have treated processes in which an electron, which is initially in its upper level emits a photon while going to its lower level. However, in quantum electrodynamics also another process is possible (fig. 1.36b). In it the electron in the lower level can emit a photon while going to the upper level. This process obviously violates the energy conservation law. However, it can be shown in quantum mechanics, in a way quite similar to Heisenberg's uncertainty relation, $\Delta x \Delta p > \hbar/2$, that a similar uncertainty relation holds for energy and time. Loosely speaking, this means that for short enough times the energy conservation law can be violated. Thus the photon can indeed be emitted but it must be reabsorbed after a sufficiently short time so that the conservation law is not violated for too long a time.

When theoretical physicists investigated the effect of these processes, also called virtual processes, they found that such processes change the energy of an electron. In the cases of a free electron or an electron of a realistic atom this energy shift turns out to be infinite! We might think that we should do away with these strange processes altogether. But then we are not able to explain energy shifts of certain atomic levels, which were observed by Lamb and Retherford. Theoretical physicists following up an idea by Bethe found an ingenious way to cope with this problem. They devised a method by which one can subtract the energy shift of a free electron from that of a bound electron in spite of the fact that both quantities were infinite. This well defined procedure is called renormalization theory. It has been also applied to other problems than the level shift.

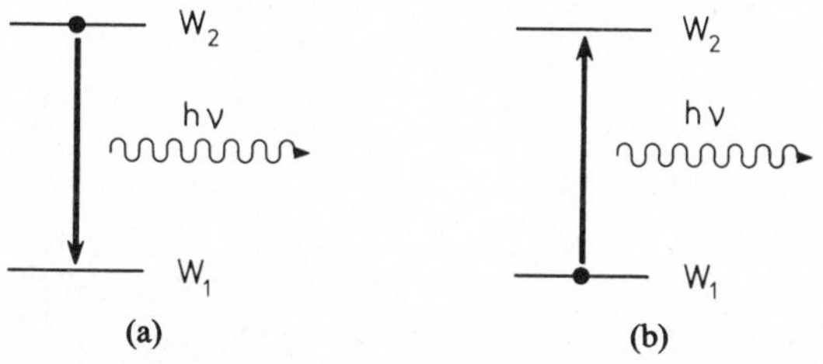

Fig. 1.36. Possible example of a real and a virtual transition in quantum electrodynamics. (a) Real transition: Initially no photon is present and the electron is in its upper state with energy W_2. It goes to its lower state with energy W_1 simultaneously emitting a photon. The energy conservation holds: $W_2 = W_1 + h\nu$. (b) Virtual transition: Initially no photon is present and the electron is in its lower state with energy W_1. It goes to its upper state with energy W_2 simultaneously emitting a photon. Clearly the energy conservation law is violated.

In all cases, quite excellent agreement between theory and the results of high-precision measurements was found.

A detailed presentation of renormalization theory is beyond the scope of our book. We will give, however, its basic ideas in section 7.8. As we shall see later, these problems can be almost entirely neglected in laser theory and nonlinear optics, so that we need not worry about them.

1.16. How this book is organized

The book is devoted to the light field, to single atoms, and to their mutual interaction. We are primarily interested in processes rather than in the treatment of stationary states. We first deepen the discussion of this introductory chapter by treating the particle and wave aspects of light including the concepts of coherence. First we will outline quantum mechanics and then study the interaction between a classical field and a "quantum mechanical" atom treating absorption and stimulated emission. We will also see how a classical field can cause a forced oscillation of electrons in an atom. This is the basis for dispersion effects. Finally we will get acquainted with an example of multiphoton absorption. The quantization of the field, which will be presented in all details, will allow us to treat the spontaneous emission as well as stimulated emission and absorption by atoms. It will lead us to a treatment of such fundamental problems as the atomic linewidth, the coherence properties of light fields and their photon statistics. We will then treat atoms which have been excited by a short pulse of light or which are steadily driven by a coherent field. This leads us to a discussion of quantum beats and the dynamic Stark effect which have been experimentally observed only recently.

The last chapter is devoted to a thorough presentation of the mathematical tools used in treating damping and fluctuations of quantum systems. These methods are indispensable for an adequate theory of the laser and for many processes of quantum optics.

1.17. Laser and nonlinear optics

In our book we shall become aware of a very strange situation. To generate fully coherent fields we must use coherently oscillating atomic or other dipoles. But coherently oscillating dipoles can be generated only by coherent fields. This leads us to a seemingly unresolvable vicious circle. As we shall see the laser has managed to solve this problem.

The laser is not only a very important new type of light source but is, as well, a rather simple system which exhibits the typical behavior of order on macroscopic scale which is often found in more complicated classical or quantum systems. The proper treatment of the laser as an open system – i.e. a system through which energy is pumped all the time – has given birth to new branches of physics and even science. It has appreciably helped to shape the field of quantum statistics of systems far from thermal equilibrium. As we now know it is a beautiful example of a self-organizing system and it has inspired a new field of science called synergetics. We shall come back to these exciting questions in the second volume.

The unique properties of laser light such as its spectral purity and its high intensity allow for a wealth of new experiments in the field of non-linear optics to which Volume 3 will be devoted. The reason for the name "non-linear optics" is easily explained. When we deal with electromagnetic fields in matter, we have to supplement Maxwell's equations with material laws. For example, the dielectric displacement $\boldsymbol{D}$ is connected with the field strength $\boldsymbol{E}$ by the relation

$$\boldsymbol{D} = \varepsilon\varepsilon_0 \boldsymbol{E} \tag{1.23}$$

where ε is the dielectric constant of the material and ε_0 the dielectric constant of vacuum. Provided ε does not depend on $\boldsymbol{E}$, (1.23) is a linear relation, which is well fulfilled in the case of light fields produced by thermal sources.

When $\boldsymbol{E}$ becomes very high, the linear relation (1.23) is no longer valid. On the purely phenomenological level of a macroscopic theory we replace (1.23) by relations of the following type

$$D = \varepsilon_0\left(\varepsilon_1 E + \varepsilon_2 E^2 + \varepsilon_3 E^3 + \dots\right) \tag{1.24}$$

where, for simplicity, we have neglected the vector character of $\boldsymbol{D}$ and $\boldsymbol{E}$.

Using such non-linear relations (1.24) in Maxwell's equations, leads to entirely new types of solutions. When we let a light wave with frequency ω impinge on a "non-linear crystal", new waves are created in its interior, associated with additional frequencies

$$\omega_2 = 2\omega \text{ (second harmonic generation)}$$

$$\omega_3 = 3\omega \text{ (third harmonic generation)}.$$

For instance when the incident light is red, in the crystal blue light may be generated.

When we shine two waves with frequencies ω and ω' on a nonlinear crystal, new waves with sum or difference frequencies are generated:

$$\omega_+ = \omega + \omega'$$

$$\omega_- = \omega - \omega'.$$

The laser can produce such high field strengths $\boldsymbol{E}$ that the non-linearities in (1.24) are substantial. Thus the laser has enabled the realization of a long felt dream of physicists, namely to pass beyond the linear range of classical optics.

Yet what is the physical reason on the microscopic level for the non-linear relation (1.24)? The answer may be given in two different ways. An approach often used generalizes the classical oscillator model, by assuming a non-linear restoring force. That is in eq. (1.1), f is assumed to depend on the displacement, q, of the oscillator

$$f = f_0 + f_1 q + f_2 q^2 + \dots . \tag{1.25}$$

While such an approach may provide us with an intuitive understanding of the underlying processes, a satisfactory theory must treat the atoms (or, more generally, matter) by means of quantum mechanics. On the other hand, the laser light field can be treated classically, which we will justify in Volume 2. In the framework of this "semi-classical" theory, we may calculate the polarization of the medium, $\boldsymbol{P}$, by means of the induced dipole moments of the individual atoms. We will demonstrate in section 4.5 by an illuminating example that these dipole moments depend in a non-linear fashion on $\boldsymbol{E}$, so that $\boldsymbol{P}$ becomes a non-linear function of $\boldsymbol{E}$. But since

$$\boldsymbol{D} = \varepsilon_0 \boldsymbol{E} + \boldsymbol{P}$$

the relation (1.24) follows.

In conclusion, let us pick another example from non-linear optics. So far, we have talked about "non-linear dispersion". There are, however, also non-linear absorption and emission effects. To explain this, let us adopt Einstein's picture of emission and absorption. In his case, the Einstein coefficients did not depend on the photon number. For instance, in absorption the transition rate of an atomic system going from the atom ground state 1 to the excited state 2 is proportional to the photon number n. But at high enough photon numbers, n, again available from the laser, this transition rate may be proportional to n^2, or n^3, etc., and in such situations one has the absorption (or emission) of more than one photon at a time by an individual atom. These processes seem to be important for isotope separation etc. We shall give a typical example of such a "multi-photon absorption" process in section 4.4.

These examples by far do not exhaust the field of nonlinear optics. They are merely meant to indicate that there exist a whole new world of fascinating non-linear phenomena.

2. The nature of light: Waves or particles?

2.1. Waves

Let us start with the idea that light can be described by waves. In the last century Maxwell (1831–1879) formulated his famous electromagnetic field equations. He showed that these equations allowed for solutions describing the propagation of electromagnetic waves. In particular it turned out that he could derive the velocity of light from the quite independent constants of electromagnetism. Thus he concluded that light is an electromagnetic wave.

Let us consider a simple example of such a wave. We identify its field amplitude with the electric field strength. We assume that the wave propagates in x-direction and that the field strength points in z-direction. We will show later on that such a wave is indeed a solution of Maxwell's equations. For the time being, however, we want to remind the reader of a few simple facts about waves. The most simple form of such a wave is (cf. fig. 2.1)

$$E(x,t) = E_0 \sin(kx - \omega t - \varphi). \tag{2.1}$$

k is the wave number which is connected with the wavelength λ by

$$k = 2\pi/\lambda. \tag{2.2}$$

The frequency ν and the circular frequency $\omega = 2\pi\nu$ are connected with the period of an oscillation T by

$$\omega = 2\pi\nu = 2\pi/T. \tag{2.3}$$

To write the expression for a plane wave in its most general form, we include the phase φ in (2.1).

Note that a shift of time t by δt is equivalent to a shift $\delta\varphi = \omega\delta t$ of the phase φ. We shall use this relation later. For this section we choose $\varphi = 0$.

It will turn out advantageous to use the complex description of waves. This description is based on the well-known relations between cosine, sine

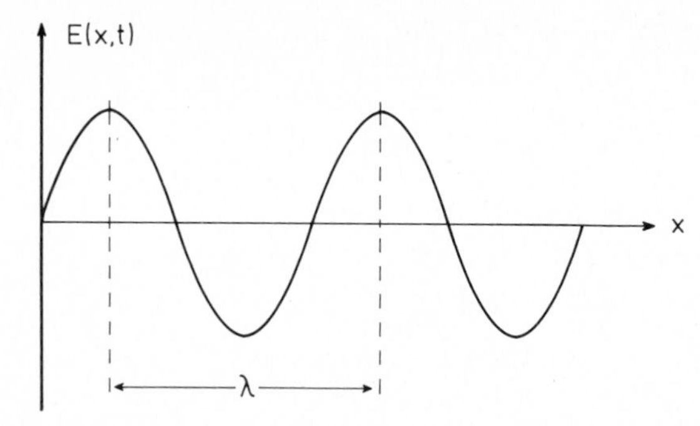

Fig. 2.1. Snapshot of the sine-wave (2.1). The field strength $E(x,t)$ is plotted at a given time versus the space coordinate x. λ is the wave length.

and the exponential function

$$\cos\alpha = \left(\tfrac{1}{2}\right)\left[\exp(\mathrm{i}\alpha) + \exp(-\mathrm{i}\alpha)\right]$$

$$\sin\alpha = \left(\tfrac{1}{2\mathrm{i}}\right)\left[\exp(\mathrm{i}\alpha) - \exp(-\mathrm{i}\alpha)\right]. \tag{2.4}$$

Using (2.4) and (2.1) we obtain

$$E(x,t) = A\exp(\mathrm{i}kx - \mathrm{i}\omega t) + A^*\exp(-\mathrm{i}kx + \mathrm{i}\omega t) \tag{2.5}$$

where the amplitude A is given by

$$A = E_0/(2\mathrm{i}). \tag{2.6}$$

For use in later chapters we will write (2.5) in the form

$$E(x,t) = E^{(+)}(x,t) + E^{(-)}(x,t) \tag{2.7}$$

where

$$E^{(+)} \propto \exp(-\mathrm{i}\omega t) \tag{2.8}$$

is called the positive frequency part and its conjugate complex $E^{(-)}$ the negative frequency part of the electric field strength. This notation sounds somewhat strange because the + sign in $E^{(+)}$ is connected with the − sign in the exponential (2.8). We will understand this notation only later when we treat the electromagnetic field quantum mechanically.

Since Maxwell's equations in vacuum are linear it follows that a superposition (fig. 2.2) of two partial waves is again a solution of Maxwell's equations. This is expressed mathematically by

$$E(x,t) = E_1(x,t) + E_2(x,t). \tag{2.9}$$

It must be noted, however, that this superposition principle can be violated when waves propagate through matter and this will indeed be an important subject of study in nonlinear optics.

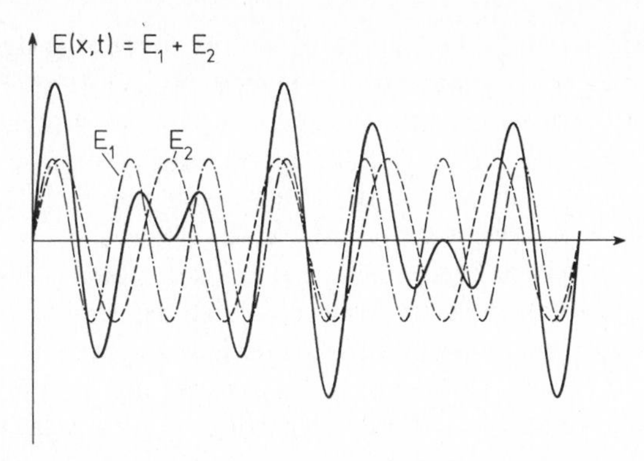

Fig. 2.2. Example of the superposition of two sinusoidal waves E_1 and E_2. E_1 is shown by a dashed dotted line. E_2 by a dashed line. The resulting superposition is shown by a solid line. Note both the regions of enhancement (positive interference) and of decrease of total amplitude (negative interference).

It is also interesting to note that at very intense fields (which are not yet available, however) one may even expect that the superposition principle breaks down for waves propagating in vacuum.

Let us, however, return to our much simpler considerations. Here the superposition principle (2.9) allows us to understand the phenomena of beats and of interference. Using for E_1 and E_2 two sine waves with different wave numbers k_j and frequencies $\omega_j, j = 1, 2$, the total wave reads

$$E_{\mathrm{tot}}(x,t) = \underbrace{A \sin(k_1 x - \omega_1 t + \varphi_1)}_{E_1} + \underbrace{A \sin(k_2 x - \omega_2 t + \varphi_2)}_{E_2}. \quad (2.10)$$

Using elementary formulas of trigonometry we cast (2.10) into the form

$$E_{\mathrm{tot}}(x,t) = \underbrace{2A \sin(\bar{k} x - \bar{\omega} t + \bar{\varphi})}_{\mathrm{I}} \underbrace{\cos(\Delta k x - \Delta\omega t + \Delta\varphi)}_{\mathrm{II}}. \quad (2.11)$$

The new wave numbers and frequencies are given by

$$\begin{aligned} \bar{k} &= (k_1 + k_2)/2, \qquad & \bar{\omega} &= (\omega_1 + \omega_2)/2, \qquad & \bar{\varphi} &= (\varphi_1 + \varphi_2)/2 \\ \Delta k &= (k_1 - k_2)/2, \qquad & \Delta\omega &= (\omega_1 - \omega_2)/2, \qquad & \Delta\varphi &= (\varphi_1 - \varphi_2)/2. \end{aligned} \quad (2.12)$$

The total wave (2.11) is represented in fig. 2.2. The first factor in (2.11), I, represents a rapidly oscillating wave in space and time. Its amplitude is slowly modulated by the second part, II (cf. fig. 2.2). Since the second function, II, can vanish at certain space–time points we obtain a complete annihilation of wave amplitudes by means of the superposition principle.

This negative interference can be easily studied by letting light propagate through a narrow slit. According to Huygen's principle each spacepoint, which is hit by a wave, becomes the starting point of spherical waves (cf. fig. 1.5). The superposition of these spherical waves gives rise to new wave fronts. A plane wave hitting a slit makes each point of the slit to the starting center of a wave with equal phase. When the waves propagate from the different points phase lags occur. Due to these phase lags it may happen that a wave amplitude maximum coincides with the minimum of another wave. In this case negative interference occurs and we shall observe dark stripes on a screen (fig. 2.3). According to diffraction theory the dark stripes occur under an angle α according to the relation

$$\sin\alpha = m\lambda/D, \qquad m = 1,2,3\dots. \tag{2.13}$$

m is an integer, λ is the wavelength of the incident light and D is the diameter of the slit.

In actual experiments using natural (thermal) light sources there are two fundamental difficulties to be taken into consideration. Light sources have a finite dimension and their atoms emit light independently of each other. Since the phases of the wave tracks emitted by the individual atoms are uncorrelated it may seem, at first sight, that the interference pattern described above does not occur when emission by many atoms takes place; quite the contrary is true. Since the phases of the individual wave tracks are uncorrelated, interference effects stemming from different wave tracks vanish when we average over the uncorrelated phases and the resulting total interference pattern is just the sum of the individual interference patterns. We shall explicitly perform such phase averages later in this book.

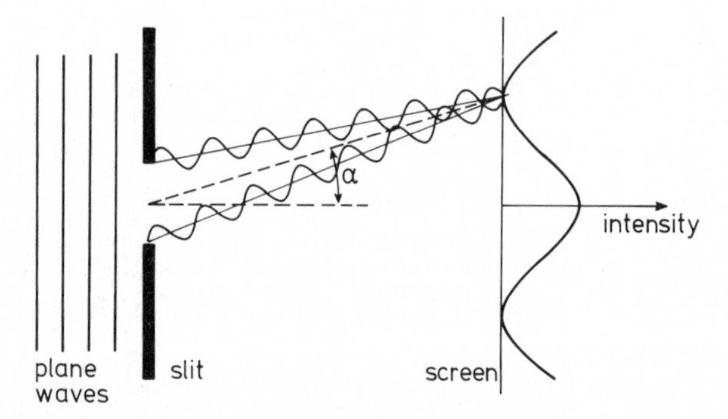

Fig. 2.3. Plane wave front is hitting a slit. Each point of the slit is a starting center of the wave with equal phases. The figure shows how the different lengths of paths cause a phase lag which results in negative interference shown by the example.

Another difficulty rests in the fact that a light source does not radiate monochromatic light but light which always possesses a certain frequency spread (being equivalent to a spread of wave lengths). As we will see later (Volume 2) the laser has made it possible to excite the field in the plane of the diffraction slit with equal phases.

A further experiment to demonstrate the wave character of light is Young's interference experiment (cf. fig. 2.4). Light emitted from a source with very narrow diameter is made parallel by a collector lens. This light then passes through two slits of a screen. We will assume that in the two slits 1 and 2 the light has the same phase.

The experimental setup can be compared with water waves where the synchronous dipping of two bars into water causes two interfering waves. Due to this interference there will be regions in which water remains at rest or other regions where the water amplitudes are particularly high. To cast this well-known phenomenon into a mathematical form we imagine that light is propagating from the two points 1 and 2 in form of spherical waves. To exhibit the most important features we neglect any polarization effects

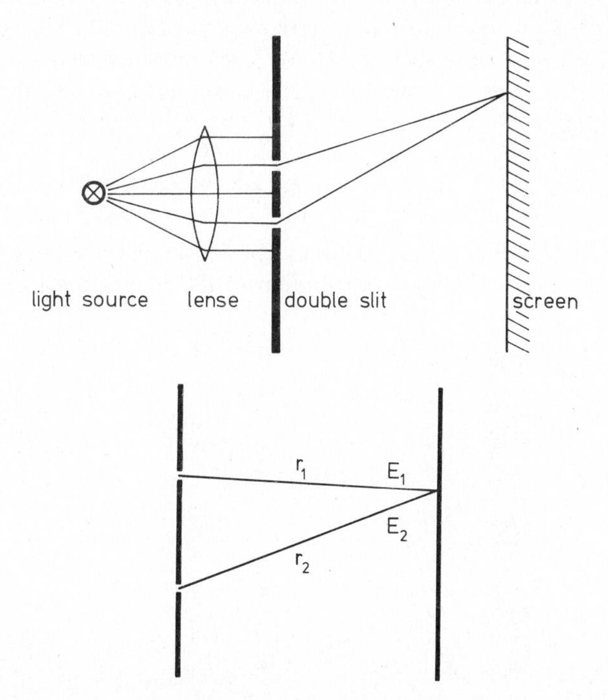

Fig. 2.4. The basic experimental arrangement of Young's double slit experiment. Light stemming from a light source is made parallel by a lens and passes through two slits. The resulting interference pattern is observed on a screen.

and represent the light wave in form of a usual spherical wave

$$E \propto (1/r)\exp[\mathrm{i}(kr - \omega t)]. \tag{2.14}$$

r is the distance between a point where the field amplitude is measured and the point where the field amplitude has been created. In our special example we have two spherical waves and we consider a point r, hit by these waves. Using the notation of fig. 2.4 we find

$$E_1^{(+)} = E_0(1/r_1)\exp[\mathrm{i}(kr_1 - \omega t)] \mathrel{\hat{=}} (1/r_1)\exp(-\mathrm{i}\omega t_1) \tag{2.15}$$

and

$$E_2^{(+)} = E_0(1/r_2)\exp[\mathrm{i}(kr_2 - \omega t)] \mathrel{\hat{=}} (1/r_2)\exp(-\mathrm{i}\omega t_2). \tag{2.16}$$

On the right-hand sides we have introduced new times t_i by the relations

$$t_i = t - r_i/c, \qquad \omega = ck. \tag{2.17}$$

These times differ from the original time t by the time $T_i = r_i/c$, which is needed by the light waves to propagate from their original space points to the point under consideration. By this time lag the field amplitude suffers a phase shift. Again we obtain an interference pattern which is represented in fig. 2.5. We shall discuss this pattern in more detail below. Let us first define, however, intensity. According to electromagnetic theory light intensity is defined by

$$\tfrac{1}{2}\left(\varepsilon_0 \boldsymbol{E}^2 + \frac{1}{\mu_0}\boldsymbol{B}^2\right)$$

where we use the MKS units. The contribution of the magnetic induction $\boldsymbol{B}$ to the intensity equals the contribution of the electric field strength, $\boldsymbol{E}$.

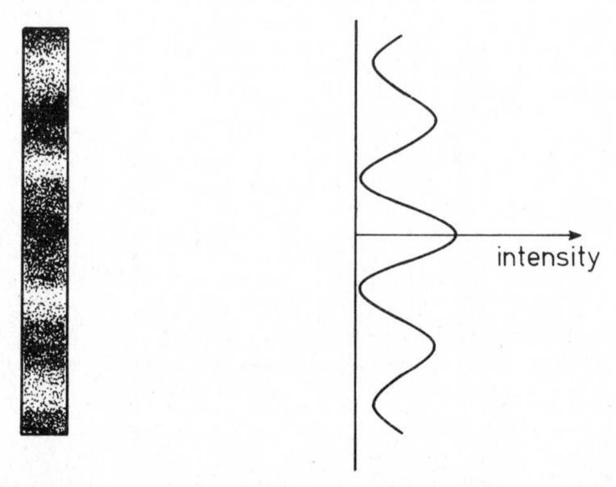

Fig. 2.5. The interference pattern of Young's double slit experiment. Left: The light intensity indicated by dark and light stripes. Right: A photogrammatic plot of the intensity (abscissa) along the screen (ordinate).

Thus it is sufficient for our purposes to use as definition of intensity the first part $\propto \boldsymbol{E}^2$. This expression can be further simplified when we use the decomposition (2.6) of E in positive and negative frequency parts, i.e. writing E in the form

$$E(x,t) = E_0^{(+)}(x)\exp[-\mathrm{i}\omega t] + E_0^{(-)}(x)\exp[\mathrm{i}\omega t]. \tag{2.17a}$$

In later applications we shall even admit that the factors $E^{(+)}$, $E^{(-)}$ may not only depend on the spatial coordinate x but also on time t. It is assumed, however, that this time dependence is much slower than that of $\exp(\mathrm{i}\omega t)$. Inserting (2.17a) into our definition of the intensity we obtain

$$I \propto E_0^{(+)^2}\exp[-2\mathrm{i}\omega t] + E_0^{(-)^2}\exp[2\mathrm{i}\omega t] + 2E_0^{(+)}E_0^{(-)}. \tag{2.18}$$

In all experiments measuring light intensity the measurement implies an average over times large compared to the period of a single oscillation. This means that we do not measure I but rather its temporal average over a time T. Thus we have to consider

$$\bar{I} = (1/T)\int_{-T/2}^{T/2} I(t)\,\mathrm{d}t \tag{2.19}$$

instead of (2.18). Inserting (2.18) into (2.19) we obtain for the first term of (2.18)

$$(2\mathrm{i}\omega T)^{-1}[\exp(\mathrm{i}\omega T) - \exp(-\mathrm{i}\omega T)]E_0^{(+)^2} \tag{2.20}$$

and a similar one for the second term of (2.18). Inserting the last term in (2.18) into the integral and neglecting the slow time dependence of E_0^+, E_0^- we obtain

$$2E_0^{(+)}E_0^{(-)}. \tag{2.21}$$

We now use the fact that the measuring time T is much bigger than the inverse frequency ω i.e.

$$T \gg \omega^{-1} \tag{2.22}$$

or written in a different way

$$\omega T \gg 1. \tag{2.23}$$

According to (2.23) the contribution (2.20) to (2.19) is much smaller than the contribution (2.21). Thus we may safely neglect the rapidly oscillating contributions which allows us to introduce the following definition of the intensity

$$I \approx \bar{I} = \langle E_0^{(+)}E_0^{(-)}\rangle \tag{2.24}$$

where we have dropped all constant factors. Our brackets will indicate a time average

$$\langle \dots \rangle = (1/T) \int_{t-T/2}^{t+T/2} \dots \mathrm{d}t. \tag{2.25}$$

This allows us later on to take care of such cases in which E_0^+, E_0^- are time dependent.

2.2. Classical coherence functions

The results of the preceding section allow us to define some important coherence functions. By means of these coherence functions we want to cast the concept of coherence into rigorous mathematical shape. In order to find out how such functions look, let us consider again Young's interference experiment in more detail. The total field strength at a space point on the screen at time t is given by

$$E_{\text{tot}}^{(+)}(t) = E^{(+)}(t - r_1/c) + E^{(+)}(t - r_2/c) \equiv E^{(+)}(1) + E^{(+)}(2). \tag{2.26}$$

By using the intensity formula (2.24) we readily obtain

$$I = \langle E^{(-)}(1)E^{(+)}(1)\rangle + \langle E^{(-)}(2)E^{(+)}(2)\rangle + \langle E^{(-)}(1)E^{(+)}(2)\rangle + \langle E^{(-)}(2)E^{(+)}(1)\rangle. \tag{2.27}$$

Using the abbreviation

$$\langle E^{(-)}(i)E^{(+)}(j)\rangle = G(i,j) \tag{2.28}$$

we write (2.27) as

$$I = G(1,1) + G(2,2) + G(1,2) + G(2,1). \tag{2.29}$$

Let us try to understand the meaning of the individual terms. To this end we assume $E^{(+)}(1)$ in the form of a spherical wave

$$E^{(+)}(1) = \text{const}(1/r_1)\exp[-\mathrm{i}\omega t_1] \tag{2.30}$$

just as in Young's experiment. The same form is, of course, adopted for E^+ (2). When we insert these expressions into (2.28) for $i = j = 1$, the two exponential functions cancel and we obtain an expression proportional $1/r_1{}^2$. A corresponding result is obtained for $G(2,2)$.

Let us now consider $G(1,2)$. Here the factor $\exp(\mathrm{i}\omega t)$ cancels out but the factor $\exp[-\mathrm{i}\omega(t_2 - t_1)]$ still remains. Thus we find

$$G(1,2) = \text{const}(1/r_1 r_2)\exp[-\mathrm{i}\omega(t_2 - t_1)]. \tag{2.31}$$

Now recall that according to its definition $t_j = t - r_j/c$. We thus recognize that varying the space point on the screen we incidentally vary t_1 and t_2.

Since the frequency of visible light is rather high, the factor $\exp[i\omega(t_2 - t_1)]$ varies relatively quickly, whereas the factors r_1^{-1} and r_2^{-1} change only very slowly. To exhibit the essentials it is therefore sufficient to consider $G(1,1)$ and $G(2,2)$ as constants and to study only the behavior of $G(1,2)$. We now turn to the case where E^+, E^- are not necessarily purely harmonic but depend on time in a more complicated way. We abbreviate $G(1,1) = G(2,2)$ by A and put

$$G(1,2) = B\exp[i\Phi]. \tag{2.32}$$

The intensity I then reads

$$I = \underbrace{2A}_{\substack{G(1,1)\\ +G(2,2)}} + \underbrace{2B\cos\Phi}_{\substack{G(1,2)\\ +G(2,1)}} . \tag{2.33}$$

The phase factor Φ varies when the point r on the screen is changed. Let us consider I as a function of the phase factor Φ. In the case $B = A$ we then obtain the shape shown in fig. 2.6, for $B < A$ the shape shown in fig. 2.7. Because an intensity can never become negative for physical reasons we have the inequality $B \leqslant A$ (which can actually be proven in a rigorous mathematical way).

When we consider the variation of an intensity on the screen a maximum of brightness follows a minimum. The contrast between bright and dark interference fringes is greater, the bigger the difference between I_{max} and I_{min} is. It is useful to normalize this difference between I_{max} and I_{min} by dividing it by $I_{max} + I_{min}$. This leads us to the definition of fringe

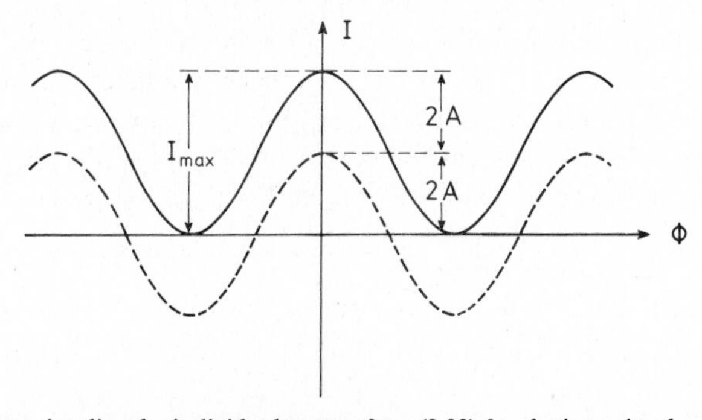

Fig. 2.6. How to visualize the individual terms of eq. (2.33) for the intensity depending on the phase lag Φ. The dashed line shows $2B\cos\Phi \equiv 2A\cos\Phi$, i.e. the case $B = A$. The intensity I is obtained from the dashed curve by shifting it into the I-direction by the amount $2A$. Evidently the minima of the solid curve lie at $I = 0$.

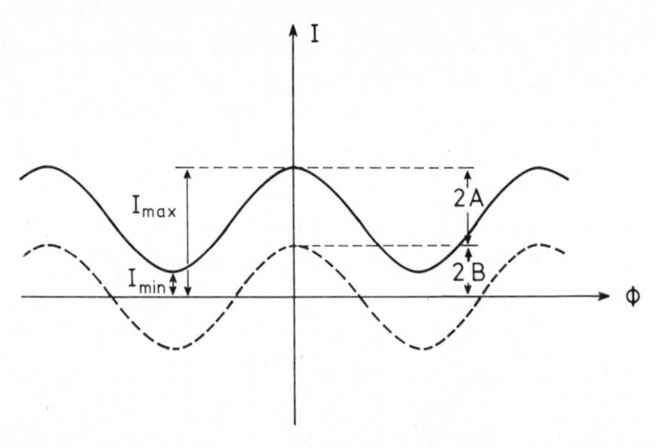

Fig. 2.7. We show what the intensity I looks like in the case that $B < 2A$. The dashed curve represents $2B\cos\Phi$. To obtain I we have to shift the dashed curve by an amount $2A$ into the direction of the I axis. The resulting total intensity is represented by the solid curve. Note that the minima of I are now shifted away from the Φ-axis. A comparison with the result of fig. 2.8 shows that complete darkness at certain points is not reached and that the contrast has become smaller.

visibility V by the relation

$$V = (I_{max} - I_{min})/(I_{max} + I_{min}). \tag{2.34}$$

In the case of our example (2.33) V is given by

$$V = B/A = |G(1,2)|/G(1,1). \tag{2.35}$$

As we will see in a minute $G(1,2)$ contains useful information about the coherence of waves. For this reason

$$G(1,2) = \langle E^{(-)}(1)E^{(+)}(2)\rangle \tag{2.36}$$

is called the mutual coherence function. Since the fringe visibility contains a normalization (division by $I_{max} + I_{min}$) it is reasonable to do the same with (2.36), i.e. dividing it by $G(1,1)$. This is, however, a somewhat overspecialized procedure because in our present example we had assumed that the first terms in (2.29) were equal. To take care of the general case we define quite generally the complex degree of mutual coherence

$$\gamma(1,2) = \frac{G(1,2)}{[G(1,1)G(2,2)]^{1/2}}. \tag{2.37}$$

As can be shown mathematically

$$|\gamma| \leqslant 1 \tag{2.38}$$

always holds.

It is most useful to compare the coherence function (2.36) [or the expression (2.37)], which we have introduced by means of a consideration of Young's double slit experiment, with the one we considered in section 1.11. To facilitate this comparison, let us use the definition (1.19), based on the positive and negative frequency parts of E. We quickly recognize that the arrangement of Young's experiment serves as a beam splitter and a delay, similar to that of fig. (1.30). Indeed, consider a plane wave front hitting the screen with the holes. Then the two holes split this wave into two new waves which are generated by means of Huygen's principle or its adequate generalization. The different path lengths of the two waves produce the delay. We may now slightly generalize the concept of coherence functions. In principle, we can measure the coherence functions of two fields E_1 and E_2 at space points $\boldsymbol{x}_1$ and $\boldsymbol{x}_2$ and times t_1 and t_2, respectively. All we have to assume is that E_1 and E_2 at these space points and times start new waves, e.g. spherical waves, which we can let interfere, for instance in a way known from Young's experiment. Because the coherence properties are "faithfully" transmitted (in vacuum) from the initial space points $\boldsymbol{x}_1$ and $\boldsymbol{x}_2$ to $\boldsymbol{x}$, we can measure

$$\langle E_1^{(-)}(\boldsymbol{x}_1, t_1) E_2^{(+)}(\boldsymbol{x}_2, t_2) \rangle \tag{2.36a}$$

where the appropriate time-average is taken.

The expression (2.36a) is a correlation function of E's with respect to space and time. Two special cases are of particular interest. When we choose $\boldsymbol{x}_1 = \boldsymbol{x}_2$, we may consider (2.36a) as function of t_1 and t_2 only. In this case (2.36a) describes temporal coherence. Similarly, we may choose $t_1 = t_2$ fixed. In this case (2.36a) refers to spatial coherence. The situation becomes particularly simple, when we can split $E_j(\boldsymbol{x}, t)$ into a time-function and a space-function

$$E_j(\boldsymbol{x}, t) = q_j(t) u_j(\boldsymbol{x})$$

where q_j may still fluctuate. In this case (2.36a) factorizes

$$(2.36a) = \langle q_j^{(-)}(t_1) q_2^{(+)}(t_2) \rangle u_1(\boldsymbol{x}_1) u_2(\boldsymbol{x}_2). \tag{2.36b}$$

In such a case the problem of calculating (2.36a) is essentially reduced to treat temporal coherence. We shall meet such a case in Volume 2 when dealing with the laser. Of course, all above expressions can be specialized to the case in which E_1 and E_2 refer to the same field: $E_1(\boldsymbol{x}, t) = E_2(\boldsymbol{x}, t)$. Now we would like to demonstrate by means of examples how the expression (2.37) for the complex degree of coherence can be used to study the coherence of light waves.

To study the coherence of a single light wave we have to split the wave into two parts. This was done in Young's interference experiments by the two slits. Furthermore we have to introduce a phase lag between the two parts and then we have to calculate (2.37) for the two thus constructed wave tracks. Let us consider to this end two examples.

(a) In the case of a pure sine wave the field is described by

$$E^{(+)} = E_0 \exp[-\mathrm{i}\omega t]. \tag{2.39}$$

Let us assume that this wave is duplicated (split) and phase lags $t^{(1)}$ or $t^{(2)}$ are applied (fig. 2.8). Inserting the resulting field amplitudes into $G(1,2)$ we find

$$G(1,2) = \frac{|E_0|^2}{T} \int_{-T/2}^{T/2} \exp\{\mathrm{i}\omega(t - t^{(1)}) - \mathrm{i}\omega(t - t^{(2)})\}\, \mathrm{d}t. \tag{2.40}$$

The integration is quite elementary and yields

$$G(1,2) = |E_0|^2 \exp\left[\mathrm{i}\omega(t^{(2)} - t^{(1)})\right]. \tag{2.41}$$

$G(1,1)$ and $G(2,2)$ can be calculated still more simply: $= |E_0|^2$. Thus the complex degree of coherence of a purely sinusoidal oscillation reads

$$\gamma(1,2) = \exp\left[\mathrm{i}\omega(t^{(2)} - t^{(1)})\right]. \tag{2.42}$$

Its modulus

$$|\gamma| = 1 \tag{2.43}$$

has attained its maximum and this maximum value is retained for all phase lags. Thus we see that a sinusoidal wave has a maximal degree of coherence.

(b) Let us consider a second example namely that of a finite wave track of length t_0 (cf. fig. 2.9). In order to determine $G(1,2)$ we duplicate (split) the wave track and introduce again a phase lag. Inserting the two corresponding expressions for the field amplitudes of the wave tracks into

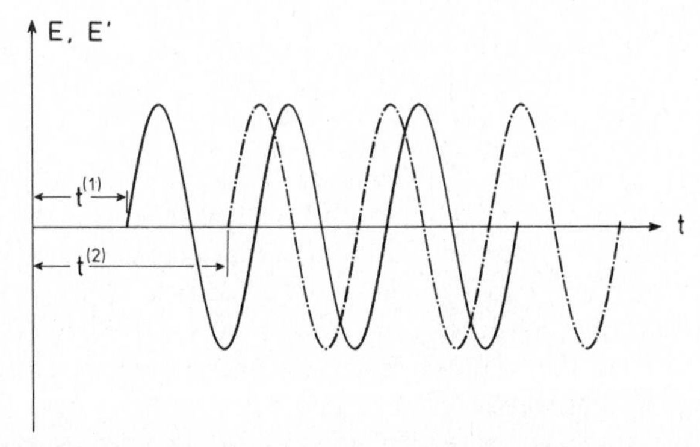

Fig. 2.8. This figure shows two waves, which are to be thought of as infinitely extended, of equal amplitudes and wavelengths but which were subjected to time lags $t^{(1)}$ and $t^{(2)}$.

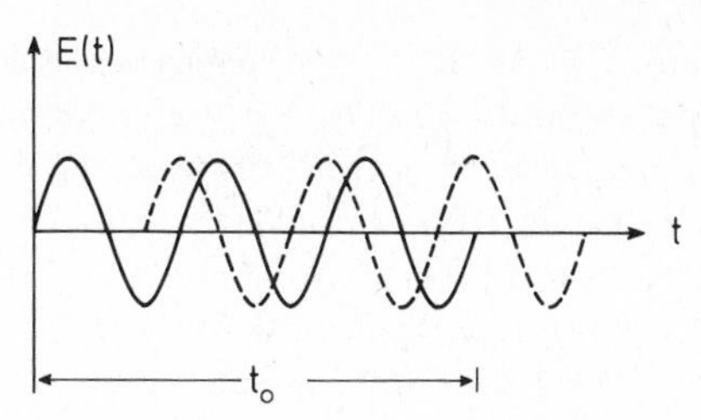

Fig. 2.9. This figure shows a wave track of finite length t_0 in time (solid line) and the same wave track but with a time lag (dashed line).

$G(1,2)$ we obtain only a nonvanishing contribution to $G(1,2)$ when the phase lag (measured in time units) is smaller than the length t_0. The result of this simple calculation, which we leave to the reader as an exercise, is represented in fig. 2.10. According to this result the degree of coherence decreases with increasing time difference. The coherence is therefore steadily decreasing. We will present more examples of realistic cases later when we have defined the quantum mechanical analogue of the coherence functions introduced above.

2.2.1. *Coherence functions of higher order*

The mutual degree of coherence (2.36) depends on two field amplitudes $E^{(-)}(1)$ and $E^{(+)}(2)$. It is, however, possible to define coherence functions of higher order in which more than two field amplitudes occur:

$$\langle E^{(-)}(1)E^{(-)}(2)\ldots E^{(-)}(n)E^{(+)}(n+1)\ldots E^{(+)}(N)\rangle$$

where $E^{(-)}(j)$ stands for $E^{(-)}(x_j, t_j)$ etc. When we discuss such correlation functions later in detail we will see that we have to discuss more complicated measurement processes which enable us to measure such

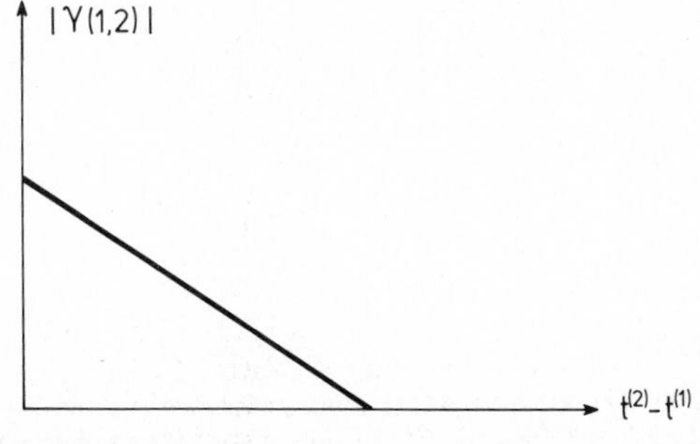

Fig. 2.10. The decay of the complex degree of coherence (2.37) as a function of the time lag.

correlation functions. The statistical properties of light fields, especially those of thermal sources and of lasers, can be described by such correlation functions. In particular, they will allow us to distinguish uniquely between laser light and light from thermal sources.

The interference patterns observed in Young's experiment seemed to be a clearcut proof of the wave-character of light. However, we know that this is only one aspect. Let us now turn to the other aspect, namely to the particle-aspect. This new development can be traced back to 1900.

2.3. Planck's radiation law

We can visualize the meaning of Planck's law in a simple way as follows. When we look into a cold furnace it appears black. When we heat up the furnace its walls first appear to glow red and later, at still higher temperatures, white. Planck's law aims at describing this change of colour or, more precisely speaking, at determining the intensity distribution of the electromagnetic field as a function of frequency. Since the atoms of the walls emit and reabsorb light, thermal equilibrium is established between the atoms and the electromagnetic field. When the radiation energy in a frequency interval $\nu, \nu + d\nu$ is measured, curves as shown in fig. 2.11 are found. Clearly the maximum of the intensity distribution is shifted to higher frequencies when the temperature is increased. This shift is expressed by Wien's law according to which the frequency shift of the intensity maximum is proportional to the increase in temperature. Let us now calculate the energy $u(\nu)\,d\nu$ of the radiation in the frequency interval $\nu, \nu + d\nu$. To this end we make use of the wave theory of light and represent $u(\nu)\,d\nu$ in

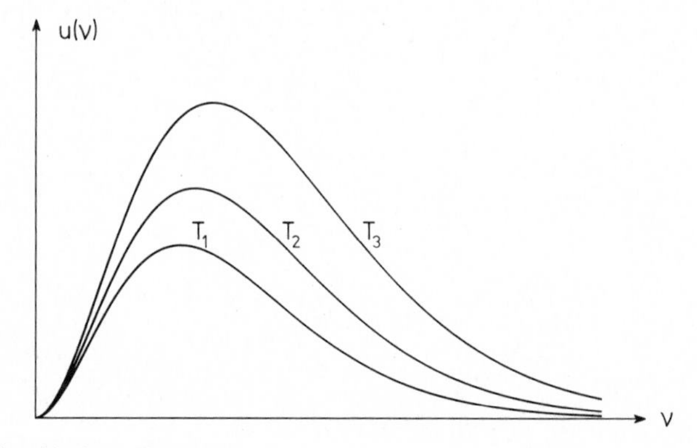

Fig. 2.11. Planck's law. The energy density $u(\nu)$ plotted versus frequency ν for different temperatures $T_1 < T_2 < T_3$. The shift of the maximum with increasing temperature is clearly visible.

the form

$$u(\nu)\,\mathrm{d}\nu = \text{number of waves in the interval } \mathrm{d}\nu \cdot \text{mean thermal energy of one wave.} \tag{2.44}$$

In order to determine the number of waves in the frequency interval $\mathrm{d}\nu$ we consider a model, namely standing waves in a metal box. The essential features of this problem can be explained when we study the electric field strength in one dimension and represent it in the form

$$E = E_0 \sin kx. \tag{2.45}$$

Since the walls of the box are assumed to be metallic, according to Maxwell's theory E must vanish at the walls (which are assumed of infinitely high conductivity). Choosing the walls perpendicular to the x-direction at the coordinate $x = 0$ and $x = L$ the sine function of (2.45) must vanish at these points. This is achieved by choosing k in (2.45) in the special form

$$k = n\pi/L, \qquad n = 1, 2, 3 \dots. \tag{2.46}$$

These considerations imply that only specific forms of E are permitted in between the walls, namely (cf. fig. 2.12)

$$E = E_0 \sin(n\pi x/L). \tag{2.47}$$

The wave number of an electromagnetic wave is connected with its frequency via the velocity of light

$$\omega = ck. \tag{2.48}$$

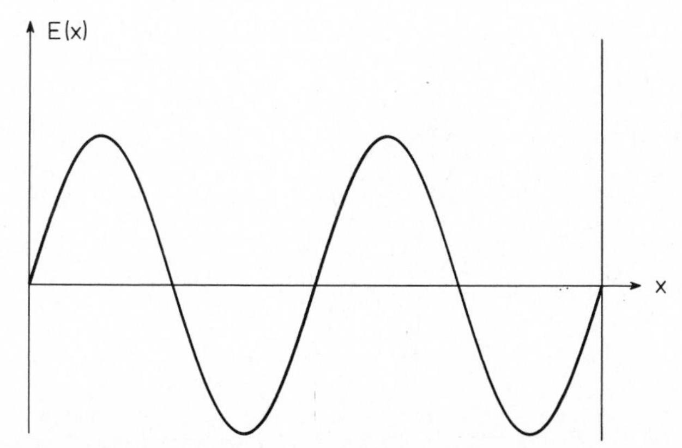

Fig. 2.12. One-dimensional example of a standing electric wave between two perfectly conducting walls.

On the other hand we have, of course,

$$\omega = 2\pi\nu. \tag{2.49}$$

From the relations (2.46), (2.48), (2.49) we obtain

$$\nu = cn/(2L). \tag{2.50}$$

Since n is an integer we can enumerate the waves. By means of (2.50) we thus obtain as number of possible waves in the interval $\mathrm{d}\nu$

$$\mathrm{d}n = (2L/c)\,\mathrm{d}\nu. \tag{2.51}$$

In this way we have solved our task to determine the number of waves in the frequency interval $\mathrm{d}\nu$, at least in one dimension. In reality we have, of course, to deal with three dimensions. As can be shown by means of Maxwell's equations the field amplitude can be again expressed by sine and cosine functions for instance in the form

$$E_z \propto \sin k_x x \sin k_y y \cos k_z z. \tag{2.52}$$

Again the wave numbers k_i are given by

$$k_i = \pi n_i/L, \qquad n_i = 1,2,3\ldots, \qquad i = x,y,z. \tag{2.53}$$

We define $k = \sqrt{k_x^2 + k_y^2 + k_z^2}$ and $n = \sqrt{n_x^2 + n_y^2 + n_z^2}$. In entire analogy of (2.50) we obtain the relation

$$\nu = cn/(2L). \tag{2.54}$$

To count the number of possible waves belonging to the frequency range $\nu, \nu + \mathrm{d}\nu$ we consider a coordinate system in which the axes correspond to the numbers n_x, n_y, n_z, respectively (fig. 2.13). Each point of this space with integer coordinates represents a possible state of the electromagnetic field oscillation. When we choose the frequency interval sufficiently large

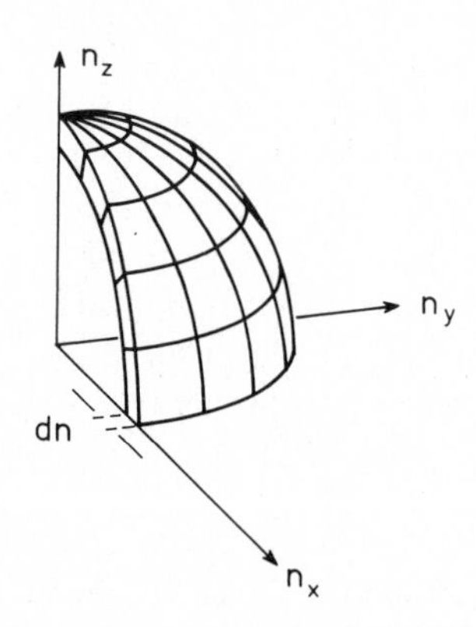

Fig. 2.13. This figure shows the section of the spherical shell in the n-space (compare text).

we may assume that the points can be counted by means of an approximation treating them as continuously distributed. The number of points in a spherical shell of thickness $\mathrm{d}n$ is given by $4\pi n^2 \mathrm{d}n$. Since all numbers n_j must be positive, this result must be divided by the number of octants, i.e. 8 and because we have two directions of polarization which have to be counted individually we must multiply our result by 2. We thus obtain as number of possible states in this spherical shell

$$\mathrm{d}N = \pi n^2 \mathrm{d}n. \tag{2.55}$$

Expressing in it n by (2.54) we obtain the number of possible waves in the frequency interval $\nu, \nu + \mathrm{d}\nu$

$$\mathrm{d}N = \frac{8\pi L^3}{c^3} \nu^2 \mathrm{d}\nu. \tag{2.56}$$

We now have to deal with the second part of our task, namely to determine the mean thermal energy of a wave. To this end we have to use formulas well known in thermodynamics. Anticipating later results we assume that the system has energy levels W_n, $n = 0, 1, 2, \ldots$. According to thermodynamics the mean thermal energy is given by

$$\overline{W} = \sum_{n=0}^{\infty} p_n W_n \tag{2.57}$$

where p_n is the probability to find the state n occupied in thermal equilibrium. We will write down p_n below.

When deriving his theory, Planck assumed that each electromagnetic wave was coupled to an atom which could be represented as an oscillator. He calculated the mean thermal energy of such an atomic oscillator and identified it with the mean energy of the specific light wave under consideration using the equipartition theorem of thermodynamics. We will see later that this assumption of material atomic oscillators is unnecessary. Indeed one may directly calculate the mean energy of an electromagnetic wave on account of quantum postulates.

Let us return, however, to Planck's ideas. In order to obtain the experimentally correct distribution function for the light intensity Planck had to make an assumption which was quite revolutionary at his time. The energy levels of an harmonic oscillator must not be continuous as it would follow from classical theory but they must be quantized according to the prescription (fig. 2.14)

$$W_n = h\nu \cdot n \tag{2.58}$$

where ν is the frequency of the light wave or equivalently the emission and absorption frequency of the corresponding atomic oscillator, h is Planck's

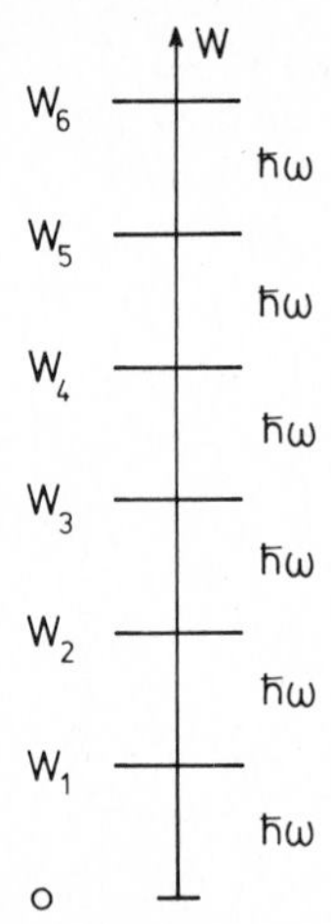

Fig. 2.14. According to Planck's fundamental idea the energy of an harmonic oscillator must be quantized. The energy levels being separated by the amount $\hbar\omega \equiv h\nu$. Note that $\omega = 2\pi\nu$ and $\hbar = h/(2\pi)$.

constant and n is an integer number $0, 1, 2, \ldots$. The occupation probability p_n was chosen by Planck to be identical with the Boltzmann distribution function of classical statistical mechanics. According to this (fig. 2.15)

$$p_n = Z^{-1} \exp[-W_n/(kT)] \tag{2.59}$$

where k is Boltzmann's constant, T absolute temperature, Z a normalization factor which normalizes the sum over the p_n to unity and is called the partition function. From the condition of normalization of the p_n's Z is

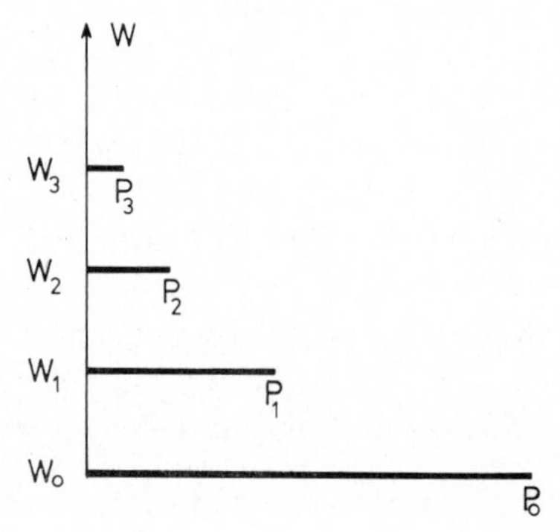

Fig. 2.15. The energy levels of the harmonic oscillator are indicated on the ordinate. Parallel to the abscissa the corresponding occupation probabilities p_n, $n = 0, 1, 2, 3$ are plotted.

easily calculated and reads

$$Z = \sum_{n=0}^{\infty} \exp[-W_n/(kT)]. \tag{2.60}$$

By inserting (2.59) and (2.60) into (2.57) the mean energy is given by

$$\overline{W} = \left[\sum_{n=0}^{\infty} W_n \exp(-\beta W_n)\right] \Big/ \left[\sum_{n=0}^{\infty} \exp(-\beta W_n)\right] \tag{2.61}$$

where we have used the abbreviation

$$\beta = 1/(kT). \tag{2.62}$$

The right-hand side of (2.61) can be written as derivative of the logarithm of the partition function Z

$$\overline{W} = -\frac{\partial}{\partial \beta} \ln(Z). \tag{2.63}$$

Due to the simple form of W_n (2.58), the partition function is a simple geometric series and can easily be evaluated

$$Z = [1 - \exp(-h\nu\beta)]^{-1}. \tag{2.64}$$

This allows us to calculate (2.63) and we obtain

$$\overline{W} = h\nu/[\exp(h\nu\beta) - 1]. \tag{2.65}$$

Now we have all the results together in order to determine the radiation energy in the interval $\nu, \nu + \mathrm{d}\nu$. By inserting (2.61) and (2.65) into (2.44) we obtain

$$u(\nu)\,\mathrm{d}\nu = \mathrm{d}N \cdot \overline{W} \tag{2.66}$$

and thus

$$u(\nu)\,\mathrm{d}\nu = (8\pi V/c^3)\nu^2\,\mathrm{d}\nu \cdot h\nu/[\exp(\beta h\nu) - 1]. \tag{2.67}$$

This formula represents the energy of the radiation field in the volume V and in the frequency interval $\nu, \nu + \mathrm{d}\nu$. When dividing (2.67) by V and $\mathrm{d}\nu$ we obtain the energy density ρ

$$\rho(\nu) = (8\pi/c^3)\nu^2 h\nu/[\exp(\beta h\nu) - 1]. \tag{2.68}$$

This is Planck's famous formula which is in excellent agreement with all experimental data (cf. fig. 2.11). In deriving this crucial formula we have learned two things. First a mathematical trick how to determine the number of field modes in a cavity and, still much more important, the idea that it is necessary to quantize energy levels. Planck's derivation was based on the idea that light is still composed of waves. In section 2.5 we shall see that there is an even much more elegant access to Planck's law namely by the assumption that light is composed of particles.

2.4. Particles of light: photons

In the preceding sections we have seen how interference experiments could convincingly prove that light consists of waves. Thus it could seem that a unique decision has been made in view of Huygen's hypothesis. There are, however, experiments which can be understood only if we assume that light is composed of particles. I mention here two well-known examples.

(a) The photoelectric effect.

When ultraviolet light impinges on metal surfaces it can free electrons. The kinetic energy of the emerging photoelectrons has been measured depending on light intensity and light frequency. According to the wave theory of light, the energy of a light wave is proportional to its intensity. Therefore one should expect that on account of energy conservation the kinetic energy of the emitted electrons increases with increasing light intensity. This is, however, not the case. With increasing light intensity more electrons are emitted but the energy of an individual electron remains unchanged (cf. fig. 2.16). On the other hand, when we vary the frequency of light, the kinetic energy of the emitted electrons changes. As Einstein has shown, this result can be easily understood by the assumption that light is composed of particles which are called photons. In this theory it is assumed that the energy of a single photon is given by $E = h\nu$, where h is Planck's constant and ν the frequency of the light wave. The kinetic energy of an electron is then connected with the photon energy by (fig. 2.17)

$$mv_{el}^2/2 = h\nu - A \tag{2.69}$$

where m is the mass of an electron, v_{el} the velocity of the emitted electron,

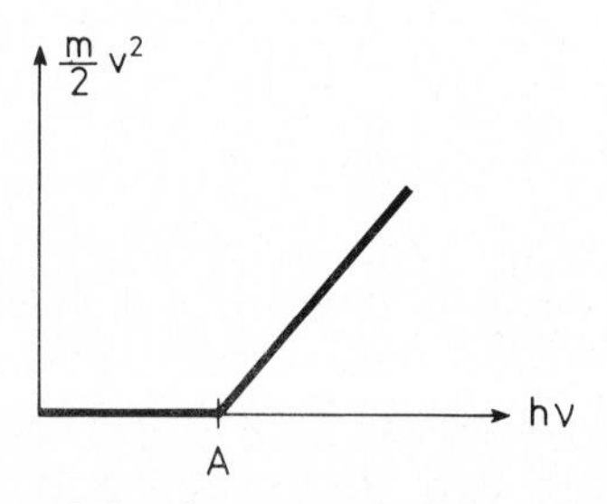

Fig. 2.16. The kinetic energy of the electrons emitted from the metal surface are plotted versus the energy of an incident light quantum. A is the work function, i.e. minimum amount of energy which is necessary to free an electron from the metal.

and A the work function. It is a potential energy which the electron has to pay when leaving the crystal (fig. 2.17). The relation (2.69) is experimentally verified in an excellent way and can be used to determine Planck's constant. By invoking the theory of relativity one can show that a photon must have also a momentum p according to the relation

$$p = h\nu/c. \tag{2.70}$$

The experimental demonstration that photons are particles, with energy $h\nu$ *and* momentum $h\nu/c$, can be done by means of the Compton effect.

(b) The Compton effect.

In this experiment a light field hits free or loosely-bound electrons. If the light field acts only as an electromagnetic wave we should expect the following effect:

The oscillating field would make the electron oscillate which would become an electric oscillator and thus be capable again of emitting light similarly to the Hertzian dipole. Accordingly the light field emitted by this electron must have the same frequency as the incident light. Thus the direction of the incident light wave can be altered (i.e. light scattering takes place) but the frequency remains unchanged.

An entirely different effect would occur if light is composed of particles. Then a photon collides with an electron (fig. 2.18). By using the usual energy and momentum conservation laws the relations

$$h\nu + W_{el} = h\nu' + W'_{el} \tag{2.71}$$

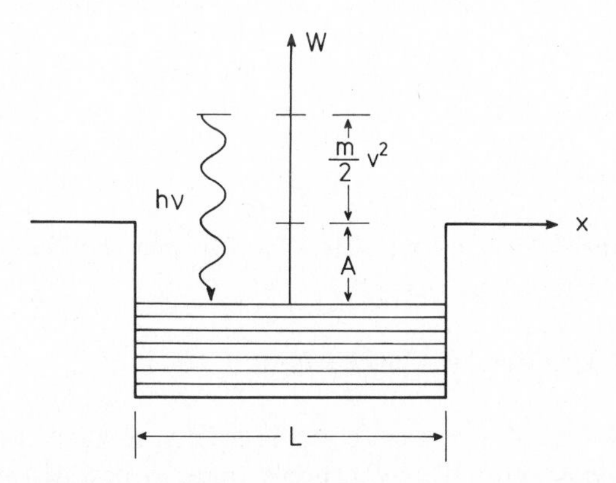

Fig. 2.17. This figure presents a simple model explaining the decomposition of the photon energy $h\nu$ into the work function A and the kinetic energy $\frac{1}{2}mv_{el}^2$. The metal of linear extension L is modelled by a box. In it each metal electron can occupy an energy level indicated by the horizontal bars within the box. The energy levels are filled by electrons up to a maximum energy. Since the electrons are bound within the metal the total energy must lie below the zero of the potential energy, which is identical in our picture with the x-axis. This model will find its justification by the quantum mechanical approach we shall discuss in section 3.6.

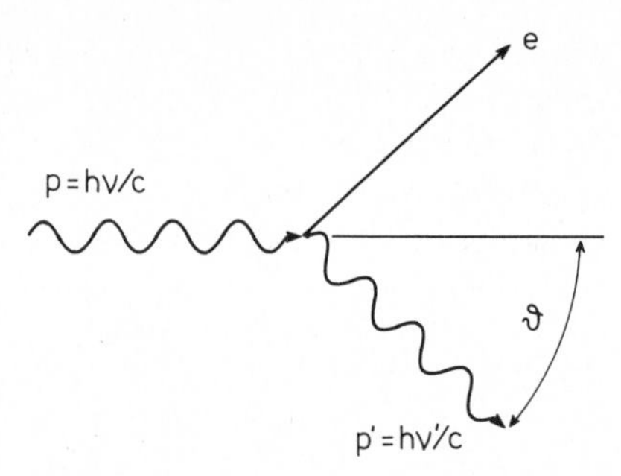

Fig. 2.18. Compton effect. Photon incident from the left hits an electron at rest. The photon is scattered by an angle ϑ as shown in figure.

and

$$\overrightarrow{h\nu\ /c} = \overrightarrow{h\nu'\ /c} + \vec{p}_{\mathrm{el}} \tag{2.72}$$

must be fulfilled. An elementary calculation yields the result that the frequency of the scattered photon is shifted with respect to that of the incident photon according to the formula

$$\nu'^{-1} - \nu^{-1} = \left(2h/m_0c^2\right)\sin^2(\vartheta/2). \tag{2.73}$$

This result is achieved if the electron is treated relativistically. Experimentally it turns out that one finds both an unshifted frequency which would point to the wave character of light (Rayleigh scattering) but in addition the Compton effect shows a well defined frequency shift in accord with (2.73).

Exercise on section 2.4

(1). Derive (2.73).

Hint: Use (2.71), (2.72) and the relativistic formula $W_{\mathrm{el}} \equiv m_0c^2$ (electron at rest), $W'_{\mathrm{el}} \equiv mc^2 = m_0c^2(1 - v^2/c^2)^{-1/2}$.

2.5. Einstein's derivation of Planck's law

In order to explain this approach let us briefly consider ideas about the structure of atoms known at Einstein's time. According to these ideas, which are mainly due to Bohr, the electrons orbit around their nuclei in atoms in well defined orbits with discrete energy levels. Let us consider two orbits with energies W_1 and W_2 (fig. 1.14) According to Bohr's hypothesis the emitted light frequency is given by $h\nu = W_2 - W_1$. Now let us consider an ensemble of atoms of which a number N_1 is in their groundstates with energy W_1 whereas a number N_2 of atoms is in their excited states with energy W_2. Following Einstein we now consider transitions between these two levels taking place under the absorption and emission of light. The number of transitions of atoms per second from the lower to the upper state, by means of absorption of light, is proportional to the number of atoms in their groundstates and to the energy density of the radiation field, ρ. Introducing the proportionality factor B_{12} we thus find

$$(\mathrm{d}N/\mathrm{d}t)_{\mathrm{abs}} = N_1 B_{12} \rho. \tag{2.74}$$

In order to be able to reproduce Planck's formula Einstein had to postulate two different emission processes (fig. 2.19):

(a) stimulated emission

The number of transitions of atoms from the excited states 2 is assumed proportional to the number of atoms N_2 and to the energy density ρ. Denoting the proportionality factor by B_{21} we obtain the corresponding transition rate

$$(\mathrm{d}N/\mathrm{d}t)_{\mathrm{em,\,in}} = N_2 B_{21} \rho. \tag{2.75}$$

(b) spontaneous emission

By this process atoms can emit light spontaneously, i.e. without presence of any lightfield. The corresponding transition rate is assumed proportional to N_2 and, using a proportionality factor, reads

$$(\mathrm{d}N/\mathrm{d}t)_{\mathrm{em,\,s}} = A N_2. \tag{2.76}$$

In thermal equilibrium where the occupation number of atoms remains constant the number of up-transitions must equal the number of down-transitions. This yields the equilibrium condition

$$N_1 B_{12} \rho = N_2 B_{21} \rho + A N_2. \tag{2.77}$$

It should be noted that this condition is necessary but not sufficient for

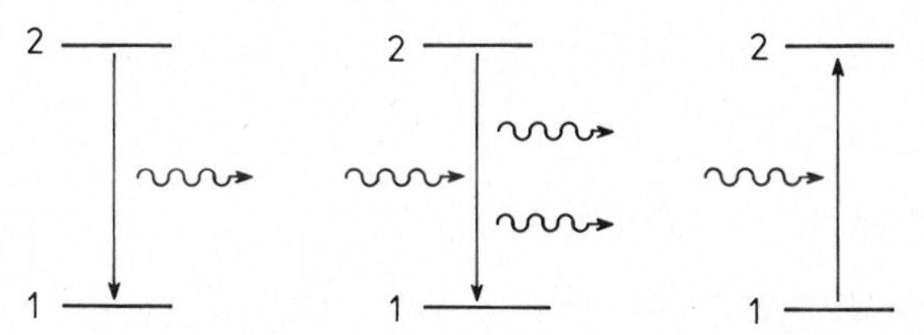

Fig. 2.19. Left: Spontaneous emission of a photon. An electron may jump from its excited state 2 into the ground state emitting a photon. Middle: Stimulated emission. A photon incident on an atom with an excited electron may cause the electron to jump to the ground level and to emit an additional photon. Similarly, n incoming photons can cause a transition of the electron of the upper to the lower level and its emission of an additional photon. Right: Absorption of one photon by the electron. The electron has initially been in its ground state and is now transferred to its upper state.

thermal equilibrium. Indeed we shall encounter later situations in which a relation (2.77) holds also for situations far from thermal equilibrium. The condition of thermal equilibrium will be introduced below [eq. (2.79)] where an explicit assumption about the N's is made. Looking at (2.77) it might seem to us that all quantities are unknown. We will see, however, that we can determine all of these quantities. Solving (2.77) for ρ yields

$$\rho = [N_1 B_{12}/(N_2 B_{21}) - 1]^{-1} \cdot A/B_{21}. \tag{2.78}$$

According to statistical mechanics the occupation numbers N_1 and N_2 are given in thermal equilibrium by

$$N_1 = \text{const}\exp[-W_1/kT]$$
$$N_2 = \text{const}\exp[-W_2/kT]. \tag{2.79}$$

In particular, we obtain

$$N_1/N_2 = \exp[-(W_1 - W_2)/kT]. \tag{2.80}$$

In the following we will make use of Bohr's postulate

$$W_2 - W_1 = h\nu \tag{2.81}$$

where ν is the frequency of the light corresponding to the transition $2 \to 1$. In order to determine the relative size of B_{12} and B_{21} we invoke the evident postulate that the energy density ρ must become infinite when temperature becomes infinite. This can be achieved only [compare (2.78) and (2.80)] if

$$T \to \infty, \rho \to \infty, \text{ i.e. } B_{12} = B_{21} \tag{2.82}$$

holds. Since we need not to distinguish between B_{12} and B_{21} we shall drop the indices. Thus (2.78) reduces to

$$\rho = [\exp(h\nu/kT) - 1]^{-1} \cdot A/B. \tag{2.83}$$

To determine the rate A/B we specialize formula (2.83) to small frequencies so that

$$h\nu \ll kT \tag{2.84}$$

holds. This case had been treated earlier in classical theory by the Rayleigh –Jeans' law which describes the density ρ by

$$\rho(\nu) = (8\pi/c^3)\nu^2 kT. \tag{2.85}$$

Expanding the exponential function in (2.83) for small exponents $h\nu/kT$ we can approximate (2.83) by

$$\rho \approx kTA/(h\nu B). \tag{2.86}$$

A comparison of this result with (2.85) yields immediately

$$A/B = (8\pi/c^3)h\nu^3. \tag{2.87}$$

Let us abbreviate the right-hand side of (2.87) by

$$D(\nu)h\nu. \tag{2.88}$$

The interpretation of $D(\nu)$ becomes evident when we think of Planck's law where $D(\nu) = 8\pi\nu^2/c^3$ denotes the number of modes in the cavity of unity volume. We shall need this quantity in a minute. Inserting (2.87) into (2.83) we obtain indeed again Planck's formula

$$\rho(\nu) = [\exp(h\nu/kT) - 1]^{-1} 8\pi\nu^2 h\nu/c^3. \tag{2.89}$$

The coefficients A and B are called Einstein's coefficients. It will be the aim of a later chapter to derive these coefficients by means of a microscopic theory. The physical meaning of A can be found very simply. According to (2.76) AN_2 is a number of spontaneous transitions per second of all the atoms. Thus A is the number of spontaneous transitions-per-second of a single atom and therefore has the dimension of the inverse of a time

$$AN_2 = N_2/\tau, \qquad A = 1/\tau. \tag{2.90}$$

τ can be interpreted as the spontaneous lifetime of an atom. By inserting (2.90) into (2.87) we can give B the form

$$B = (h\nu D(\nu)\tau)^{-1}. \tag{2.91}$$

The steps performed above can be generalized to non-equilibrium situations and thus play a basic role in modern laser theory. Here we confine ourselves to indicating some simple generalizations. We will come back to this point in much greater detail in Volume 2 dealing with laser theory.

We write the energy density of the radiation field in the following somewhat different form

$$\rho(\nu)\mathrm{d}\nu = \underbrace{(8\pi\nu^2\mathrm{d}\nu/c^3)}_{\mathrm{I}}\cdot\underbrace{h\nu}_{\mathrm{II}}\cdot\underbrace{[\exp(h\nu/kT) - 1]^{-1}}_{n(\nu)}. \tag{2.92}$$

The first factor I is again the number of field modes in the interval $\nu, \nu + \mathrm{d}\nu$ in the unit volume. The second factor II is the energy of the single photon. The last factor can be interpreted as the number of photons occupying individual waves. We now extend Einsteins considerations to a non-equilibrium case. We assume that each atomic transition creates or annihilates a photon. Then the following rate equation must hold. The temporal increase of photon number is given by stimulated and spontaneous emission. This rate is decreased by the corresponding absorption rate. By denoting the number of photons by $\bar{n}$ we obtain by means of (2.74)–(2.76) the prototype of a rate equation for photon production

$$\mathrm{d}\bar{n}/\mathrm{d}t = (N_2 - N_1)B\rho + AN_2. \tag{2.93}$$

Now let us analyze this photon number $\bar{n}$. To this end we assume that an atom emits photons in a certain frequency range $\Delta\nu$ or absorbs photons out of this frequency range. The total number of photons $\bar{n}$ is then given by the number of possible modes in the volume V multiplied by the number of photons of an individual mode. The number of modes has been determined in the previous section. Again, using the abbreviation (2.88) we obtain

$$\bar{n} = V \cdot D(\nu)\Delta\nu \cdot n(\nu). \tag{2.94}$$

We now wish to transcribe eq. (2.93) into an equation for the photon number $n(\nu)$. By using the relations (2.91), (2.92), and (2.94) we are led to the equation

$$\mathrm{d}n/\mathrm{d}t = (N_2 - N_1)wn + wN_2 \tag{2.95}$$

where we have used the abbreviation

$$w = (VD(\nu)\Delta\nu\tau)^{-1}. \tag{2.96}$$

As we will see later in the frame of laser theory, (2.95) is based on the assumption that the photons of the frequency interval $\Delta\nu$ are all produced or absorbed at exactly the same rate. This is not quite the case. Nevertheless (2.95) is a good starting point to develop the basic ideas of a laser theory.

When we look back at this chapter we are still confronted with a puzzle. On the one hand, there is convincing evidence that light behaves like waves, on the other hand, there is evidence, not less convincing, that light consists of particles. As we have indicated in chapter 1, it was left to quantum theory to reconcile this apparent contradiction and we shall come back to this question in chapter 5 and subsequent chapters.

3. The nature of matter. Particles or waves?

3.1. A wave equation for matter: The Schrödinger equation

According to chapter 2, light does not only appear as waves but also as particles. To a wave with frequency ν photons each with an energy $W = h\nu$ are attributed. Using further that $c = \lambda\nu$ we can cast formula (2.70) into the form $p = h/\lambda$ telling us that a photon, attributed to a wavelength λ, carries a momentum p. For later purposes we can cast such relations into a more convenient form. Just recall that

$$2\pi\nu = \omega \tag{3.1}$$

and

$$2\pi/\lambda = k \tag{3.2}$$

(ω = circular frequency and k = wave number). Furthermore we introduce an abbreviation for Planck's constant divided by 2π, namely

$$\hbar = h/2\pi. \tag{3.3}$$

This allows us to write the original relations

$$W = h\nu \tag{3.4a}$$

and

$$p = h/\lambda \tag{3.5a}$$

in the form

$$W = \hbar\omega \tag{3.4}$$

and

$$p = \hbar k. \tag{3.5}$$

Originally, in the theory of light, we read these two formulas from right to left, i.e. we attributed particle properties to waves. De Broglie had the ingenious idea to attribute wave character to particles by reading eqs. (3.4) and (3.5) from left to right. So as I represent it here it sounds of course trivial, however, note that his idea was to think of particles, for instance electrons, as waves when everybody was convinced that these were merely particles. According to his hypothesis, a particle with energy W and momentum p should have the possibility to manifest itself as a wave. This hypothesis was fully substantiated by electron diffraction at crystal surfaces (fig. 3.1). As we know from diffraction theory, we can expect the diffraction pattern only if the wavelength is of the same order of magnitude as the dimension of the diffraction grating. In the case of electrons, the required grating is of the order of the lattice distance of crystals.

Historically, it is worth mentioning that the experimenters Davisson and Germer had done some of these experiments even before De Broglie's hypothesis was established, but they could not interpret their results. Let us try to cast this hypothesis into a more mathematical form. The simplest example of a wave is given in the form

$$\exp[ikx - i\omega t] \tag{3.6}$$

as we have seen before in chapter 2. Invoking the superposition principle for them we can immediately explain diffraction experiments. As we know from many other examples in physics, expressions describing waves are always solutions of certain wave equations. This suggests that we should look for an equation to which (3.6) is a solution. Since we deal with free particles we have to require that energy and momentum are connected by the relation

$$W = p^2/(2m). \tag{3.7}$$

On the other hand, W and p can be expressed by ω and k according to

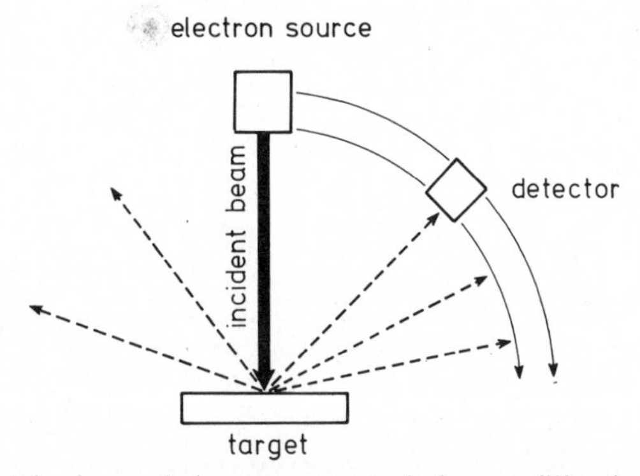

Fig. 3.1. Experimental scheme of the arrangement of electron diffraction experiment by Davisson and Germer. When the detector is put in different positions as indicated in this figure, diffraction minima and maxima are found.

(3.4) and (3.5). Thus as a result of (3.7), a relation between k and ω follows. When we wish to derive a general wave equation we have to secure such a relation. As we know from wave equations they contain differentiations with respect to space and time. Therefore we are led to differentiate the wave function (3.6) with respect to time. By multiplying the result with $\mathrm{i}\hbar$ we obtain

$$\mathrm{i}\hbar\frac{\partial}{\partial t}\exp[\mathrm{i}kx - \mathrm{i}\omega t] = \hbar\omega\exp[\mathrm{i}kx - \mathrm{i}\omega t] \tag{3.8}$$

and according to (3.4)

$$\mathrm{i}\hbar\frac{\partial}{\partial t}\exp[\mathrm{i}kx - \mathrm{i}\omega t] = W\exp[\mathrm{i}kx - \mathrm{i}\omega t]. \tag{3.9}$$

Of course, we have multiplied (3.8) with $\mathrm{i}\hbar$ to obtain just the energy W. We proceed in a similar way to obtain the expression $p^2/2m$ which occurs on the right hand side of eq. (3.7). Differentiating the exponential function (3.6) twice with respect to x we obtain the factor $(-k^2)$. To obtain p^2 and eventually $p^2/2m$ we have still to multiply $(-k^2)$ by $-\hbar^2/2m$. This yields the following equations

$$-\frac{\hbar^2}{2m}\frac{\partial^2}{\partial x^2}\exp[\mathrm{i}kx - \mathrm{i}\omega t]$$
$$= \frac{\hbar^2k^2}{2m}\exp[\mathrm{i}kx - \mathrm{i}\omega t] = \frac{p^2}{2m}\exp[\mathrm{i}kx - \mathrm{i}\omega t]. \tag{3.10}$$

By equating (3.9) and the right-hand side of (3.10) and dividing both sides by (3.6), we obtain the relation (3.7) as we had wished it. On the other hand, the expressions in (3.10) were obtained in a general way by differentiation processes so that we might equally well equate the left-hand side of (3.8) and the left-hand side of (3.10). Writing more generally instead of the known function (3.6) a wave function $\psi(x,t)$, which is still to be determined, we find eq.

$$-\frac{\hbar^2}{2m}\frac{\partial^2}{\partial x^2}\psi(x,t) = \mathrm{i}\hbar\frac{\partial}{\partial t}\psi(x,t). \tag{3.11}$$

This is indeed the famous Schrödinger equation for the special case of a freely moving particle. Of course, to some readers this derivation of the Schrödinger equation may look somewhat heuristic. The reader should keep in mind, however, that all fundamental equations of physics can be derived only in an inductive way. On account of more or less evident facts one has to develop a certain hypothesis and to cast it in mathematical form. Then one has to draw as many conclusions of this formulation as possible and to compare them with experimental results. In the case of the Schrödinger equation and the more general Schrödinger equation of a particle under the influence of external fields these predictions have been substantiated very thoroughly. One limitation should be observed, however: the Schrödinger equation refers to particles moving in the nonrelativistic domain. We will use the Schrödinger equation in our book because most of the important effects in lasers and non-linear optics can be treated by it. For the fully relativistic case, the Dirac equation must be considered.

Within the heuristic spirit of our derivation it is quite simple to derive a Schrödinger equation for a particle moving in a force field. We assume that this force field can be derived from a potential V. When the force is conservative, in classical physics the energy conservation law holds. We introduce the momentum p which is connected with the velocity $\mathrm{d}x/\mathrm{d}t$ of a particle by $p = m\,\mathrm{d}x/\mathrm{d}t$. The sum of kinetic and potential energy then yields the constant energy W

$$p^2/(2m) + V = W. \tag{3.12}$$

The expression on the left-hand side is known in mechanics as the Hamiltonian

$$H = p^2/(2m) + V. \tag{3.13}$$

When V is constant, the relation (3.12) can be obtained from a Schrödinger equation by adding the expression

$$V\psi \tag{3.14}$$

to the left-hand side of (3.11). It has turned out that this Schrödinger equation remains valid even when V depends on space x and time t. This yields the Schrödinger equation

$$\left[-\frac{\hbar^2}{2m}\frac{\partial^2}{\partial x^2} + V(x)\right]\psi(x,t) = \mathrm{i}\hbar\frac{\partial}{\partial t}\psi(x,t). \tag{3.15}$$

In general we have to deal not only with one-dimensional but three-dimensional motions. To extend the Schrödinger equation (3.15) to this case we write (3.7) in the form

$$W = \left(p_x^2 + p_y^2 + p_z^2\right)/(2m). \tag{3.16}$$

The sum over the squares of the momentum components is to be replaced by a sum over second derivatives with respect to x, y, z in order to obtain the correct wave equation

$$\frac{\partial^2}{\partial x^2} \Rightarrow \frac{\partial^2}{\partial x^2} + \frac{\partial^2}{\partial y^2} + \frac{\partial^2}{\partial z^2} \equiv \Delta. \tag{3.17}$$

The Δ symbol is called Laplace-operator. Often the resulting Schrödinger equation is written in a somewhat formal way

$$H\psi = \mathrm{i}\hbar \frac{\partial}{\partial t}\psi. \tag{3.18}$$

In it H is the Hamiltonian

$$H = -\frac{\hbar^2}{2m}\Delta + V(\boldsymbol{x}). \tag{3.19}$$

In writing the Schrödinger equation in the form of eq. (3.18), (3.19) must be understood in such a way that V stemming from H must be multiplied with ψ and that in addition the Laplace-operator acts on ψ.

Provided V is time independent we can derive the time independent Schrödinger equation. In such a case the expression in brackets on the l.h.s. of eq. (3.15) is an operator containing x and acting only on x, while the r.h.s. contains the operator $\partial/\partial t$ acting only on t. As is shown in mathematics in such a case one may find the solutions $\psi(x,t)$ in the form of a product,

$$\psi(x,t) = f(t)\psi(x) \tag{3.20}$$

where f depends only on t and $\psi(x)$ on the r.h.s. of (3.20) only on x. After inserting this ansatz into (3.15), we can split the resulting equation into two equations, one for $f(t)$ alone, and one for $\psi(x)$ alone. The equation for $f(t)$ reads $i\hbar(\partial f/\partial t) = Wf$, where W is a still undetermined constant (cf. exercise 4 on this section). The solution of this equation reads

$$f = \exp[-\mathrm{i}Wt/\hbar].$$

The same procedure can be applied to the Schrödinger equation (3.18) containing the Hamiltonian (3.19). We therefore try the ansatz

$$\psi(\boldsymbol{x},t) = \exp(-\mathrm{i}Wt/\hbar)\psi(\boldsymbol{x}). \tag{3.20a}$$

Since the differentiation, with respect to time, acts on the r.h.s. of (3.18) only on the exponential function we obtain on the r.h.s. of (3.18) the factor W. Then on both sides of (3.18) we are left with the exponential factors $\exp(-\mathrm{i}Wt/\hbar)$. Dividing both sides by this factor leaves us with the time independent Schrödinger equation

$$H\psi = W\psi. \tag{3.21}$$

Since H has the dimension of energy, W must have this dimension also. We will get acquainted with solutions of equations (3.21) in section 3.3 and we shall present further examples in sections 3.4 and 3.7. As we will discover we find solutions of the Schrödinger equation (3.21) which correspond to closed orbits in classical mechanics. The basic new features in the wave theory of matter will be that for bound states we obtain a discrete sequence of energy levels W quite in agreement with Planck's original hypothesis. We shall distinguish these discrete energy levels by indices and write $W_1, W_2, \ldots$. To these energy levels there belong wave functions which we denote correspondingly by indices $\psi_1, \psi_2, \ldots$.

Since we deal with mechanics which treats the motion of particles and since this motion is quantized we call this theory "quantum mechanics". For its understanding the correct interpretation of the wave function ψ is crucial. In contrast to conventional wave theories it has turned out that ψ can be correctly interpreted only in a statistical sense. According to this interpretation $|\psi(\boldsymbol{x})|^2 \, \mathrm{d}V$ gives the probability of finding the particle (electron) around the space point $\boldsymbol{x}$ in the volume $\mathrm{d}V$ (fig. 3.2).

The reason for this interpretation can easily be seen. Consider an experiment analogous to Young's double slit experiment (cf. section 1.10), but now done with an electron beam. Or, more realistically, we may think of the electron diffraction experiment by Davisson and Germer which we discussed at the beginning of this section. Let us assume that we try to measure the scattered electrons by a set of individual localized electron counters which we have arranged around the sample (similarly to fig. 3.1). When we lower the intensity of the electron beam so much that only one electron is present at a time, the following happens: Only one of the counters clicks, and we are not able to predict which one. However, when we repeat this experiment very often and plot the number of events versus the position of the individual counters, we obtain a curve which coincides with the one given by $|\psi(\boldsymbol{x})|^2$, except for a constant proportionality factor. ψ is the wave-function determined by the Schrödinger equation. Since we can cover all positions by the counters by shifting them, $\boldsymbol{x}$ is indeed the continuous space variable. Clearly, $|\psi(\boldsymbol{x})|^2 \, \mathrm{d}V$ is a measure for the probability to find the electron around the space point $\boldsymbol{x}$ in the volume $\mathrm{d}V$, as we have just postulated it.

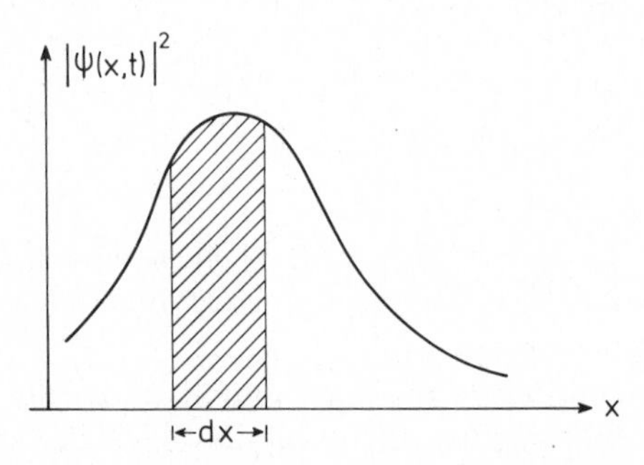

Fig. 3.2. $|\psi(x,t)|^2$ is plotted versus x. The shaded area shows the probability $|\psi(x,t)|^2 \, \mathrm{d}x$, i.e. the probability of finding the electron within the interval $x \ldots x + \mathrm{d}x$.

As we will find again and again in this book, the statistical interpretation of the wave function (and also of other quantities which we will define below) is the price we have to pay for the reconciliation of the wave and particle pictures.

The interpretation of $|\psi|^2 \mathrm{d}V$ as a probability implies a normalization condition. According to fundamental postulates of probability theory the sum (or the integral) over the individual probabilities to find the particle at a certain point $\boldsymbol{x}$ must be normalized to unity. We shall use this normalization condition

$$\int |\psi(\boldsymbol{x})|^2 \mathrm{d}V = 1 \tag{3.22}$$

henceforth. Furthermore one may show that wave functions with different indices are orthogonal to each other, i.e. they obey the relations

$$\int \psi_m^*(\boldsymbol{x})\psi_n(\boldsymbol{x})\,\mathrm{d}V = \begin{cases} 1, & \text{for } m = n \\ 0, & \text{for } m \neq n. \end{cases} \tag{3.23}$$

Here we have included a case $m = n$ which corresponds to (3.22). This orthogonality property (3.23) is quite analogous to the orthogonality of 2 vectors whose scalar product vanishes.

Exercises on section 3.1

(1) Determine the normalized solutions $\varphi(x)$ of the time-independent Schrödinger equation with

$$H = -\frac{\hbar^2}{2m}\frac{\partial^2}{\partial x^2} \qquad (W \text{ real!})$$

where $\varphi(x)$ obeys the periodic boundary condition. L is length of periodicity interval.

$$\varphi(x + L) = \varphi(x).$$

Show that

$$\int_0^L \varphi_n^*(x)\varphi_m(x)\,\mathrm{d}x = \delta_{n,m}$$

where

$$\varphi_n(x) = \frac{1}{\sqrt{L}}\exp(\mathrm{i}kx), \qquad k = \frac{2\pi n}{L}, \qquad n = \text{integer}.$$

(2) Treat the problem of exercise 1 in three dimensions, where

$$\varphi(x+L,y,z)=\varphi(x,y,z)$$
$$\varphi(x,y+L,z)=\varphi(x,y,z)$$
$$\varphi(x,y,z+L)=\varphi(x,y,z).$$

(3) Treat the following one-dimensional problem. A free particle is enclosed in a box whose walls at $x = 0$ and $x = L$ cannot be penetrated by the particle. What are the energies and wave functions? Check the orthogonality of the wave functions.
Hint: Solve the free particle Schrödinger equation for $0 \leqslant x \leqslant L$ and subject the wave functions to the condition $\psi(0) = \psi(L) = 0$. Use a superposition of the form

$$\psi(x) = a\exp(\mathrm{i}kx) + b\exp(-\mathrm{i}kx).$$

(4) Show that the time-dependent Schrödinger equation (3.15) can be split into

$$\mathrm{i}\hbar\partial f/\partial t = Wf \qquad \text{and} \qquad H\psi(x) = W\psi(x)$$

by means of $\psi(x,t) = f(t)\psi(x)$.
Hint: Transform the resulting equation into

$$\frac{\mathrm{i}\hbar\partial f/\partial t}{f(t)} = \frac{H\psi(x)}{\psi(x)}. \qquad (*)$$

Since the l.h.s. depends only on t, the r.h.s. only on x, but x and t are independent variables, $(*)$ implies that each side can be equal only to a constant, which we call W.

(5) Show that

$$\psi(x,t) = \sum_{n=1}^{\infty} c_n \exp(-\mathrm{i}W_n t/\hbar)\psi_n(x) \qquad (*)$$

where the c_n's are constant coefficients and the ψ_n's are solutions of the time-independent Schrödinger equation (3.21), i.e.

$$H\psi_n = W_n\psi_n \qquad (**)$$

is a solution of the time dependent Schrödinger equation (3.18).
Hint: Insert $(*)$ into (3.18) and use $(**)$.

3.2. Measurements in quantum mechanics and expectation values

The statistical interpretation of the wave function has many important consequences and the reader is well advised to recall this interpretation again and again into his memory. When we measure the space coordinate of a particle it can be found at that point or another one. This is strongly reminiscent of rolling dice. When we wish to make predictions when rolling dice or on other stochastic processes we can make such predictions only with respect to probabilities or to average values. For instance, if we throw a die very many times, we can determine the mean number of spots. Let us denote the probability of obtaining a certain number of spots n when throwing a die once by P_n. Then the mean number of spots is defined by the sum over the number of spots $n = 1,\ldots,6$ multiplied by the probability P_n

$$\bar{n} = \sum_n nP_n. \tag{3.24}$$

This average is strongly reminiscent of the calculation of the center of gravity in mechanics, n corresponding to the coordinate and P_n to the masses. In a similar interpretation (3.24) means nothing but weighting the individual measured values n.

Let us apply this idea to quantum mechanics and let us start with measuring the particle coordinate. The probability of finding a particle at space point $\boldsymbol{x}$ in the volume $\mathrm{d}V$ was given by $|\psi(\boldsymbol{x})|^2\,\mathrm{d}V$. This expression plays the same role as P_n in the case of dice. The measured space coordinate $\boldsymbol{x}$ of the particle now corresponds to the number of spots n. This allows for direct translation of (3.24) and thus for a definition of the average value of the space coordinate $\boldsymbol{x}$ of a "particle" in quantum mechanics.

$$\bar{\boldsymbol{x}} = \int \boldsymbol{x}|\psi(\boldsymbol{x})|^2\,\mathrm{d}V. \tag{3.25}$$

The statistical interpretation of quantum mechanics holds not only for the space coordinate but equally well for the momentum. The rule for calculating the average value of the momentum reads

$$\bar{p}_x = \int \psi^*(\boldsymbol{x})(\hbar/\mathrm{i})\frac{\partial}{\partial x}\psi(\boldsymbol{x})\,\mathrm{d}V. \tag{3.26}$$

For its derivation we must refer the reader to the usual textbooks in quantum mechanics.

Quite generally we can formulate rules for the transition from classical mechanics to quantum mechanics. In classical mechanics we are dealing with certain measurable quantities, often called observables, such as coordinate or momentum of a particle or its energy, etc. As we have seen in section 3.1 and here again, operators are attributed to these observables. These quantities are called operators because they operate in a well defined way on the wave functions. For instance, the space operator x multiples the wave function by it. The momentum operator $(\hbar/\mathrm{i})\,\mathrm{d}/\mathrm{d}x$ differentiates the subsequent wave function, etc. In order to be able to make predictions about the outcome of measurements one has to form average (mean) values by the prescription (3.25) and (3.26).

These mean values make a prediction about the experimental average values when the experiment is repeated very often. Therefore these mean values are also called expectation values. Let us supplement this scheme by the example of the kinetic or the potential energy. This leads us to table 3.1.

Table 3.1

One-dimensional case

Observable	Operator	Expectation values
Space coordinate x	x	$\int \psi^*(x) x \psi(x)\,\mathrm{d}V$
Momentum p_x	$\frac{\hbar}{i}\frac{\partial}{\partial x}$	$\int \psi^*(x) \frac{\hbar}{i}\frac{\partial}{\partial x}\psi(x)\,\mathrm{d}V$
Kinetic energy $p^2/2m$	$-\frac{\hbar^2}{2m}\frac{\partial^2}{\partial x^2}$	$\int \psi^*(x) \frac{-\hbar^2}{2m}\frac{\partial^2}{\partial x^2}\psi(x)\,\mathrm{d}V$
Potential energy $V(x)$	$V(x)$	$\int \psi^*(x) V(x) \psi(x)\,\mathrm{d}V$
Total energy $p^2/2m + V(x)$	$\frac{-\hbar^2}{2m}\frac{\partial^2}{\partial x^2} + V(x)$	$\int \psi^*(x)\{\frac{-\hbar^2}{2m}\frac{\partial^2}{\partial x^2} + V(x)\}\psi(x)\,\mathrm{d}V$

3.2.1. Dirac notation

In the more modern literature, both in texts and original scientific publications, wave functions and expectation values are often represented in still another way. As we will see in the next sections we have not only to deal with expressions of the form

$$\int \psi^*(x)x\psi(x)\,\mathrm{d}V \tag{3.27a}$$

but also with expressions of the form

$$\int \psi_m^*(x)x\psi_n(x)\,\mathrm{d}V \tag{3.27}$$

and similar expressions containing other operators Ω instead of x

$$\int \psi_m^*(x)\Omega\psi_n(x)\,\mathrm{d}V. \tag{3.28}$$

The other notation mentioned above reads

$$\int \psi_m^*(x)\Omega\psi_n(x)\,\mathrm{d}V = \langle m|\Omega|n\rangle. \tag{3.29}$$

The English physicist Dirac has split this expression $\langle m|\Omega|n\rangle$ into $\langle m|$ and $|n\rangle$. Making a pun he also split the word "bracket" into "bra" and "ket" which led him to the bra-ket notation

$$\begin{array}{ccc} \text{bra} & \text{and} & \text{ket} \\ \updownarrow & & \updownarrow \\ \langle m| & & |n\rangle. \end{array} \tag{3.30}$$

In order to go over from Schrödinger's notation to Dirac's notation one has to make the replacements*

$$\begin{gathered} \psi_n(x) \Rightarrow |n\rangle \\ \int \mathrm{d}V\psi_m^*(x) \Rightarrow \langle m| \\ \Omega \Rightarrow \Omega. \end{gathered} \tag{3.31}$$

Another common abbreviation for (3.29) is Ω_{mn}. When we let run m and n from one to N we may arrange the Ω_{mn}'s in a square array or, in other words, in the matrix

$$(\Omega_{mn}) = \begin{bmatrix} \Omega_{11} & \Omega_{12}\cdots & \cdots\Omega_{1N} \\ \Omega_{21} & \Omega_{22}\cdots & \cdots\Omega_{2N} \\ \vdots & & \vdots \\ \vdots & & \vdots \\ \Omega_{N1} & \Omega_{N2}\cdots & \cdots\Omega_{NN} \end{bmatrix}. \tag{3.32}$$

*Still, another notation is $\int \varphi^*(x)\Omega\chi(x)\,\mathrm{d}V = \langle\varphi|\Omega|\chi\rangle$.

Ω_{mn} are therefore called matrix elements. For later use we need the notion of a Hermitian matrix and a Hermitian operator. A matrix (Ω_{mn}) is called Hermitian, if its elements fulfill the relation

$$\Omega_{nm} = \Omega^*_{mn}, \tag{3.32a}$$

i.e. we obtain the conjugate complex of Ω_{mn} when we exchange the indices m and n. According to (3.29) an operator Ω can be connected with $\Omega_{mn} \equiv \langle m|\Omega|n\rangle$. An operator Ω whose matrix elements fulfill (3.32b), is called a Hermitian operator.

In the final part of this section we want to complete our presentation of the formal frame of quantum mechanics. This part is somewhat more difficult and can be skipped until the reader begins chapter 5 and the later chapters.

3.2.2. Equations for the determination of the wave function ψ. The measuring operator

Our considerations of the present section were based on the assumption that the wave function $\psi(\boldsymbol{x})$ is already given. In section 3.1 we were acquainted with the Schrödinger equation as an equation to determine $\psi(\boldsymbol{x})$. We now want to show that aside from the Schrödinger equation there are other equations by which we can determine $\psi(\boldsymbol{x})$. (For simplicity we again consider the one-dimensional case.) In quantum theory, the space-dependent part of the wave function describing the motion of a free particle with momentum $\hbar k$ is given by

$$\exp[\mathrm{i}kx]. \tag{3.33}$$

On the other hand, we just came across (compare table 3.1) the operator which corresponds to the momentum p_x, namely $(\hbar/\mathrm{i})\,\mathrm{d}/\mathrm{d}x$. When we apply this operator to the wave function (3.33) we readily obtain

$$\frac{\hbar}{\mathrm{i}}\frac{\mathrm{d}}{\mathrm{d}x}\exp[\mathrm{i}kx] = \hbar k \exp[\mathrm{i}kx]. \tag{3.34}$$

Similarly when we apply the operator of the kinetic energy to the same wave function we obtain

$$-\frac{\hbar^2}{2m}\frac{\mathrm{d}^2}{\mathrm{d}x^2}\exp[\mathrm{i}kx] = \frac{\hbar^2k^2}{2m}\exp[\mathrm{i}kx]. \tag{3.35}$$

The factor in front of the r.h.s. is just the classical kinetic energy

$$p^2/2m. \tag{3.36}$$

From these cases we learn that there are operators Ω (for instance $\Omega = (\hbar/\mathrm{i})\,\mathrm{d}/\mathrm{d}x$) which, when applied to a wave function ψ, yield the wave function multiplied by the measured value, for instance $\hbar k$. Thus we are led to consider equations of the form

$$\Omega\psi_l(x) = \omega_l\psi_l(x) \tag{3.37*}$$

where the index l distinguishes the different functions ψ which obey equation (3.37). The constants ω_l are called eigenvalues. According to the table 3.1, we may attribute a quantum mechanical operator which is called Ω to each classical observable. As is shown in quantum mechanics, these operators Ω are Hermitian. According to mathematical theorems, the eq. (3.37) allows for a complete set of eigenfunctions ψ_l with real eigenvalues ω_l. According to the fundamental postulate of quantum mechanics the eigenvalues ω_l are precisely those values which can be measured experimentally. For instance when we identify Ω with the momentum operator, the ω_l's are just the values of the momentum we can measure, namely $\hbar k$. Similarly when we use the Hamiltonian (3.19), i.e. the energy operator, the ω_l's are just the quantized energies W_l we can measure.

A fundamental problem in quantum mechanics is whether we can measure two different physical quantities simultaneously with absolute precision. We have seen in the introductory chapter (namely section 1.9) that in general by measuring a certain quantity, say the location of the particle, we simultaneously make it impossible to measure the momentum precisely. We now want to derive a criterium which allows us to decide whether we can measure two observables simultaneously with absolute precision. To this end we introduce the measuring operators $\Omega^{(1)}$ and $\Omega^{(2)}$ just corresponding to these two observables. According to the fundamental postulate just introduced above we must require that the same wave function ψ fulfills the two "measuring equations"

$$\Omega^{(1)}\psi(x) = \omega^{(1)}\psi(x) \tag{3.38}$$

and

$$\Omega^{(2)}\psi(x) = \omega^{(2)}\psi(x). \tag{3.39}$$

We can easily find a condition for $\Omega^{(1)}$ and $\Omega^{(2)}$ so that the simultaneous solution of (3.38) and (3.39) is possible. To this end we let the operator $\Omega^{(2)}$ act on both sides from the left on eq. (3.38) and similarly the operator $\Omega^{(1)}$ on both sides of eq. (3.39). Then in the equation resulting from (3.38) we make again use of eq. (3.39) and vice versa. Subtracting the two resulting

*It is shown in mathematics that these equations must be supplemented by appropriate boundary conditions on ψ. In our following discussion it is assumed that those conditions are defined, and fulfilled by ψ_l. For explicit examples consult the exercises at the end of the section.

equations from each other we immediately find the following equation

$$(\Omega^{(2)}\Omega^{(1)} - \Omega^{(1)}\Omega^{(2)})\psi = (\omega^{(2)}\omega^{(1)} - \omega^{(1)}\omega^{(2)})\psi = 0. \tag{3.40}$$

Since the left-hand side of (3.40) vanishes for a complete set of ψ's, one has introduced a shorthand writing

$$\Omega^{(1)}\Omega^{(2)} - \Omega^{(2)}\Omega^{(1)} = 0. \tag{3.41}$$

To check such a relation one has always to think that the total expression on the l.h.s. of (3.41) must be applied to an arbitrary wave function. We shall study the significance of the relation (3.41) in some exercises on this section. Let us assume that the relation (3.41) is fulfilled. Then one shows in mathematics that it is always possible to find the wave functions $\psi(\boldsymbol{x})$ so that the eqs. (3.38) and (3.39) are simultaneously fulfilled. We may summarize these results as follows: The relation (3.41) is necessary and sufficient that the observables to which $\Omega^{(1)}$ and $\Omega^{(2)}$ are attributed can be measured simultaneously with absolute precision.

3.2.3. The relation between equation (3.37) and the time-dependent Schrödinger equation

From our above considerations, it seems to follow that the Hamiltonian (3.19) plays the same role as any other measuring operator, for instance the momentum operator. This is indeed true where the equation (3.37) is concerned. For $\Omega = H$ (3.19), eq. (3.37) is just the time-independent Schrödinger equation. We must bear in mind, however, that H appears in the time-dependent Schrödinger equation also. This equation plays an extra role, because it determines in any case the temporal development of $\psi(\boldsymbol{x}, t)$. The connection of (3.37), including the time-independent Schrödinger equation, and the time-dependent Schrödinger equation can be seen as follows: We first choose an observable we want to measure, and the corresponding operator Ω. Then we solve (3.37) (or perform the measurement). We choose a certain $\psi_{l_0}(\boldsymbol{x})$ (or the wave function belonging to the measured value ω_{l_0}). This $\psi_{l_0}(\boldsymbol{x})$ serves as the initial condition for the solution of the time-dependent Schrödinger equation at the initial time t_0 (= time of measurement) so that

$$\psi(\boldsymbol{x}, t_0) = \psi_{l_0}(\boldsymbol{x}).$$

Therefore, after the measurement was made, the further time-evolution of the wave-function is uniquely determined by the time-dependent Schrödinger equation.

Exercises on section 3.2

(1) To get used to the bra and ket notation, show

$$\langle n|x|m\rangle = \langle m|x|n\rangle^* \qquad (\alpha)$$

$$\left\langle n\left|\frac{\hbar}{\mathrm{i}}\frac{\partial}{\partial x}\right|m\right\rangle = \left\langle m\left|\frac{\hbar}{\mathrm{i}}\frac{\partial}{\partial x}\right|n\right\rangle^* \qquad (\beta)$$

$$\left\langle n\left|-\frac{\hbar^2}{2m}\frac{\partial^2}{\partial x^2}\right|m\right\rangle = \left\langle m\left|-\frac{\hbar^2}{2m}\frac{\partial^2}{\partial x^2}\right|n\right\rangle^* \qquad (\gamma)$$

$$\langle n|V(\boldsymbol{x})|m\rangle = \langle m|V(\boldsymbol{x})|n\rangle^* \qquad (\delta)$$

for arbitrary $\langle n|$, $|m\rangle$.
Hints: (α) Use

$$\langle n|x|m\rangle = \int_{-\infty}^{\infty} \psi_n^*(\boldsymbol{x})x\psi_m(\boldsymbol{x})\,\mathrm{d}V.$$

(β) Use

$$\left\langle n\left|\frac{\hbar}{\mathrm{i}}\frac{\partial}{\partial x}\right|m\right\rangle = \int_{-\infty}^{\infty} \psi_n^*(\boldsymbol{x})\frac{\hbar}{\mathrm{i}}\frac{\partial}{\partial x}\psi_m(\boldsymbol{x})\,\mathrm{d}V$$

and integrate by parts. It is assumed that $|\psi_n|^2 \to 0$ for $|\boldsymbol{x}| \to \pm\infty$. ($\gamma$) analogous to ($\beta$). ($\delta$) analogous to ($\alpha$).

(2) Verify the following properties of bras and kets.
Let $\varphi = c\psi$, where c is a (complex) constant, and Ω an arbitrary operator which can also be $\equiv 1$. Then

$$\langle\chi|\Omega|\varphi\rangle \equiv \langle\chi|\Omega|c\psi\rangle = c\langle\chi|\Omega|\psi\rangle$$

$$\langle\varphi|\Omega|\chi\rangle \equiv \langle c\psi|\Omega|\chi\rangle = c^*\langle\psi|\Omega|\chi\rangle.$$

Let $\varphi = c_1\psi_1 + c_2\psi_2$, then

$$\langle\chi|\Omega|\varphi\rangle \equiv \langle\chi|\Omega|c_1\psi_1 + c_2\psi_2\rangle = c_1\langle\chi|\Omega|\psi_1\rangle + c_2\langle\chi|\Omega|\psi_2\rangle$$

and

$$\langle\varphi|\Omega|\chi\rangle \equiv \langle c_1\psi_1 + c_2\psi_2|\Omega|\chi\rangle = c_1^*\langle\psi_1|\Omega|\chi\rangle + c_2^*\langle\psi_2|\Omega|\chi\rangle.$$

Hint: Use each time the integral defining $\langle\varphi|\Omega|\chi\rangle$, that is $\langle\varphi|\Omega|\chi\rangle = \int\varphi^*\Omega\chi\,\mathrm{d}V$ (cf. footnote on page 73).

(3) Calculate

$$\int_0^L |\psi_n(x)|^2 x \, \mathrm{d}x$$

$$\int_0^L \psi_n^*(x) \frac{\hbar}{\mathrm{i}} \frac{\partial}{\partial x} \psi_n(x) \, \mathrm{d}x$$

$$\int_0^L \psi_n^*(x) \left(-\frac{\hbar^2}{2m} \frac{\partial^2}{\partial x^2} \right) \psi_n(x) \, \mathrm{d}x$$

for

$$\psi_n(x) = \frac{1}{\sqrt{L}} \exp(\mathrm{i}kx), \qquad k = \frac{2\pi n}{L}, \qquad n = 1, 2 \ldots .$$

(4) In classical mechanics the vector of angular momentum is defined by

$$\mathfrak{L}_x = yp_z - zp_y$$

$$\mathfrak{L}_y = zp_x - xp_z$$

$$\mathfrak{L}_z = xp_y - yp_x .$$

Translate these quantities into quantum mechanical operators.
Hint: Use the rule

$$\boldsymbol{p} = \frac{\hbar}{\mathrm{i}} \nabla = \frac{\hbar}{\mathrm{i}} \left(\frac{\partial}{\partial x}, \frac{\partial}{\partial y}, \frac{\partial}{\partial z} \right).$$

(5) The following exercise intends to clarify the meaning of the expansion coefficients c_n in the wave-packet of exercise 5 on section 3.1.
(a) Show that (*) of exercise 5 on section 3.1 is normalized provided

$$\sum_{n=0}^{\infty} |c_n|^2 = 1. \tag{$*$}$$

(b) Show that

$$\langle \psi | H | \psi \rangle = \sum_{n=0}^{\infty} W_n |c_n|^2 \tag{$**$}$$

(for hints see below.) This relation is extremely important, because it allows us to interpret the meaning of $|c_n|^2$. The l.h.s. is the expectation ($\equiv$ mean) value of the energy operator ($\equiv$ Hamiltonian H). On the r.h.s. is a sum over the energies W_n of the states (wave functions) ψ_n. Thus (* *) can be interpreted as

$$\overline{W} = \sum_{n=0}^{\infty} W_n |c_n|^2 .$$

Now compare the formation of this expression with (3.24). Clearly, W_n

corresponds to the number of spots that can be found in principle. But in addition $|c_n|^2$ corresponds to P_n, including the normalization condition

$$\sum_n P_n = 1.$$

(Compare (∗).) This gives us the clue to the interpretation of the meaning of $|c_n|^2$: $|c_n|^2$ is the probability to find the quantum system in the state "n" when we measure the energy.

Hints: (a) Insert (5∗ of exercise 3.1) into $\langle\psi|\psi\rangle$, multiply term by term and use

$$\langle\psi_n|\psi_m\rangle = \delta_{nm}. \tag{$***$}$$

(b) Insert (5∗ of exercise 3.1) into $\langle\psi|W|\psi\rangle$, multiply term by term, use $H\psi_n = W_n\psi_n$ and (∗ ∗ ∗).

(6) *Commutation relations*: We use the following abbreviation for the commutator of two operators Ω_1, Ω_2:

$$\Omega_1\Omega_2 - \Omega_2\Omega_1 = [\Omega_1, \Omega_2].$$

Prove

$$\left[\frac{\hbar}{\mathrm{i}}\frac{\partial}{\partial x}, -\frac{\hbar^2}{2m}\frac{\partial^2}{\partial x^2}\right] = 0$$

$$\left[\frac{\hbar}{\mathrm{i}}\frac{\partial}{\partial x}, \left(-\frac{\hbar^2}{2m}\frac{\partial^2}{\partial x^2} + V(x)\right)\right] = \frac{\hbar}{\mathrm{i}}\frac{\partial V}{\partial x}$$

$$\left[\left(-\frac{\hbar^2}{2m}\frac{\partial^2}{\partial x^2} + V(x)\right), x\right] = -\frac{\hbar^2}{m}\frac{\partial}{\partial x} \equiv \frac{\hbar}{m\mathrm{i}}p_x$$

where the abbreviation $p_x = (\hbar/\mathrm{i})\partial/\partial x$ was used.

Calculate

$$\left[-\frac{\hbar^2}{2m}\frac{\partial^2}{\partial x^2}, \left(-\frac{\hbar^2}{2m}\frac{\partial^2}{\partial x^2} + V(x)\right)\right] = ?$$

$$\left[\left(-\frac{\hbar^2}{2m}\Delta + V(\boldsymbol{x})\right), \boldsymbol{x}\right] = -\frac{\hbar^2}{m}\nabla \equiv \frac{\hbar}{m\mathrm{i}}\boldsymbol{p}.$$

Hints: Insert after [· · ·] an arbitrary wave function φ and perform the differentiations.

(7) Let Ω be a measuring operator and ψ_l the eigenfunctions of eq. (3.37). Put $\psi = \Sigma c_l\psi_l$ and show that

$$\langle\psi|\Omega|\psi\rangle = \sum |c_l|^2\omega_l.$$

Hint: Same as for exercise 5.

3.3. The harmonic oscillator

Consider the motion of a particle of mass m bound elastically to its equilibrium position (fig. 3.3). As particle coordinate q we take the displacement from the equilibrium position. Denoting the force constant by f the classical equation of motion reads

$$m\frac{d^2q}{dt^2} = -fq. \tag{3.42}$$

The particle momentum p is connected with its velocity by

$$p = m\frac{dq}{dt} \tag{3.43}$$

or solving for dq/dt the relation reads

$$\frac{d}{dt}q = p/m. \tag{3.44}$$

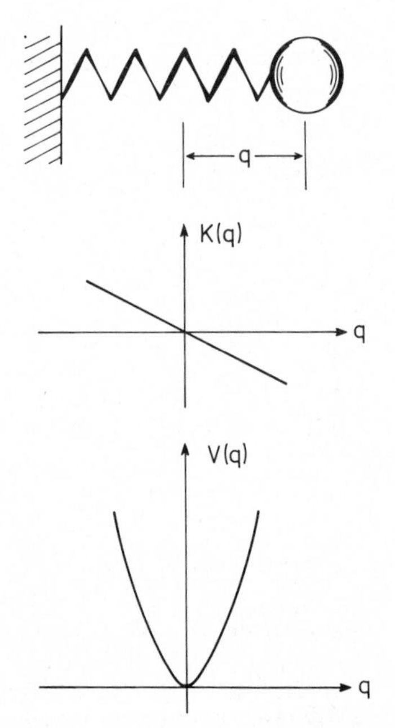

Fig. 3.3. The harmonic oscillator. *Upper part*: Mechanical model of a ball with mass m displaced by an amount q from its equilibrium position. *Middle part*: The elastic force $K(q)$ as a function of the displacement coordinate q (Hooke's law). *Lower part*: The potential energy $V(q)$ as a function of q.

Differentiating (3.43) with respect to time and using (3.42) we obtain

$$\frac{\mathrm{d}}{\mathrm{d}t}p = -f \cdot q. \tag{3.45}$$

The equations (3.44) and (3.45) can be cast in an especially elegant form using the Hamiltonian:

$$H = \frac{p^2}{2m} + \frac{f}{2}q^2. \tag{3.46}$$

The Hamiltonian can be obtained as sum of a kinetic and potential energy. The potential energy V is obtained from the force F by means of the usual relation

$$F = -\frac{\partial V}{\partial q} = -f \cdot q \tag{3.47}$$

and reads

$$V = \frac{f}{2}q^2. \tag{3.48}$$

Differentiating (3.46) with respect to p we obtain the right hand side of equation (3.44) and differentiating it with respect to q we obtain except for the minus sign the right hand side of (3.45). This allows us to write (3.44) and (3.45) in the form

$$\frac{\mathrm{d}}{\mathrm{d}t}q = \frac{\partial H}{\partial p} \tag{3.49}$$

$$\frac{\mathrm{d}}{\mathrm{d}t}p = -\frac{\partial H}{\partial q}. \tag{3.50}$$

This is an example of the Hamiltonian equations. We will make use of them when quantizing the light field. From the classical harmonic oscillator it is known that its oscillation frequency ω is connected with the force constant by

$$f = m\omega^2. \tag{3.51}$$

With (3.51) the expression (3.46) acquires the form

$$H = \frac{p^2}{2m} + \frac{m\omega^2}{2}q^2. \tag{3.52}$$

We enter the realm of quantum mechanics when we replace the momentum p by an operator

$$p = \frac{\hbar}{\mathrm{i}}\frac{\mathrm{d}}{\mathrm{d}q}. \tag{3.53}$$

p and q satisfy the commutation relation

$$pq - qp = \frac{\hbar}{\mathrm{i}}. \tag{3.54}$$

The best way to understand this relation is by thinking of both sides of (3.54) as applied to an arbitrary wave function ψ and using the explicit representation (3.53)

$$\frac{\hbar}{\mathrm{i}}\frac{\mathrm{d}}{\mathrm{d}q}(q\psi) - q\frac{\hbar}{\mathrm{i}}\frac{\mathrm{d}}{\mathrm{d}q}\psi = \frac{\hbar}{\mathrm{i}}\psi. \tag{3.55}$$

Differentiating the first term of (3.55) by the product rule, we obtain $(\hbar/\mathrm{i})\psi + (\hbar/\mathrm{i})q\,\mathrm{d}\psi/\mathrm{d}q$. By subtracting the second term as indicated in (3.55), we do, in fact obtain precisely the right hand side of (3.55). As (3.55) holds for any differentiable function, we may treat (3.54) as an identity. Although the basis of (3.54) may appear rather trivial, commutation relations are of fundamental importance to many problems in quantum optics. Substituting (3.53) into (3.52), we obtain the Hamiltonian operator of the harmonic oscillator. With it the Schrödinger equation becomes

$$\left\{-\frac{\hbar^2}{2m}\frac{\mathrm{d}^2}{\mathrm{d}q^2} + \frac{m\omega^2}{2}q^2\right\}\psi(q) = W\psi(q). \tag{3.56}$$

To treat (3.56) further, let us introduce a dimensionless coordinate ξ defined by

$$q = \sqrt{\frac{\hbar}{m\omega}}\,\xi. \tag{3.57}$$

We further put

$$\psi(q) \equiv \psi\left(\sqrt{\frac{\hbar}{m\omega}}\,\xi\right) = \varphi(\xi) \tag{3.57a}$$

so that the Schrödinger equation (3.56) takes the form

$$\frac{\hbar\omega}{2}\left\{-\frac{\mathrm{d}^2}{\mathrm{d}\xi^2} + \xi^2\right\}\varphi(\xi) = W\varphi(\xi). \tag{3.58}$$

For later use, we write its Hamiltonian in the form

$$\tfrac{1}{2}\hbar\omega(\pi^2 + \xi^2) \tag{3.59}$$

where π has the meaning of a momentum. If we were now able to treat the operators ξ and $\mathrm{d}/\mathrm{d}\xi$ as ordinary numbers, it might well appear that the operator on the left hand side of (3.58)

$$\left(-\frac{d^2}{d\xi^2}+\xi^2\right)$$

could be regarded as an expression of the type

$$(-\alpha^2+\beta^2) \tag{3.60}$$

which could then be factorized in the form

$$(-\alpha+\beta)(\alpha+\beta). \tag{3.61}$$

Let us therefore, just as an experiment, replace the left-hand side of (3.58) by

$$\hbar\omega\frac{1}{\sqrt{2}}\left(-\frac{d}{d\xi}+\xi\right)\left(\frac{d}{d\xi}+\xi\right)\frac{1}{\sqrt{2}}\varphi. \tag{3.62}$$

By multiplying out the brackets, but taking care to keep the operators ξ and $d/d\xi$ in the right order, we obtain

$$(3.62)=\underbrace{\frac{\hbar\omega}{2}\left\{-\frac{d^2}{d\xi^2}+\xi^2\right\}\varphi(\xi)}_{\text{I}}\underbrace{-\frac{\hbar\omega}{2}\left\{\frac{d}{d\xi}\xi-\xi\frac{d}{d\xi}\right\}\varphi.}_{\text{II}}$$

Here the first expression is the same as the left-hand side of (3.58) as we should wish. By the commutation relation

$$\frac{d}{d\xi}\xi-\xi\frac{d}{d\xi}=1 \tag{3.63}$$

the second term reduces to

$$\text{II}=-\frac{\hbar\omega}{2}\varphi. \tag{3.64}$$

[The commutation relation (3.63) was derived directly from the earlier one (3.54) by substituting ξ for q using (3.57).] The expressions in brackets (3.62) are again operators; let us write them at first purely formally as follows:

$$\frac{1}{\sqrt{2}}\left(-\frac{d}{d\xi}+\xi\right)=b^+ \tag{3.65}$$

$$\frac{1}{\sqrt{2}}\left(\frac{d}{d\xi}+\xi\right)=b. \tag{3.66}$$

As (3.62) and the left-hand side of (3.58) differ by the term (3.64), we introduce the displaced energy

$$W' = W - \tfrac{1}{2}\hbar\omega. \tag{3.67}$$

Then the Schrödinger equation (3.58) can finally be replaced by

$$\hbar\omega b^+ b\varphi = W'\varphi. \tag{3.68}$$

To derive a commutation relation for b and b^+, let us construct the term $bb^+ - b^+ b$ and substitute the operators according to (3.65) and (3.66). Using (3.63), we then obtain the basic commutation relation

$$bb^+ - b^+ b = 1. \tag{3.69}$$

For reasons which will emerge later, we will call operators satisfying condition (3.69) Bose operators.

We will now show that, with the aid of operators b^+, b and commutation relation (3.69), we can construct eigenstates of the harmonic oscillator. Here we start with the assumption (compare exercise 2 at the end of this section) that the energy W of the quantum mechanical oscillator is always positive, and certainly has a lower bound. Let us denote the state with the lowest energy value W_0' by φ_0. When multiplying equation

$$\hbar\omega b^+ b\varphi_0 = W_0'\varphi_0 \tag{3.70}$$

from the left by the operator b, we obtain

$$\hbar\omega(bb^+)b\varphi_0 = W_0' b\varphi_0. \tag{3.71}$$

By using the commutation relation (3.69), let us replace bb^+ by $(1 + b^+ b)$ on the left-hand side of (3.71), so that

$$\hbar\omega\{1 + b^+ b\}b\varphi_0 = W_0' b\varphi_0. \tag{3.72}$$

Finally, taking $\hbar\omega b\varphi_0$ to the right hand side, we obtain

$$\hbar\omega b^+ b(b\varphi_0) = (W_0' - \hbar\omega)b\varphi_0. \tag{3.73}$$

However this equation shows that $b\varphi_0$ is a new eigenfunction with the eigenvalue $W_0' - \hbar\omega$, contrary to our assumption that φ_0 is the lowest state. This contradiction can be resolved only if

$$b\varphi_0 = 0. \tag{3.74}$$

This equation will be used from now on to define the ground state φ_0.

Let us try to solve (3.68) using nothing but algebra. With this aim in mind, let us multiply equation (3.68) by b^+ from the left:

$$b^+(\hbar\omega b^+ b)\varphi = W' b^+\varphi. \tag{3.75}$$

Using commutation relation (3.69) and the steps just indicated, we obtain

$$\hbar\omega b^+ b(b^+\varphi) = (W' + \hbar\omega)b^+\varphi. \tag{3.76}$$

Clearly, if φ is an eigenfunction of (3.68), $b^+ \varphi$ is also an eigenfunction, and its eigenvalue is larger than the eigenvalue of the former function by an amount $\hbar\omega$. By applying the operator b^+ n times, we obtain the nth excited eigenfunction

$$\varphi_n = (b^+)^n \varphi_0. \tag{3.77}$$

Since equation (3.68) is homogeneous in φ, there remains, as usual with eigenfunctions, a free constant coefficient. We will choose its value to normalize φ_n and from now on we will define normalization in the dimensionless coordinate ξ by

$$\int_{-\infty}^{\infty} \varphi_n^*(\xi)\varphi_n(\xi)\,\mathrm{d}\xi = 1.$$

As we shall show below, the normalization factor can be determined purely algebraically. For the moment, let us anticipate the result, and write the normalized eigenfunction in the form

$$\varphi_n = \frac{1}{\sqrt{n!}}(b^+)^n \varphi_0. \tag{3.78}$$

What are the corresponding eigenvalues? To obtain the value for $n = 0$, we multiply (3.74) by b^+ and compare this with the general equation (3.68). Obviously $W_0' = 0$. As n-fold application of b^+ to φ_0 increases the energy W_0' n times by $\hbar\omega$, the energy value corresponding to (3.78) is

$$W' = n\hbar\omega \tag{3.79}$$

or, in terms of the original energy scale on which (3.46) was based (cf. fig. 3.4),

$$W = \hbar\omega\left(n + \tfrac{1}{2}\right). \tag{3.80}$$

(3.79) may be interpreted such that in the nth state there are n energy quanta of magnitude $\hbar\omega$. By applying b^+ to φ_n, the number of quanta is increased by one, i.e. one additional quantum is created, and therefore b^+ is called a creation operator. When b is applied to φ [see (3.71) to (3.73)] a quantum is annihilated, so that b is called annihilation operator (see fig. 3.5).

The formalism of the creation and annihilation operators b^+, b will be of fundamental importance when we quantize the light field. In the context of finding an explicit example for wave functions and energy levels we now want to show that we can easily derive the wave functions explicitly. To this end we start from equation (3.74)

$$b\varphi_0 = 0 \tag{3.81}$$

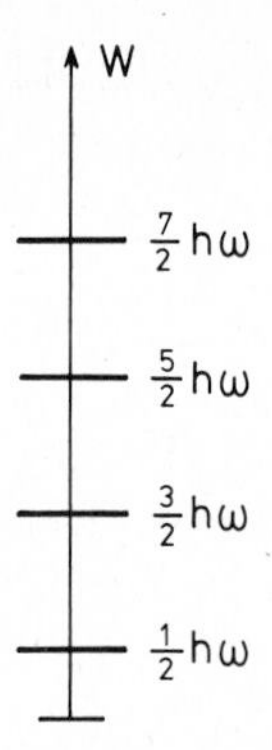

Fig. 3.4. The energy levels of the quantum mechanical harmonic oscillator.

and insert the explicit expression for b into it

$$\frac{1}{\sqrt{2}}\left(\frac{d}{d\xi}+\xi\right)\varphi_0=0. \tag{3.82}$$

This is a first-order differential equation. As one verifies immediately its solution reads

$$\varphi_0=\mathcal{N}\exp\left[-\xi^2/2\right]. \tag{3.83}$$

The factor $\mathcal{N}$ serves to normalize this function and is given by

$$\mathcal{N}=(\pi)^{-1/4}. \tag{3.84}$$

Thus our procedure allows us to derive simply and explicitly the wave function of the groundstate. The next excited wave function can be

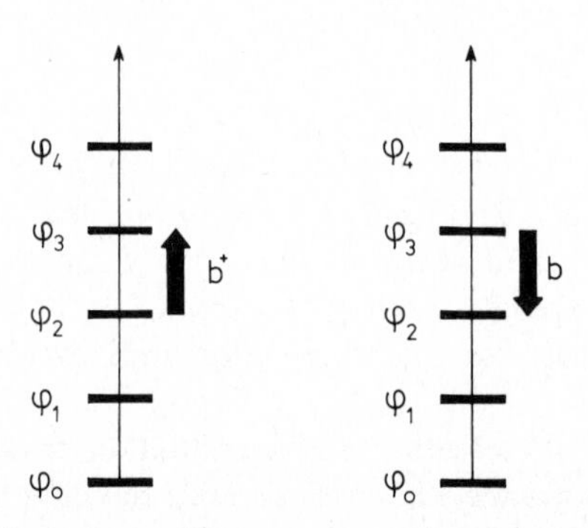

Fig. 3.5. *Left*: The creation operator b^+ causes a transition one step up in the ladder of the energy levels with their corresponding wave functions. *Right*: The annihilation operator b causes a transition one step downwards on the ladder of the energy levels with their corresponding wave functions.

obtained according to (3.78) by applying b^+ on φ_0. We thus obtain

$$\varphi_1 = b^+ \varphi_0 = \frac{1}{\sqrt{2}}\left(-\frac{d}{d\xi} + \xi\right)\mathfrak{N}\exp[-\xi^2/2] \tag{3.85}$$

or after performing the differentiation

$$\varphi_1 = \sqrt{2}\,\mathfrak{N}\xi\exp[-\xi^2/2]. \tag{3.86}$$

The normalization factor is again given by (3.84). This procedure can be continued in an explicit way. The next example reads

$$\varphi_2 = \frac{\mathfrak{N}}{\sqrt{2}}(2\xi^2 - 1)\exp[-\xi^2/2]. \tag{3.87}$$

The first 4 wave functions of the harmonic oscillator are exhibited in fig. 3.6. Quite generally, when we continue our procedure, we find φ_n in the general form

$$\varphi_n = H_n(\xi)\exp[-\xi^2/2]. \tag{3.88}$$

H_n is called a Hermitian polynomial.

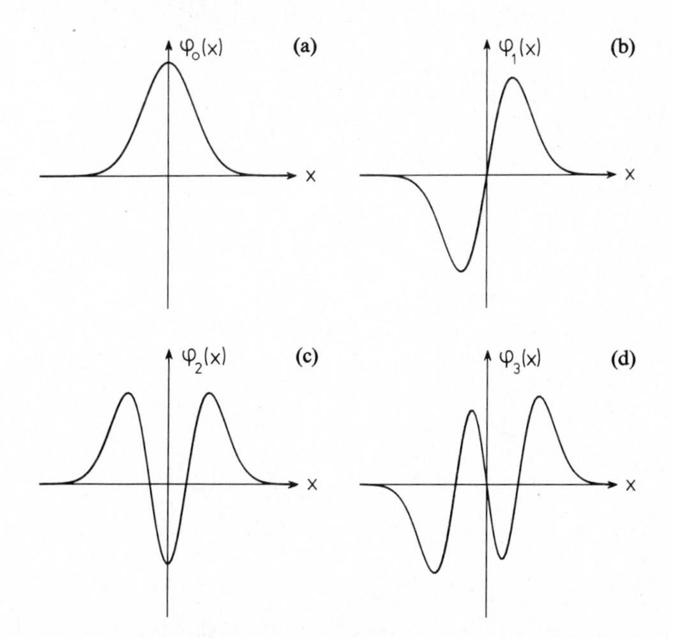

Fig. 3.6.a–d. *Upper left*: The oscillator wave function of the ground state $n = 0$. *Upper right*: The first excited state, $n = 1$. *Lower left*: The second excited state, $n = 2$. *Lower right*: The third excited state, $n = 3$. Note the increasing number of nodes with increasing n.

The following exercises are indispensable for an understanding of many aspects of quantum optics as will transpire in the course of this book. Therefore the reader is advised to pay particular attention to these exercises.

Exercises on section 3.3

(1) Prove $bb^+ - b^+ b = 1$.
Hint: Write $(bb^+ - b^+ b)\varphi(\xi)$ and use the definition of

$$b, b^+ \text{ by } \xi, \frac{\partial}{\partial \xi}.$$

(2) Prove

$$\int_{-\infty}^{\infty} \varphi^*(\xi)\left[-\frac{\partial^2}{\partial \xi^2} + \xi^2\right]\varphi(\xi)\, \mathrm{d}\xi \geqslant 0.$$

Hint: Assume that $|\varphi(\xi)| \to 0$ for $\xi \to \pm\infty$ and integrate

$$\int_{-\infty}^{\infty} \varphi^*(\xi)\left(-\frac{\partial^2}{\partial \xi^2}\right)\varphi(\xi)\, \mathrm{d}\xi$$

by parts, which yields

$$\int_{-\infty}^{\infty} \left(\frac{\partial \varphi^*(\xi)}{\partial \xi}\right)\left(\frac{\partial \varphi(\xi)}{\partial \xi}\right) \mathrm{d}\xi \geqslant 0.$$

(3) Wave packets, an example
To solve the time-dependent Schrödinger equation

$$\tfrac{1}{2}\left(-\frac{\partial^2}{\partial \xi^2} + \xi^2\right)\varphi(\xi, t) = \mathrm{i}\hbar\frac{\partial \varphi(\xi, t)}{\partial t} \qquad (*)$$

make the ansatz

$$\varphi(\xi, t) = \underbrace{c_0\frac{1}{\sqrt{\pi}}\exp(-\xi^2/2)}_{\varphi_0} + \underbrace{c_1\sqrt{\frac{2}{\pi}}\,\xi\exp(-\xi^2/2)}_{\varphi_1}\exp(-\mathrm{i}\omega t). \qquad (**)$$

(a) Show that $(**)$ fulfills the Schrödinger equation $(*)$.
(b) Show that $\int \varphi_0^* \varphi_1\, \mathrm{d}\xi = 0$. (The integral can be explicitly evaluated.)
(c) Show that $|c_0|^2 + |c_1|^2$ must be $= 1$ so that $\int |\varphi|^2 \mathrm{d}\xi = 1$.

(d) Choose $c_0 = c_1 = 1/\sqrt{2}$ and discuss the time-development of the real part of ($* *$).
Hint: $\mathrm{Re}(\exp(\mathrm{i}\omega t)) = \cos \omega t$; consult figs. 3.7.

(4) In the above section we have seen how to solve the Schrödinger equation of the harmonic oscillator by purely algebraic methods. In this exercise we want to learn how to calculate matrix elements by purely algebraic methods, avoiding any explicit integration. We note that we may transcribe any matrix element formed of ξ and $\partial/\partial\xi$ into those containing b^+, b.
Example:

$$\langle \varphi_m | \xi | \varphi_n \rangle \equiv \int_{-\infty}^{\infty} \varphi_m^*(\xi)\xi\varphi_n(\xi)\,\mathrm{d}\xi$$

can be transformed into those containing b^+, b by means of the transformations

$$\xi = \frac{1}{\sqrt{2}}(b + b^+) \qquad \frac{1}{\mathrm{i}}\frac{\partial}{\partial \xi} = \frac{1}{\mathrm{i}\sqrt{2}}(b - b^+).$$

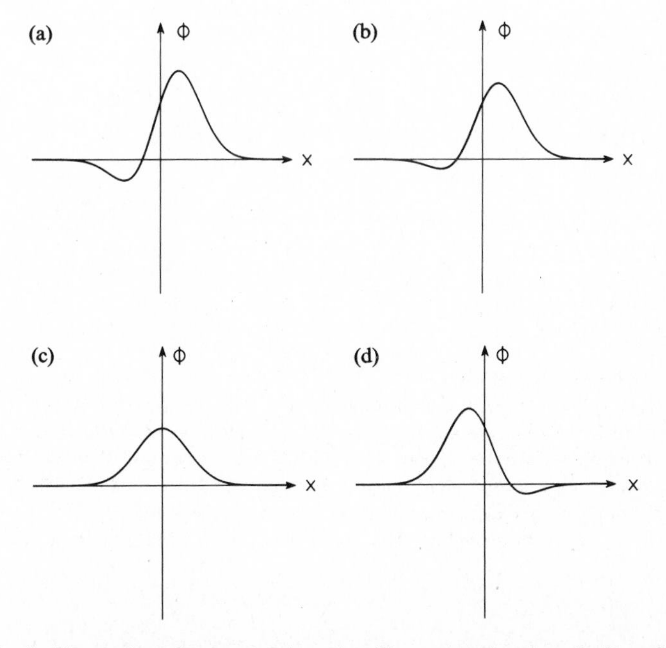

Fig. 3.7 (a–j). The motion of a wave packet of oscillator wave functions during a full cycle. Wave packets were first introduced by Schrödinger. The sequence of figures shows the motion of the wave packet ($* *$) of exercise 3 of this section. We have chosen $c_0 = c_1 = \frac{1}{2}$.

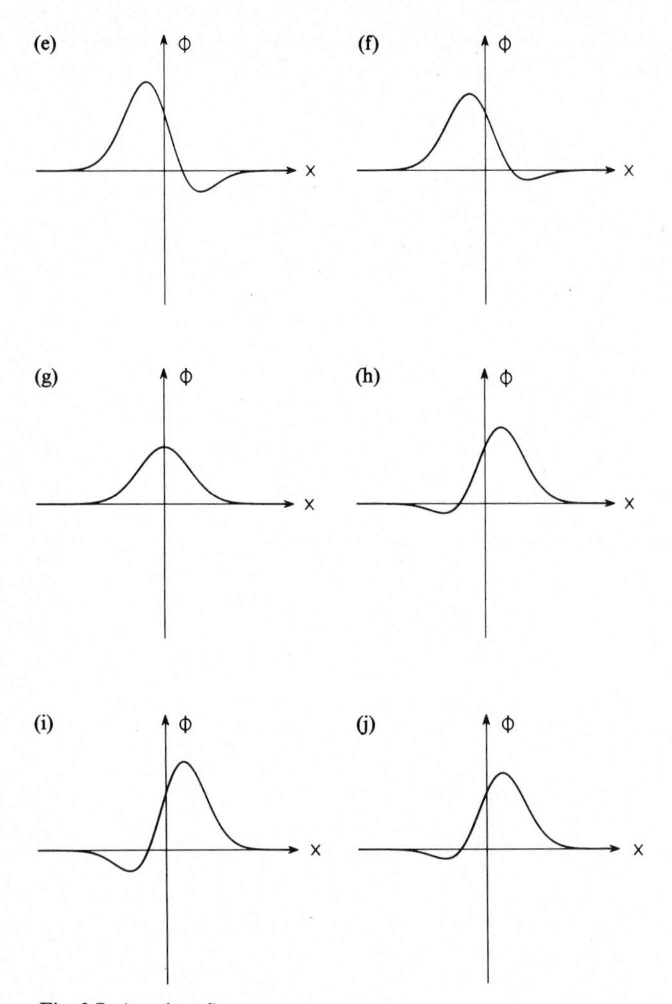

Fig. 3.7. (continued)
The individual figures show the appearance of the total wave packet $\mathrm{Re}(\varphi(x,t))$ for the following time sequence: $t_n = n\pi/(4\omega)$, $n = 0, 1, \ldots, 9$. Later we will see that wave packets of this and a still more general type play a crucial role in quantum optics. As our figures show the wave packet reaches its original shape after the period $T = t_8 = 2\pi/\omega$. This is a unique feature of harmonic oscillators. The reader should be warned that in most other quantum systems wave packets are spreading more and more in space when time is increasing.

We now formulate the basic rules, where φ and χ are functions of ξ which vanish for $\xi \to \pm\infty$. Prove the following rules

$$\langle \varphi | b^+ \chi \rangle = \langle b\varphi | \chi \rangle \tag{α}$$

$$\langle \varphi | b\chi \rangle = \langle b^+ \varphi | \chi \rangle \tag{β}$$

$$\langle c^* b^+ b\varphi | \chi \rangle = \langle \varphi | c b^+ b\chi \rangle \tag{γ}$$

$$\langle \varphi_0 | b^+ \varphi_0 \rangle = 0 \tag{δ}$$

$$\langle \varphi_0 | \varphi_n \rangle = 0, \qquad n \geqslant 1. \tag{ε}$$

(φ_n: oscillator wave functions, $n \geqslant 0$).
Hints: We treat (α) as an example. Write the l.h.s. as integral

$$\int_{-\infty}^{\infty} \varphi^*(\xi) \frac{1}{\sqrt{2}} \left(-\frac{\partial}{\partial \xi} + \xi \right) \chi(\xi) \, d\xi. \tag{$*$}$$

Use partial integration

$$\int_{-\infty}^{\infty} \varphi^*(\xi) \left(-\frac{\partial}{\partial \xi} \right) \chi(\xi) \, d\xi = \int_{-\infty}^{\infty} \left(\frac{\partial}{\partial \xi} \varphi^*(\xi) \right) \chi(\xi) \, d\xi$$

and write

$$\int_{-\infty}^{\infty} \varphi^* \xi \chi \, d\xi = \int_{-\infty}^{\infty} (\xi \varphi^*) \chi \, d\xi.$$

With these tricks it follows

$$(*) = \int_{-\infty}^{\infty} \left[\frac{1}{\sqrt{2}} \left(\frac{\partial}{\partial \xi} + \xi \right) \varphi \right]^* \chi \, d\xi,$$

i.e. the r.h.s. of (α). (β) and (γ) can be proven similarly. Perform the steps! (δ): Use (α) and $b\varphi_0 = 0$; (ε): use (α), write

$$\varphi_n = \frac{1}{\sqrt{n!}} (b^+)^n \varphi_0 = b^+ \underbrace{\frac{1}{\sqrt{n!}} (b^+)^{n-1} \varphi_0}_{= \chi}.$$

(5) Calculate $\langle n | \xi | m \rangle$ and $\langle n | \pi | m \rangle$, $\pi = (1/\mathrm{i})(\partial/\partial \xi)$ for the first 3 oscillator wavefunctions, $m, n = 0, 1, 2$.
Hint: Remember the bra-ket notation of section 3.2, use (3.83), (3.84), (3.86), (3.87). Integrate by parts and use

$$\int_{-\infty}^{\infty} \exp[-\xi^2] \, d\xi = \sqrt{\pi}.$$

(6) Assume that $\langle\varphi_m|\varphi_n\rangle = \delta_{m,n}$, where the φ's are oscillator wave functions. Show:

$$\langle\varphi_m|b^+|\varphi_n\rangle = \sqrt{n+1}\,\delta_{m,n+1}$$

$$\langle\varphi_m|b|\varphi_n\rangle = \sqrt{n}\,\delta_{m,n-1}$$

(i.e. especially

$$\langle\varphi_n|b^+|\varphi_n\rangle = \langle\varphi_n|b|\varphi_n\rangle = 0)$$

without using any integration!

Hints: Use

$$\varphi_n = \frac{1}{\sqrt{n!}}(b^+)^n\varphi_0$$

and (α) or (β) of exercise 4 above.

(7) This is a somewhat more difficult exercise: Prove

$$\langle\varphi_m|\varphi_n\rangle = \delta_{m,n}$$

using

$$\varphi_l = \frac{1}{\sqrt{l!}}(b^+)^l\varphi_0, \qquad l = m, n, \qquad b\varphi_0 = 0.$$

Hints: Use (6.α) if $n \leqslant m$ and $b\varphi_n = \sqrt{n}\,\varphi_{n-1}$, $n = 0, 1, 2, \ldots$.
Use (6.β) if $n \geqslant m$.
Use in both cases $\langle\varphi_0|\varphi_0\rangle = 1$.

(8) Oscillating dipole moments

Let the oscillator be charged with charge e. $e\xi$ is the classical dipole moment. In quantum mechanics its expectation value is given by

$$\vartheta = \int_{-\infty}^{\infty} \varphi^*(\xi, t)(e\xi)\varphi(\xi, t)\,\mathrm{d}\xi.$$

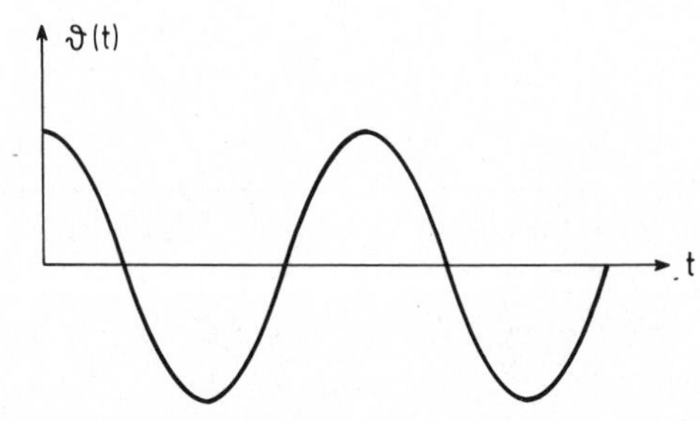

Fig. 3.8. Example of an oscillating dipole moment caused by a wave packet of the harmonic oscillator.

Evaluate ϑ using the following oscillator wave functions:
(a) $\varphi = \varphi_0(\xi)$
(b) $\varphi = \varphi_1(\xi)\exp(-\mathrm{i}\omega t)$
(c) $\varphi = c_0\varphi_0(\xi) + c_1\varphi_1(\xi)\exp(-\mathrm{i}\omega t)$, where $|c_0|^2 + |c_1|^2 = 1$, $c_0 \neq 0$, $c_1 \neq 0$ (i.e. a wave packet). (See, e.g. fig. 3.8.)

3.4. The hydrogen atom

It is not the purpose of this book to develop the quantum theory of matter in all details. We only wish to exhibit those features that are important for an understanding of the processes going on between light and matter. As it will transpire later, to understand the basic processes it is not necessary to know all the details about wave functions and energy levels. Therefore we will confine our discussions of this and the following sections to some general features of quantum systems.

A still rather simple quantum system consists of an electron bound to a nucleus, say to a proton. Since the mass of a proton is about 2000 times bigger than that of an electron, in a good approximation we can neglect the motion of the proton. We consider a nucleus with charge Ze. Its Coulomb interaction energy with the electron is given by $-Ze^2/(4\pi\varepsilon_0 r)$, where ε_0 is the dielectric constant of vacuum in MKS-units, and r is the distance between the electron and the nucleus. The Schrödinger equation reads (m_0: electron mass) (cf. also fig. 3.9)

$$\left\{-\frac{\hbar^2}{2m_0}\left(\frac{\partial^2}{\partial x^2} + \frac{\partial^2}{\partial y^2} + \frac{\partial^2}{\partial z^2}\right) - \frac{Ze^2}{4\pi\varepsilon_0 r}\right\}\psi(\boldsymbol{x}) = W\psi(\boldsymbol{x}). \tag{3.89}$$

The wave functions depend on the three coordinates x, y, z. It turns out that there are two types of solutions to (3.89). The first type corresponds to closed orbits or bound states in classical mechanics and possesses a discrete energy spectrum. The other type of motion corresponds to scattering states in which the electron comes from infinity and after being scattered by the nucleus again travels to infinity. The corresponding energy spectrum is continuous. The wave functions of the bound state are characterized by three quantum numbers n, l, m. n is called the principal quantum number and runs from $1, 2, \ldots$ till infinity. It labels the energy levels of the electron in the hydrogen atom. l is called the angular momentum quantum number and takes on the values $l = 0, 1, 2, \ldots, n-1$. m is called the magnetic quantum number. It denotes the direction of the

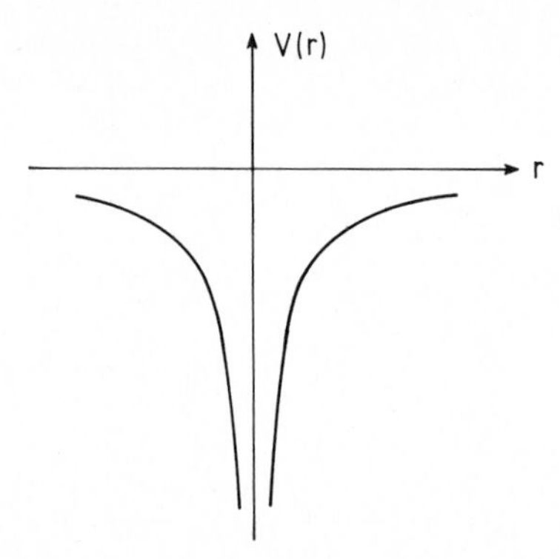

Fig. 3.9. The potential energy $V(r)$ of an electron bound to a positively charged nucleus as a function of its distance r from the nucleus.

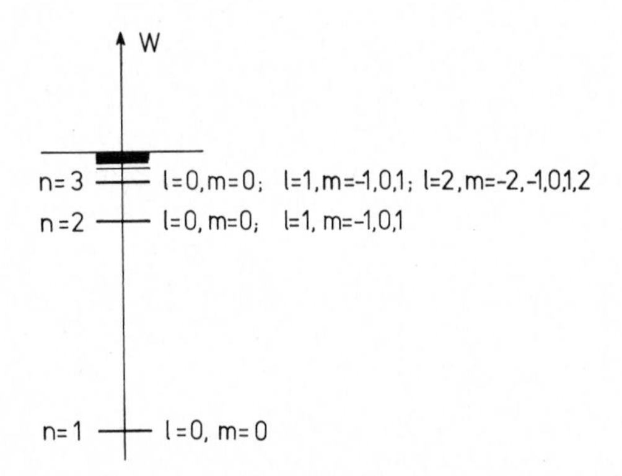

Fig. 3.10. The first three energy levels of the hydrogen atom together with their quantum numbers. The position of the abscissa indicates the zero of the potential energy where ionisation occurs.

angular momentum vector in space and it takes on the values $m = -l, -l+1, \ldots, l$. The energy of the hydrogen atom is given by (cf. fig. 3.10)

$$W_n = -\frac{m_0 e^4}{8h^2 \varepsilon_0^2} \frac{1}{n^2}. \tag{3.90}$$

It is remarkable that quite different quantum numbers and thus different kinds of wave functions can be attached to the same energy level W_n. We call such energy levels degenerate.

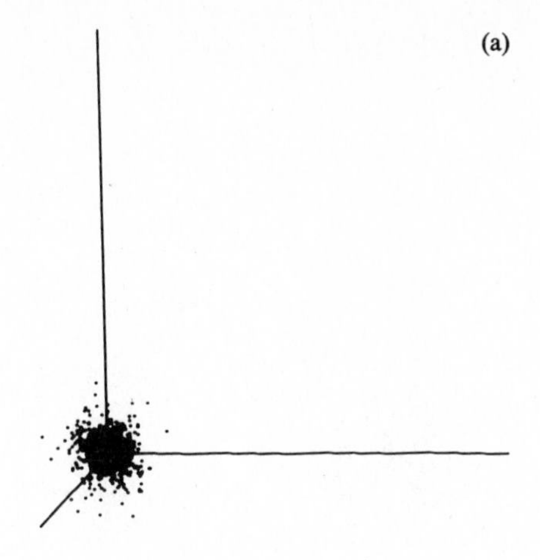

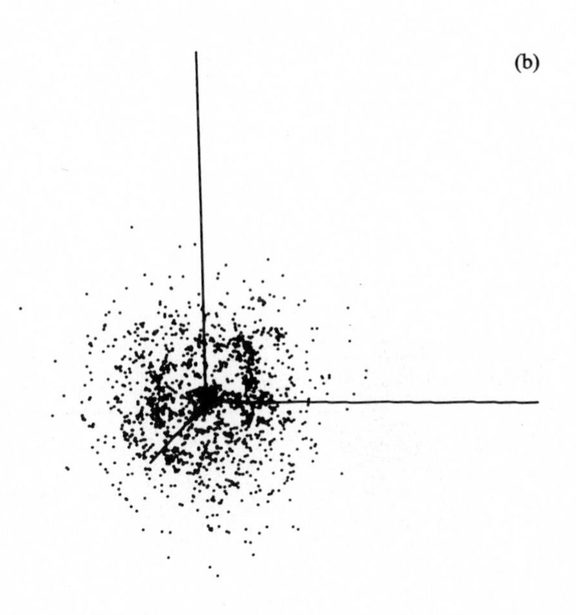

Fig. 3.11. Electron density distributions for the first eigenfunctions of the hydrogen atom. The density of dots is a measure for the probability density $|\psi(\boldsymbol{x})|^2$ of the electron in its quantum state n, l, m (computer plot). Functions with $l = 0$ are called s-functions, those with $l = 1$ p-functions and those with $l = 2$ d-functions.
(a) $n = 1$, $l = 0$, $m = 0$ (1s-function). (b) $n = 2$, $l = 0$, $m = 0$ (2s-function). (Continued).

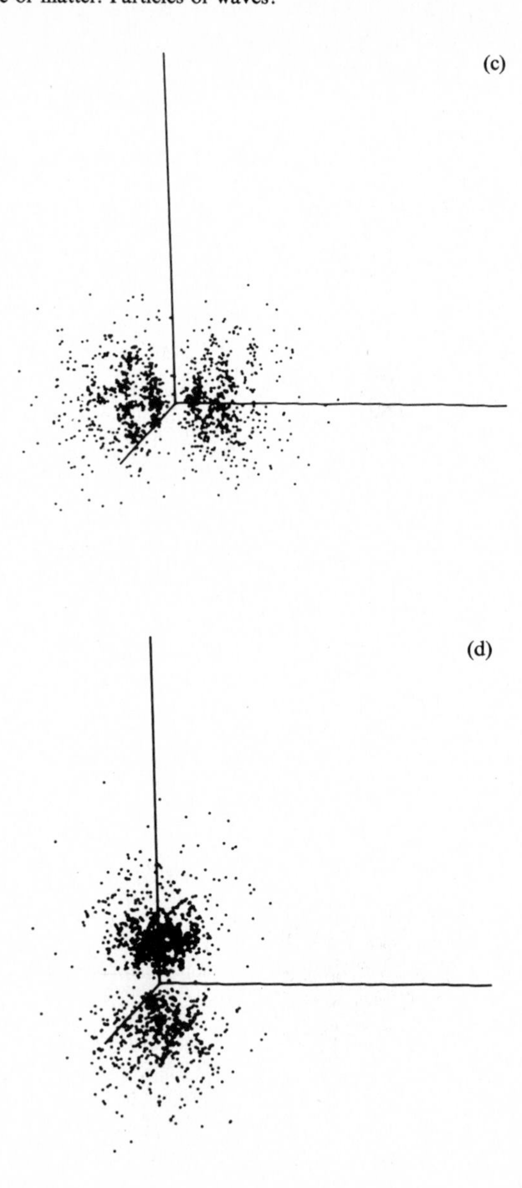

Fig. 3.11. (Continued.) (c) Because ψ is complex for $m = \pm 1$, often a linear combination of wave functions with $m = \pm 1$ and $m = -1$ is used so that ψ becomes real. Here we used $\psi = (-i/\sqrt{2})(\psi_{2,1,1} - \psi_{2,1,-1})$, i.e. $n = 1$, $l = 1$ are kept as quantum numbers. (d) Another example of a $2p$-function: $n = 2$, $l = 1$, $m = 0$.

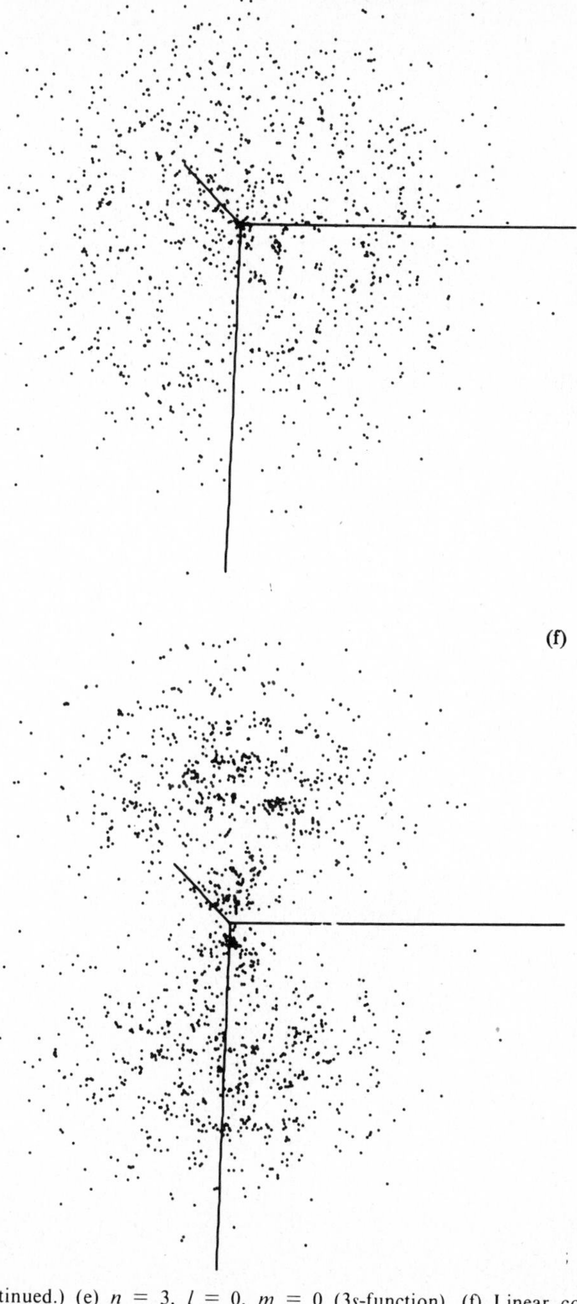

Fig. 3.11. (Continued.) (e) $n = 3$, $l = 0$, $m = 0$ (3*s*-function). (f) Linear combination $\psi = \left(-i/\sqrt{2}\right)(\psi_{3,1,1} - \psi_{3,1,-1})$.

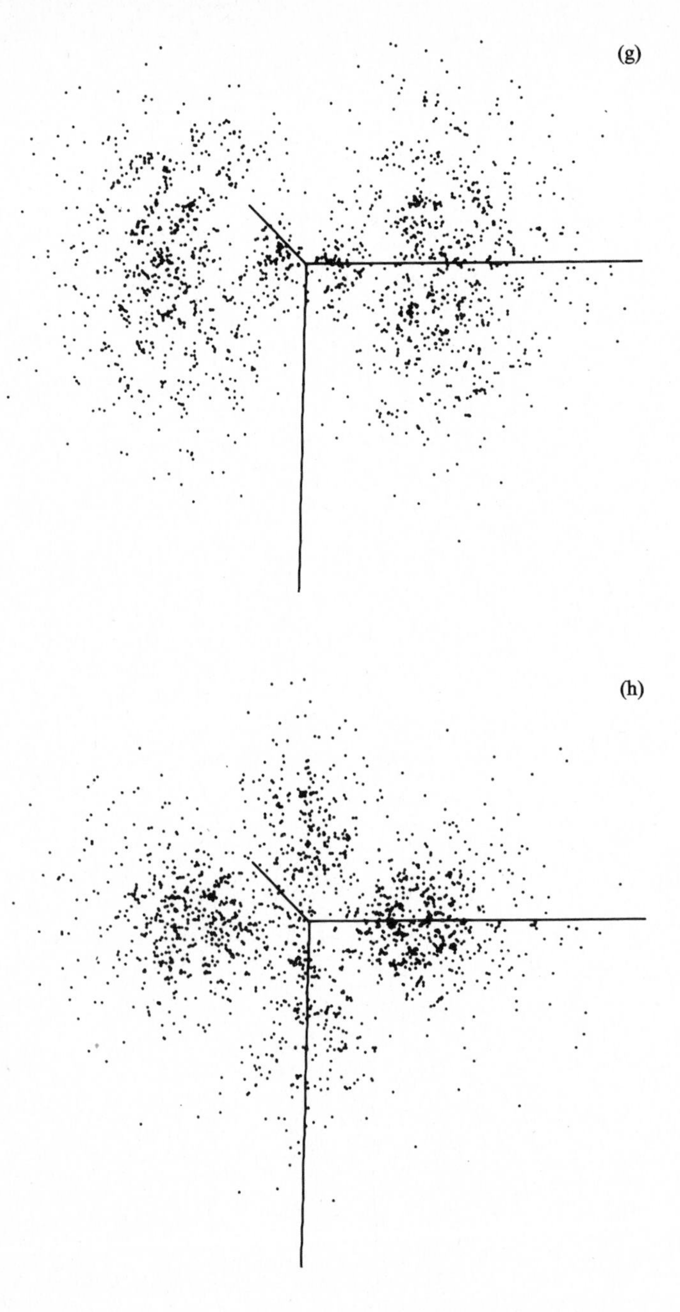

Fig. 3.11. (Continued.) (g) $n = 3$, $l = 1$, $m = 0$ (3p-function). (h) $n = 3$, $l = 2$, $m = 0$ (3d-function).

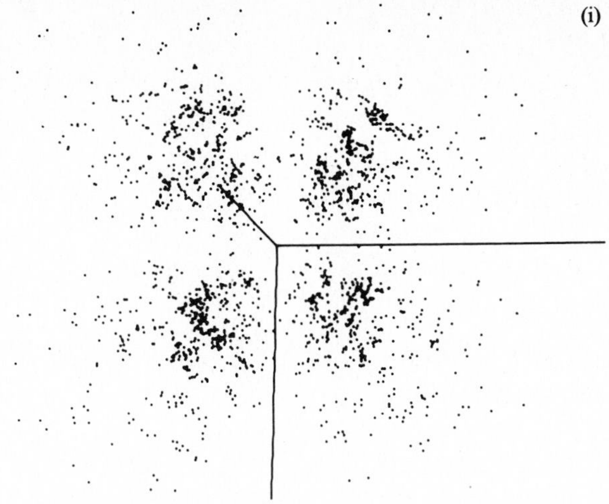

Fig. 3.11. (Continued.) (i) Linear combination $\psi = (-i/\sqrt{2})(\psi_{3,2,1} - \psi_{3,2,-1})$.

Exercises on section 3.4

(1) The hydrogen atom
We introduce spherical polar coordinates (c.f. fig. 3.12)

$$x = r \sin\vartheta \cos\varphi, \quad y = r \sin\vartheta \sin\varphi, \quad z = r \cos\vartheta.$$

Volume element $dV = dx\,dy\,dz = r^2\,dr \sin\vartheta\,d\vartheta\,d\varphi$. Then the Laplacian is given by

$$\Delta = \frac{\partial^2}{\partial r^2} + \frac{2}{r}\frac{\partial}{\partial r} + \frac{1}{r^2}\left[\frac{1}{\sin^2\vartheta}\frac{\partial^2}{\partial\varphi^2} + \frac{\partial^2}{\partial\vartheta^2} + \cot\vartheta\frac{\partial}{\partial\vartheta}\right].$$

The Hamiltonian of the electron of the hydrogen atom reads

$$H = -\frac{\hbar^2}{2m}\Delta - \frac{e^2}{4\pi\varepsilon_0 r}.$$

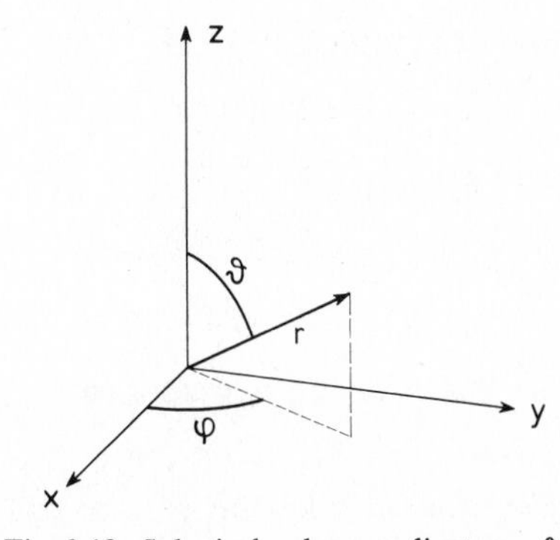

Fig. 3.12. Spherical polar coordinates r, ϑ, φ.

The solutions of the corresponding time-independent Schrödinger equation

$$H\psi_{n,l,m} = W_n\psi_{n,l,m} \tag{$*$}$$

are, as indicated, characterized by the quantum numbers n, l, m. We give a few explicit examples

$$n = 1, l = 0, m = 0: \qquad \psi_{1,0,0} = \frac{a_0^{-3/2}}{\sqrt{\pi}} \exp[-\rho]$$

$$n = 2, l = 0, m = 0: \qquad \psi_{2,0,0} = \frac{a_0^{-3/2}}{\sqrt{8\pi}} \exp\left[-\tfrac{1}{2}\rho\right]\left(1 - \tfrac{1}{2}\rho\right)$$

$$n = 2, l = 1, m = 0: \qquad \psi_{2,1,0} = \frac{a_0^{-3/2}}{4\sqrt{2\pi}} \exp\left[-\tfrac{1}{2}\rho\right]\rho\cos\vartheta \tag{$**$}$$

$$n = 2, l = 1, m = \pm 1: \qquad \psi_{2,1,\pm 1} = \frac{a_0^{-3/2}}{8\sqrt{\pi}} \exp\left[-\tfrac{1}{2}\rho\right]\rho\sin\vartheta\exp[\pm i\varphi$$

where the normalized distance $\rho = r/a_0$, with

$$a_0 = 4\pi\varepsilon_0\hbar^2/(me^2)$$

(Bohr radius) has been used.

(a) Show that the functions $\psi_{n,l,m}$ given in $(**)$ solve the Schrödinger equation $(*)$ with the energies

$$W_n = -\frac{e^4 m}{32\pi^2\hbar^2\varepsilon_0^2}\frac{1}{n^2}.$$

Show that the $\psi_{n,l,m}$ are orthogonal and normalized.

(b) Show that the only nonvanishing dipole matrix elements

$$\vartheta_{n,l,m;n',l',m'} = \int \psi^*_{n,l,m}[e\mathbf{x}]\psi_{n',l',m'}\,\mathrm{d}V$$

are given by the following components of ϑ

$$\vartheta^{(z)}_{1,0,0;2,1,0} \approx 0.745 a_0 e$$

$$\vartheta^{(z)}_{2,0,0;2,1,0} = 3 a_0 e$$

$$\vartheta^{(x)}_{1,0,0;2,1,\pm 1} \pm i\vartheta^{(y)}_{1,0,0;2,1,\pm 1} \approx \pm 1.053 a_0 e$$

$$\vartheta^{(x)}_{2,0,0;2,1,\pm 1} \pm i\vartheta^{(y)}_{2,0,0;2,1,\pm 1} \approx \pm 4.243 a_0 e.$$

This gives the selection rules for the z-component

$$\Delta m \equiv m - m' = 0, \qquad \Delta l \equiv l - l' = \pm 1.$$

The corresponding selection rules for the x-, y-components read

$$\Delta m \equiv m - m' = \pm 1, \qquad \Delta l \equiv l - l' = \pm 1.$$

Hints:
(a) Insert the wave functions (* *) into (*) and perform the differentiations.
(b) In performing the integrations, make use of a representation of factorials

$$\int_0^\infty \xi^n \exp[-\xi]\,\mathrm{d}\xi = n!$$

(2) Calculate

$$\vartheta = \int \psi^* e x \psi \,\mathrm{d}V$$

where ψ is the wave packet

$$\psi(\boldsymbol{x}, t) = \frac{1}{\sqrt{2}}\left(\psi_{1,0,0}(\boldsymbol{x}) \exp[-\mathrm{i}W_1 t/\hbar] + \psi_{2,1,0}(\boldsymbol{x}) \exp[-\mathrm{i}W_2 t/\hbar]\right)$$

of the hydrogen wave functions of exercise (1).

(3) Prove the following commutation relations for the components of the angular momentum operator $\mathfrak{L}$:

$$[\mathfrak{L}_x, \mathfrak{L}_y] = \mathrm{i}\hbar\mathfrak{L}_z$$

$$[\mathfrak{L}_y, \mathfrak{L}_z] = \mathrm{i}\hbar\mathfrak{L}_x$$

$$[\mathfrak{L}_z, \mathfrak{L}_x] = \mathrm{i}\hbar\mathfrak{L}_y$$

$$[\mathfrak{L}^2, \mathfrak{L}_x] = [\mathfrak{L}^2, \mathfrak{L}_y] = [\mathfrak{L}^2, \mathfrak{L}_z] = 0.$$

Hint: Apply the commutators and $\mathfrak{L}_x$, $\mathfrak{L}_y$, $\mathfrak{L}_z$, respectively, to an arbitrary wave function $\psi(x, y, z)$. Use the explicit representation of $\mathfrak{L}_x$, $\mathfrak{L}_y$, $\mathfrak{L}_z$ of exercise 4 on 3.2 and perform the differentiations.

(4) Prove

$$[\mathfrak{L}_z, H] = 0$$

for

$$H = -\frac{\hbar^2}{2m}\Delta + V(r).$$

It follows that

$$H, \mathfrak{L}^2, \mathfrak{L}_z$$

are operators which commute pairwise. Discuss what this means for simultaneous measurability.

Hint: Proceed as in the foregoing exercise and use

$$r^2 = x^2 + y^2 + z^2.$$

3.5. Some other quantum systems

In all other atoms than the hydrogen atoms there are several electrons in a single atom. This gives rise to a highly complicated problem due to the mutual interaction between the different electrons. However, it has proved useful to lump all the interactions between the electrons together into a single potential function in which an individual electron moves. In this procedure one first assumes a given distribution of charges of the electrons and then calculates the effective potential of an electron in the field of all

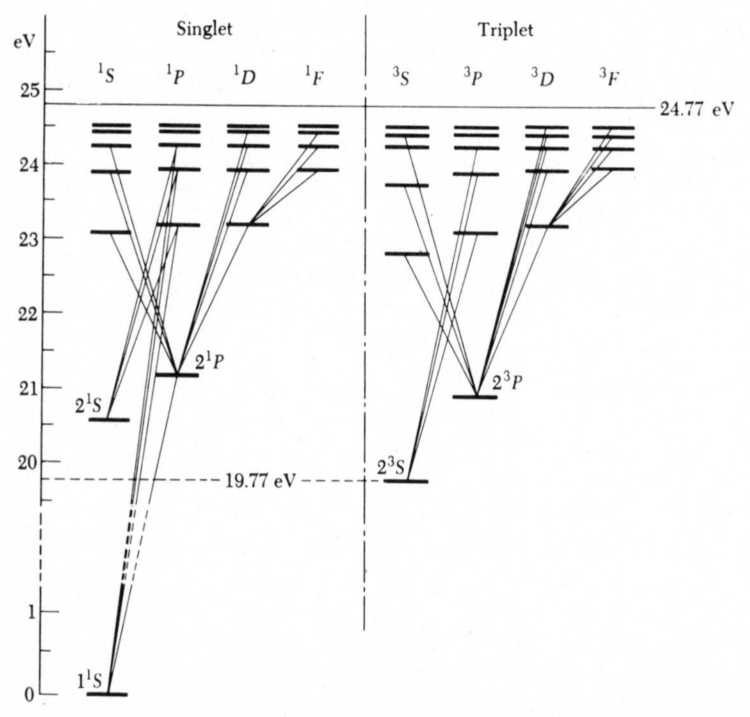

Fig. 3.13. Energy levels of the helium atom. In the singlet and triplet states, the spins of the electrons are anti-parallel and parallel, respectively. The letters *S*, *P*, *D*, *F* designate the total orbital angular momentum of the electrons. The left superscript 1 or 3 designates the multiplicity (singlet or triplet). The nonexistence of a triplet 1S state is a direct consequence of the Pauli exclusion principle: Two electrons with parallel spin cannot occupy the same quantum state. The big numbers 1 and 2 indicate the bigger of the two principal numbers n, n' of the two electrons.

the other charges and of the nucleus. One then solves the Schrödinger equation and finds in this way a new set of wave functions for the electrons. These wave functions then determine new charge distributions and the procedure can be repeated. In this way when continuing this approach approximate wave functions and energy levels can be determined. This procedure is called the Hartree procedure and when certain symmetry properties of the total wave function are observed, the Hartree–Fock procedure.

To understand the structure of atoms it must be assumed that each electron possesses an internal degree of freedom, called spin. The spin can adopt only two quantum states. With its aid it is possible to formulate Pauli's principle. It says that a state characterized by the quantum numbers n, l, m and the spin quantum number can be occupied, at maximum, by a single electron.

Since the effective Hartree potential in general differs from the Coulomb potential (3.89) the energy levels of the hydrogen atoms are shifted and in particular the degenerate levels can split into sublevels for different values of l. Furthermore, in addition to the Coulomb interaction a number of other interactions are effective in atoms, for instance magnetic interactions. These give rise to further splittings and shifts of energy levels. Turning the argument around we can state that a careful measurement of energy levels can give us important hints on the fundamental forces acting in atoms. This is why spectroscopic investigations are so important for atomic physics. An example of an energy level diagram is given in fig. 3.13.

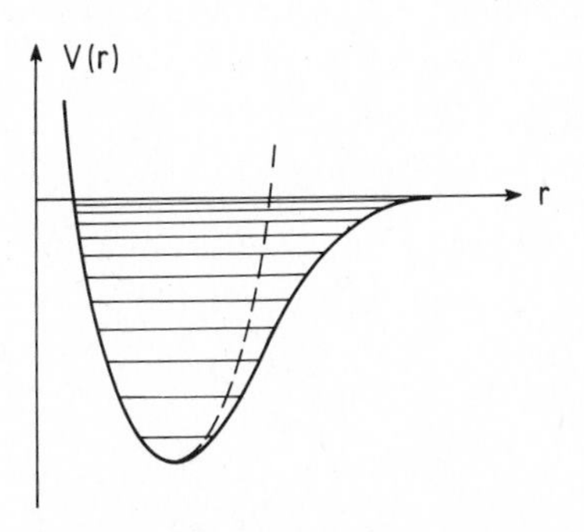

Fig. 3.14. This figure shows the potential binding energy of two atoms at a distance r between their centers of gravity within a two-atomic molecule. The atoms can oscillate. Their relative motion is described by the Schrödinger equation of a quantum mechanical oscillator. Since the potential energy $V(r)$ is not harmonic (compare the dashed curve), especially the energy levels lying higher are no more equidistant as in the purely harmonic case. We have thus an example for the energy levels of a quantum mechanical anharmonic oscillator. Indicated are only the levels belonging to bound states.

As we know from chemistry atoms can be bound together to form molecules. Their quantum states are not only determined by the electrons, but also by the nuclei, contributing with their own degrees of freedom. The nuclei (together with their electrons) can perform oscillations as well as rotations. According to quantum theory these oscillations (vibrations) and rotations must again be quantized. While very often the low lying oscillatory levels are determined by an oscillator potential and are therefore equidistant, at higher excitation levels deviations may occur (cf. fig. 3.14). We shall discuss some explicit examples in the second volume dealing with laser theory.

3.6. Electrons in crystalline solids

In quite a number of cases it is possible to form crystalline solids by putting many atoms of the same or of few different kinds together. Again it is not the purpose of our introductory text to derive the whole quantum theory of electrons in solids. But the most important features, which we shall use later on, can be quite easily seen.

Again we have to deal with a very difficult many body problem due to the many electrons and nuclei present in a solid. In a first approximation it is assumed that the nuclei form a rigid lattice. In the spirit of the Hartree–Fock approximation the electron–electron interactions are replaced by an effective potential field $V(\boldsymbol{x})$ in which a single electron moves (single electron model). Since we deal with a regular lattice the potential V is periodic with the lattice constants (cf. fig. 3.15). As a consequence of this periodicity it can be shown that quite independently of the explicit form of V the wave function must have the form (cf. fig. 3.16).

$$\psi_{\boldsymbol{k},n} = \exp(\mathrm{i}\boldsymbol{k}\boldsymbol{x})u_{\boldsymbol{k},n}(\boldsymbol{x}). \tag{3.91}$$

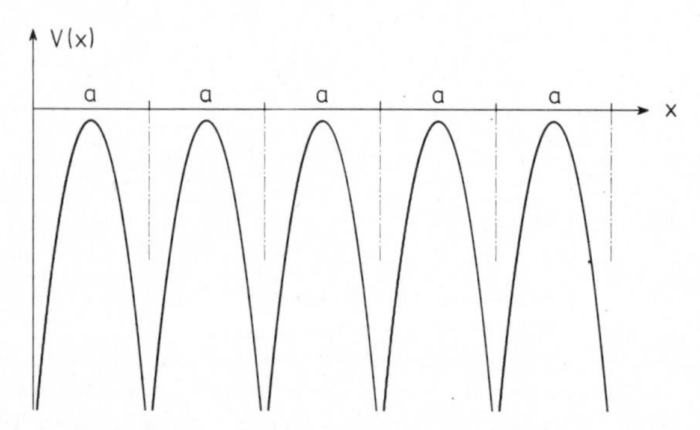

Fig. 3.15. A one-dimensional plot of the periodic potential which an electron finds in a crystal.

In it the function u has the same periodicity as the potential function V. Thus an electron in a periodic lattice has a wave function which is strongly reminiscent of that of a free electron in vacuum. The difference lies in the periodic modulation factor u. $\boldsymbol{k}$ plays, of course, the role of a wave number. As is shown in solid state physics, the energy levels depend on essentially two indices namely the continuous wave vector $\boldsymbol{k}$ and a discrete quantum number n. The energy levels are now grouped into bands which are separated by gaps. Examples are shown in fig. 3.17. In any finite crystal these energy levels are not entirely continuous but have a very narrow spacing. That is in other words, they are still discrete. These discrete energy levels can then be filled up by electrons which have to obey Pauli's principle. If the upper most band is entirely filled up, an electric field cannot cause motion of the electrons and we are dealing with an insulator. When the upper most band is not entirely filled an electric field can cause a current and we have a metal.

An interesting intermediate case is provided by semiconductors. In them, in their ground state, the "valence" band is completely filled, and the next band, known as the conduction band, empty. However, when the energy gap is small enough, thermal excitation may bring electrons from the valence band into the conduction band and electric conduction becomes possible. Electrons can also be brought from the valence band to the conduction band by optical excitation (internal photo-effect giving rise to photoconductivity).

The optical and electrical (and other) properties of solids can be strongly influenced by "impurity" atoms. For instance, impurity atoms in an

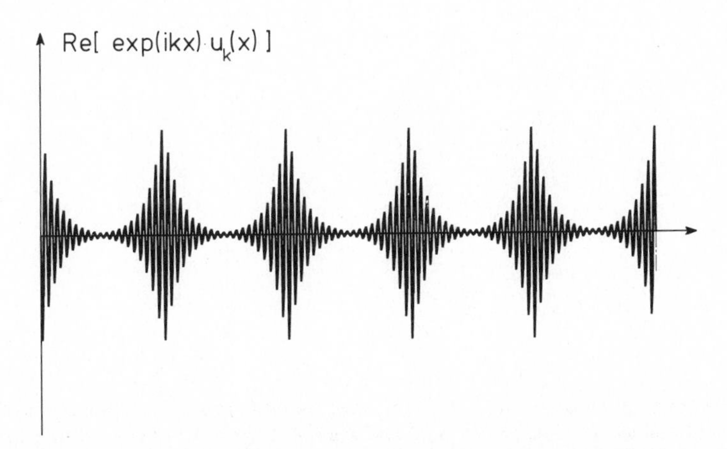

Fig. 3.16. The real part of the Bloch wave function is plotted versus the space coordinate x. In this example the electron is periodically concentrated around the individual nuclei on the lattice.

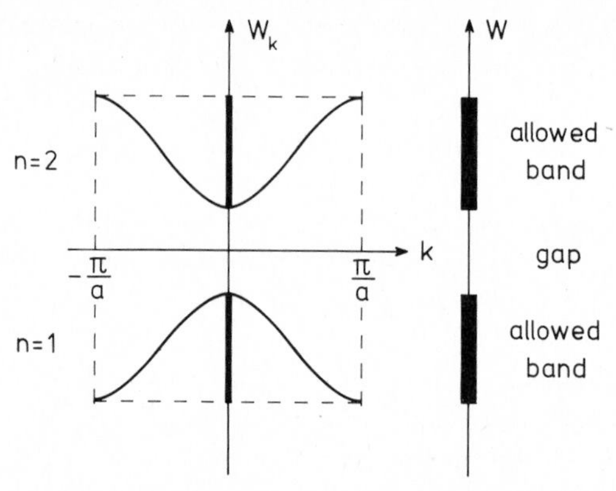

Fig. 3.17. Energy level diagram of an electron in the periodic potential of a lattice. *Left part*: The energy W_k of the electron is plotted versus the wave vector k occurring in the Bloch wave function (3.91). The curve can be periodically extended along the k-axis. a is the lattice constant of the one-dimensional crystal considered here. *Right part*: When we project the energy levels on the energy axis W we find the allowed bands which are interrupted by forbidden zones also called gaps.

insulator introduce new electronic levels below the conduction band (or above the valence band). From these levels, electrons can be thermally excited to the conduction band leading again to a certain type of semiconductors.

Some of such impurity atoms have an energy spectrum closely resembling the hydrogen spectrum, but with a screened Coulomb potential

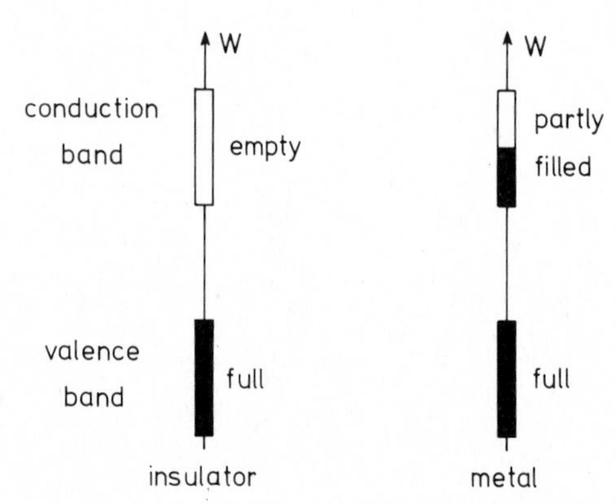

Fig. 3.18. This figure shows how the energy band model allows for an explanation of the difference between an insulator (left part) and a metal. In an insulator the valence band is completely filled up by electrons, whereas the conduction band is entirely empty. In a metal the conduction band is partly filled by electrons.

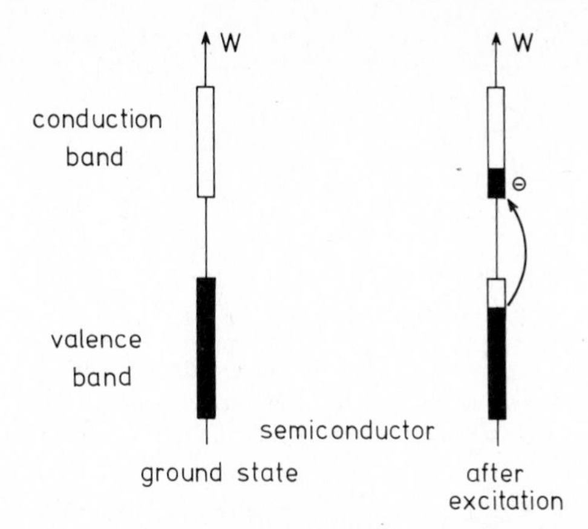

Fig. 3.19. Application of the energy band model to semiconductors. At sufficiently low temperatures the semiconductor is in its ground state. The valence band is entirely filled up, the conduction band is empty and no electrical conduction is possible. By thermal excitation or by shining light on a semiconductor electrons can be excited from the valence band to the conduction band which is then partly filled. This then allows for an electrical conduction.

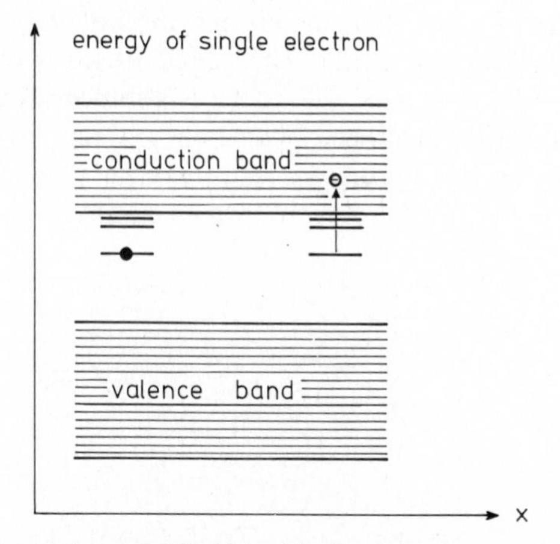

Fig. 3.20. An extension of the energy band model when impurity centers are present. We plot the energy of a single electron versus the x axis. Note that such a plot must be taken cum grano salis due to Heisenberg's uncertainty relation. Nevertheless it gives us good visualization of the whole situation. The horizontal bars show the individual energy levels of the electron in the perfect crystal. Due to impurities localized states are introduced which lie below the conduction band or above the valence band. In the left part of this picture the electron of the impurity is sitting in its ground state, the right part shows the situation where the electron has been excited by light or thermal excitation from its ground state into a state of the conduction band.

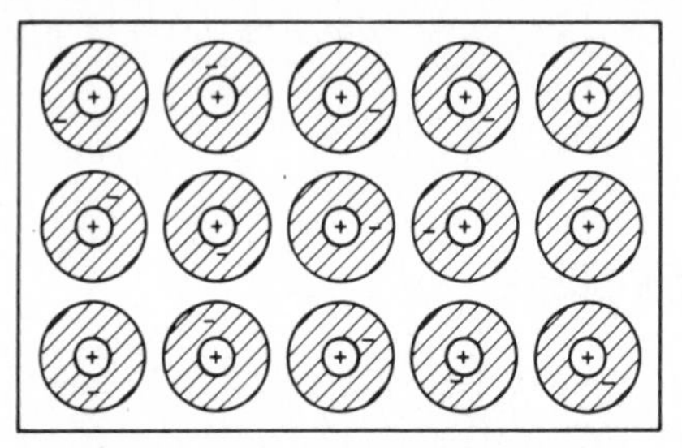

Fig. 3.21. To describe the ground state or an insulator of semiconductor one may use instead of the energy band model a model referring to localized electronic states. Here each nucleus indicated by a plus sign is surrounded by negatively charged clouds of electrons.

$-(e^2Z)/(4\pi\varepsilon\varepsilon_0 r)$ where $\varepsilon > 1$ is the dielectric constant of the crystal. As a consequence, the spacing between the energy levels is narrower than in the hydrogen atom.

We will come back to all of these energy level diagrams later when dealing with laser action of semiconductors and with nonlinear optical properties of solids.

The optical spectrum of semiconductors without or even with impurities exhibits a number of lines which cannot be explained in the single electron model. To get an understanding of this new kind of states let us consider a semiconductor in its groundstate. The whole lattice is electrically neutral because the positive charge of the nuclei is compensated by the negative charge of the individual electrons. When we now remove an electron from an atom by exciting it to another band it leaves a positive charge behind it by which it is attracted. As may be shown in more detail not only the

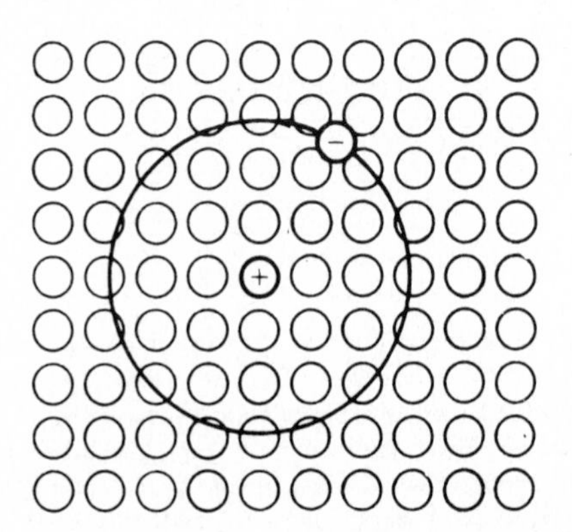

Fig. 3.22. Model of an exciton. One electron is kicked away of its individual atom leaving a positively charged hole behind it. It now circles in an excited state around the positive hole.

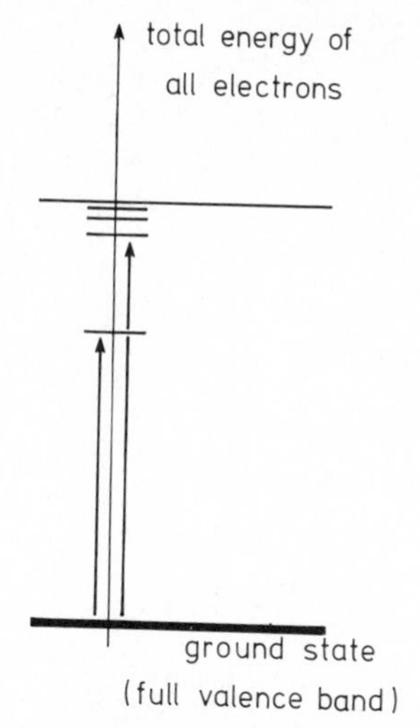

Fig. 3.23. This figure represents the energy levels of an exciton at rest. We have plotted here the total energy of all electrons of a crystal. The state in which all electrons fill up the valence band states completely is indicated here as the energy of the ground state. The most upper level shows the minimum energy required to lift an electron from the valence band into the condition band. Below this level we find the energy levels of the bound states of the excitons.

electron but also the remaining positive charge (= "hole") can now move freely in the crystal similar to particles but with certain effective masses m_1 and m_2. Furthermore a Coulomb-force, which is screened by the dielectric constant ε of the crystal, acts between these two particles. Thus we see that new excited states of the crystal become possible where the two particles give rise to a hydrogenlike energy scheme (cf. fig. 3.23). Such an electron–hole pair is called exciton. When a crystal is highly excited, especially optically, many excitons can be produced which in certain crystals can form exciton molecules (biexcitons) or a new kind of state called electron–hole-droplets.

Exercises on section 3.6

(1) Subject the Bloch wave function to the periodicity condition (one-dimensional example) $\psi(x+L) = \psi(x)$, where $L = Na$, N is integer, a is lattice constant.

Show:

$$k = \frac{2\pi m}{L} = \frac{2\pi m}{Na}, \qquad m \text{ integer.} \tag{$*$}$$

Show:

$$\int_0^L \mathrm{d}x\,\psi_{k,n}^*(x)\frac{\hbar}{\mathrm{i}}\frac{\mathrm{d}}{\mathrm{d}x}\psi_{k',n}(x) = 0, \qquad \text{for } k \neq k'.$$

Hint: Write

$$\int_0^L \mathrm{d}x(\cdots) = \sum_{l=0}^{N-1}\int_{la}^{(l+1)a}(\cdots)\,\mathrm{d}x.$$

Further hint: Show that

$$\int_{la}^{(l+1)a}(\cdots)\,\mathrm{d}x = \exp[\mathrm{i}(-k+k')la]\int_0^a(\cdots)\,\mathrm{d}x$$

and use

$$\sum_{l=0}^{N-1}\exp[\mathrm{i}(k'-k)la] = \begin{cases} N, & \text{for } k' = k \\ \dfrac{1-\exp[\mathrm{i}(k'-k)Na]}{1-\exp[\mathrm{i}(k'-k)a]}, & \text{for } k \neq k'. \end{cases}$$

Show that

$$\sum_{l=0}^{N-1}\exp[\mathrm{i}(k'-k)la] = 0, \qquad \text{for } k \neq k'.$$

Hint: Use (*).

(2) Show that the matrix element

$$\int_0^L \mathrm{d}x\,\psi_{k,n}^*(x)\exp[\mathrm{i}qx]\frac{\hbar}{\mathrm{i}}\frac{\mathrm{d}}{\mathrm{d}x}\psi_{k',n}(x) = 0, \qquad \text{unless } k = k' + q.$$

Generalize the result to 3 dimensions.
Hint: Same as for exercise above.
Later we will see that this matrix element decides which optical transitions between Bloch states are possible. The result given is called: k-selection rule.

3.7. Nuclei

We quite briefly touch on other quantum systems, namely nuclei, which may be of importance in future laser experiments. Nuclei are composed of neutrons and protons (nucleons) which both have masses about 2000 times the electron mass. While neutrons carry no charge, protons have a charge opposite to that of the electrons. Though the forces within nuclei are the "nuclear forces" (and not only Coulomb forces) the whole formulation of

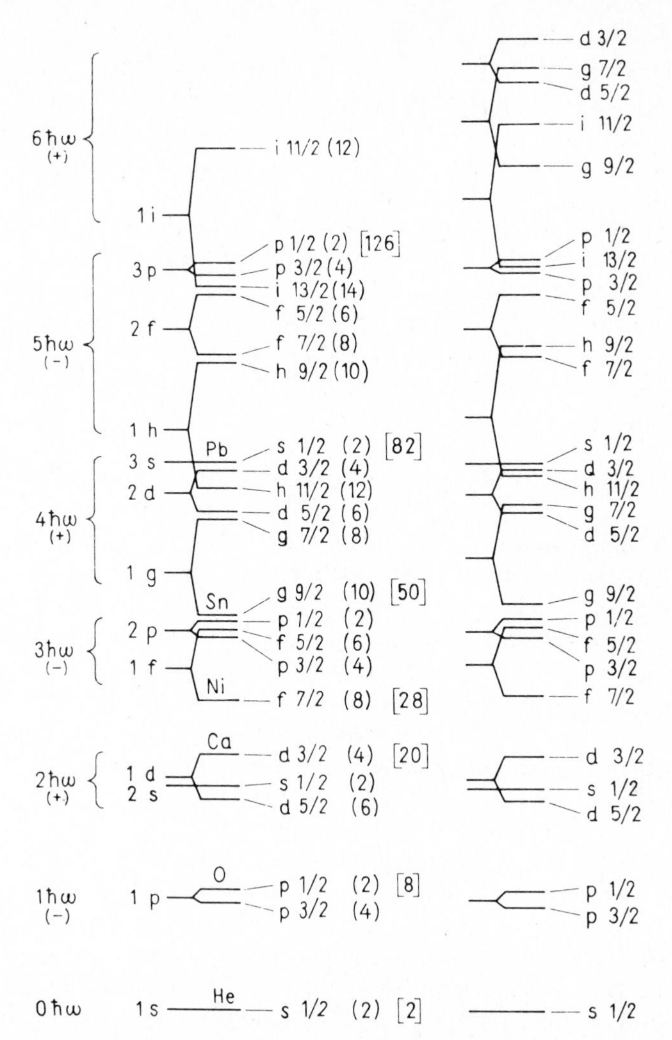

Fig. 3.24. Energy scheme for protons (left) and neutrons (right) according to the shell model. The interaction of one nucleon with all others is assumed to give rise to an effective potential for that nucleon. A simple harmonic oscillator potential would yield the degenerate levels as indicated in the left column. Due to spin–orbit-coupling this degeneracy is removed as indicated. The possible population numbers of the individual levels are given in brackets, the numbers of all possible particles in the nucleus are given in square brackets (up to a certain occupied level). (After Klingenberg, P.; Rev. Mod. Phys. **24** (1952), 63).

quantum theory seems to be still valid. A good deal of the energy level structure can be understood in the frame of the Hartree–Fock approach. A single nucleon is considered which moves in the force field generated by all other nucleons. It turns out, however, that there is a strong coupling between the angular momentum and the spin (see section 3.8) of the nucleons, i.e. a strong "LS-coupling" (cf. fig. 3.24).

Furthermore nuclei can perform collective motions similar to oscillations and rotations of liquid droplets. What is important in the context of our later applications is that the basic concepts of laser theory apply equally well to transitions between quantum states of nuclei. The essential difference lies in the fact that the photons emitted by these transitions are of very short wavelength and lie in the γ-ray region.

3.8. Quantum theory of electron and proton spin

The electron possesses not only its three translational degrees of freedom but in addition an internal degree of freedom. Since it can be visualized by a spinning of the electron around its own axis (whatever this means) this new degree of freedom is called spin. Some other elementary particles, such as protons, also possess a spin. Since the spin has the physical meaning of angular momentum it is a vector $\boldsymbol{s}$ with three spatial components s_x, s_y, s_z. When the electron (or proton) spin is measured in a preferential direction, say parallel to the z-axis, only two values of s_z were found, namely $\frac{1}{2}\hbar$ and $-\frac{1}{2}\hbar$. This preferential direction is in most cases generated by an external constant magnetic field.

In order to develop an adequate formalism we attribute two different states and thus two wave functions to the situation "spin up" or "spin down." We call these two functions $\varphi_\uparrow$ and $\varphi_\downarrow$, respectively. Taking the formalism of quantum theory as developed in our previous sections seriously we have to describe a measurement quantum theoretically by the application of an operator s_z on the corresponding wave function. We choose the wave function in such a way that the application of the measuring operator on the wave function yields the corresponding measured value. Since we have only two measured values, namely $\frac{1}{2}\hbar$ and $-\frac{1}{2}\hbar$ we thus expect the relations

$$s_z\varphi_\uparrow = \tfrac{1}{2}\hbar\varphi_\uparrow \qquad s_z\varphi_\downarrow = -\tfrac{1}{2}\hbar\varphi_\downarrow . \tag{3.92, 93}$$

We now look for a formalism which yields the relations (3.92) and (3.93) so to speak automatically. It has turned out that this goal can be achieved by means of matrices. We write down the result and then will verify it. We chose s_z in the form

$$s_z = \frac{\hbar}{2}\begin{pmatrix} 1 & 0 \\ 0 & -1 \end{pmatrix} \tag{3.94}$$

and the spin functions in the form

$$\varphi_\uparrow = \begin{pmatrix} 1 \\ 0 \end{pmatrix}, \quad \varphi_\downarrow = \begin{pmatrix} 0 \\ 1 \end{pmatrix}. \tag{3.95}$$

The multiplication of the matrix s_z and the spin functions $\varphi_\downarrow, \varphi_\uparrow$ obeys the rule of matrix multiplication. We remind the reader of that definition for two general matrices

$$\begin{pmatrix} a & b \\ c & d \end{pmatrix}\begin{pmatrix} x \\ y \end{pmatrix} = \begin{pmatrix} ax + by \\ cx + dy \end{pmatrix}. \tag{3.96}$$

Using this relation one verifies immediately that (3.94) and (3.95) lead immediately to the relations (3.92) and (3.93). The most general spin function is obtained by the superposition of $\varphi_\uparrow$ and $\varphi_\downarrow$ with coefficients a, b.

$$\varphi = a\varphi_\uparrow + b\varphi_\downarrow \equiv \begin{pmatrix} a \\ b \end{pmatrix}. \tag{3.97}$$

To obtain the normalization condition we have still to introduce scalar products. With φ_1 and φ_2 given in the general forms

$$\varphi_1 = \begin{pmatrix} a_1 \\ b_1 \end{pmatrix} \tag{3.98}$$

and

$$\varphi_2 = \begin{pmatrix} a_2 \\ b_2 \end{pmatrix}. \tag{3.99}$$

We define the scalar product by

$$\langle \varphi_1 | \varphi_2 \rangle \equiv (a_1^*, b_1^*)\begin{pmatrix} a_2 \\ b_2 \end{pmatrix} = a_1^* a_2 + b_1^* b_2. \tag{3.100}$$

The normalization condition for a function φ reads

$$\langle \varphi | \varphi \rangle = |a|^2 + |b|^2 = 1. \tag{3.101}$$

Once we have found an explicit representation for s_z it is not difficult, at least in principle, to determine the explicit form of s_x and s_y. We will not give the details here because they don't yield much physical insight. What one does to find the explicit form of s_x and s_y is the following:
One first writes down two general 2×2 matrices for s_x and s_y. Then one invokes the fact that s_y, s_x are components of angular momentum. In

quantum theory it is shown quite generally that the components of any angular momentum obey commutation relations (compare exercise 3 on section 3.4). s_x, s_y and s_z are then subjected to these commutation relations. The resulting equations for the unknown matrices can then be solved. It turns out that the solution is determined within a certain arbitrariness which leaves some freedom of choice of a particular simple form but has no bearing on any physics as can be shown in detail. For our purposes the following choice is favorable

$$s_x = \tfrac{1}{2}\hbar\begin{pmatrix} 0 & 1 \\ 1 & 0 \end{pmatrix} \tag{3.102}$$

$$s_y = \tfrac{1}{2}\hbar\begin{pmatrix} 0 & -\mathrm{i} \\ \mathrm{i} & 0 \end{pmatrix} \tag{3.103}$$

We now wish to formulate a Schrödinger equation for the spin. It is experimentally known that with the electron or proton spin a magnetic moment is connected according to (MKS-units)

$$\mu = \frac{e\hbar}{2m} \tag{3.104}$$

where e is the elementary charge, m the mass of the particle. If m is the electron mass, the magnetic moment (3.104) is also called Bohr's magneton. Since the magnetic moment is a vector parallel (or anti-parallel) to the spin we write more generally

$$\boldsymbol{\mu} = \frac{e}{m}\boldsymbol{s}. \tag{3.105}$$

The factor $\frac{1}{2}\hbar$ is now contained in the spin operator $\boldsymbol{s}$. We now proceed quite in analogy to the derivation of the Schrödinger equation for the electron, i.e. we start from a classical expression. The energy of the spin in a spatially homogeneous magnetic field with induction $\boldsymbol{B}$ is according to electrodynamics given by

$$-\boldsymbol{\mu}\cdot\boldsymbol{B}. \tag{3.106}$$

We now wish to let the expression (3.106) become an operator analogous to the Hamiltonian operator in the conventional Schrödinger equation of section 3.1. We know already how this can be achieved, where $\boldsymbol{s}$ is, as we have seen, now an operator. Inserting (3.105) into (3.106) allows us immediately to formulate an equation for the electron spin

$$-\frac{e}{m}\boldsymbol{B}\cdot\boldsymbol{s}\varphi = W\varphi. \tag{3.107}$$

When we chose the magnetic field $\boldsymbol{b}$ in preferential direction along the z-axis we have

$$\boldsymbol{B} = (0, 0, B_z). \tag{3.108}$$

In this case, the left-hand side of (3.104) coincides up to a numerical factor, with the left hand sides of (3.92) and (3.93). In other words, the functions (3.95) are just eigenfunctions to (3.107) with the corresponding eigenvalues

$$W = \pm \frac{e\hbar}{2m} B_z . \tag{3.109}$$

The energy of a spin in a constant magnetic field in z-direction is just our expression given by a classical theory of the interaction of a parallel or antiparallel spin. Similarly to the ordinary Schrödinger equation we can now also formulate a time-dependent Schrödinger equation

$$-\frac{e}{m} \boldsymbol{B} \cdot \boldsymbol{s} \varphi = \mathrm{i} \hbar \dot{\varphi} . \tag{3.110}$$

The time-dependent equation has to be applied in particular when the magnetic field is time-dependent. Let us here consider a time-dependent solution in a constant magnetic field. One then immediately verifies that, for instance, the following wave packet

$$\varphi = 2^{-1/2} \left[\exp(\mathrm{i}\omega_0 t/2) \varphi_\downarrow + \exp(-\mathrm{i}\omega_0 t/2) \varphi_\uparrow \right] \tag{3.111}$$

is a solution of (3.110). The meaning of the wave packet (3.111) as well as of other spin functions can best be explored by determining expectation values for the components of the spin operator. These expectation values are defined by

$$\langle s_j \rangle \equiv \langle \varphi | s_j | \varphi \rangle , \qquad \varphi = \begin{pmatrix} a \\ b \end{pmatrix}, \qquad j = x, y, z$$

$$\langle s_j \rangle \equiv (a, b)^* s_j \begin{pmatrix} a \\ b \end{pmatrix} \tag{3.112}$$

where we have to represent s_j by the corresponding matrices (3.102), (3.103), (3.94). A short calculation which we leave to the reader as an exercise yields:

$$\langle s_z \rangle = 0 \tag{3.113}$$

$$\langle s_x \rangle = \tfrac{1}{2} \hbar \cos \omega_0 t \tag{3.114}$$

$$\langle s_y \rangle = \tfrac{1}{2} \hbar \sin \omega_0 t . \tag{3.115}$$

The three components of the expectation value of $\boldsymbol{s}$ form again a vector. Its motion is exhibited in fig. 3.25. Evidently in this case the spin rotates around the z-axis in the x–y plane. We have thus an example of a free precession of the spin. When we choose a more general wave packet φ (compare exercise) the spin still rotates around the z-axis but now in a different plane (compare fig. 3.26). The spin thus behaves quite similarly to a spinning top.

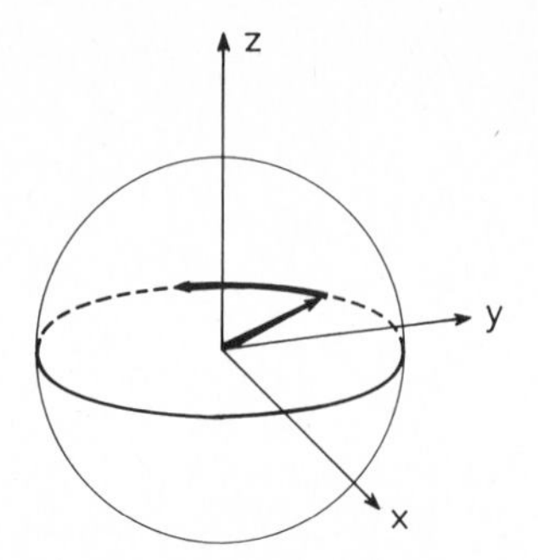

Fig. 3.25. The motion of the expectation values of $\boldsymbol{s}$ for the wave packet (3.111) (cf. (3.113)–(3.115)).

As we will see later, this analogy can be carried even further namely when other components of the magnetic field are non-vanishing.

Exercise on section 3.8

(1) Choose

$$\varphi = c_1 \varphi_\downarrow \exp(\mathrm{i}\,\omega_0 t/2) + c_2 \varphi_\uparrow \exp(-\mathrm{i}\,\omega_0 t/2)$$

where

$$|c_1|^2 + |c_2|^2 = 1 .$$

Calculate:

$$\langle s_x \rangle \equiv \langle \varphi | s_x | \varphi \rangle$$
$$\langle s_y \rangle \equiv \langle \varphi | s_y | \varphi \rangle$$
$$\langle s_z \rangle \equiv \langle \varphi | s_z | \varphi \rangle .$$

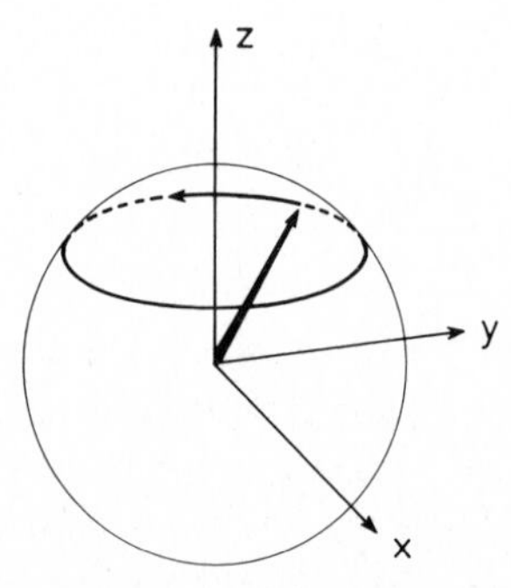

Fig. 3.26. The motion of the expectation value of the spin $\boldsymbol{s}$ calculated by means of the wave function of exercise 1 of the section 3.8.

Show that $\langle \boldsymbol{s} \rangle = (\langle s_x \rangle, \langle s_y \rangle, \langle s_z \rangle)$ fullfills the equations

$$\frac{\mathrm{d}}{\mathrm{d}t}\langle s_x \rangle = \mu \langle s_y \rangle B_z$$

$$\frac{\mathrm{d}}{\mathrm{d}t}\langle s_y \rangle = -\mu \langle s_x \rangle B_z$$

$$\frac{\mathrm{d}}{\mathrm{d}t}\langle s_z \rangle = 0$$

(cf. also fig. 3.26) or, in short

$$\frac{\mathrm{d}}{\mathrm{d}t}\langle \boldsymbol{s} \rangle = \mu \langle \boldsymbol{s} \rangle \times \boldsymbol{B}, \qquad \boldsymbol{B} = (0, 0, B_z).$$

Hint: Use the rule for matrix multiplication and the explicit matrix representations of s_x, s_y, s_z given in the text.

(2) Prove

$$s_x s_y - s_y s_x = i\hbar s_z$$

$$s_y s_z - s_z s_y = i\hbar s_x$$

$$s_z s_x - s_x s_z = i\hbar s_y$$

and

$$s_x^2 = s_y^2 = s_z^2 = \tfrac{1}{4}\hbar^2 \begin{pmatrix} 1 & 0 \\ 0 & 1 \end{pmatrix}.$$

4. Response of quantum systems to classical electromagnetic oscillations

4.1. An example. A two-level atom exposed to an oscillating electric field

To start with an explicit example let us consider an electron of an atom. The motion of this electron is described by a Schrödinger equation of the form

$$H_0\psi = \mathrm{i}\hbar\frac{\mathrm{d}\psi}{\mathrm{d}t}. \tag{4.1}$$

We assume that (4.1) allows for a series of solutions describing stationary states, which in a classical interpretation means that the electron moves in certain orbits. Quantum mechanically speaking, its motion is described by wave functions with time independent probability distributions $|\psi(\boldsymbol{x})|^2$. We have attached an index 0 to the Hamiltonian to indicate that we are dealing with unperturbed motion.

Now let us subject the atom to an external field, for instance a monochromatic light wave or an incoherent light field. Then already in classical physics we have to expect that the electron is thrown out of its original orbit and will perform new kinds of motion. We expect a similar behavior in quantum mechanics. Fortunately there are a number of cases of practical interest in which the new kind of motion, or more precisely speaking, the new kind of wave function can be determined rather simply. To demonstrate this procedure let us consider a system with only two energy levels. This sounds rather strange because we know that electrons in atoms possess an infinity of energy levels. However, we will see below that in a number of cases, when an electron interacts with light, only two energy levels are of central importance and all others can be neglected. This is, for instance, the case when the difference between the two energy levels under consideration is quite different from the differences between any other energy levels. A further condition for this assumption is that the perturbation is not too strong.

Now let us turn to our problem. We have to solve the Schrödinger equation

$$i\hbar\frac{d\psi}{dt} = H\psi \tag{4.2}$$

where the Hamiltonian

$$H = H_0 + H^{\mathrm{p}} \tag{4.2a}$$

consists of the Hamiltonian H_0 of the unperturbed motion and the Hamiltonian caused by the external field, H^{p}. Let us assume that we know the stationary solutions and energies of the time independent unperturbed Schrödinger equation

$$H_0\varphi_n = W_n\varphi_n . \tag{4.2b}$$

Since we assume that only two levels, i.e. $n = 1$ and $n = 2$ are relevant the most general form of the solution of (4.2) must be chosen in the form of a superposition of the two unperturbed solutions ($n = 1, 2$). This leads us to the ansatz

$$\psi = c_1(t)\varphi_1 + c_2(t)\varphi_2 \tag{4.3}$$

where we admit that the coefficients c_1, c_2 are still time dependent. It will be now our task to derive equations for the still unknown coefficients c_1, c_2. To this end we insert (4.3) into (4.2). We may use the fact that φ_j obeys equation (4.2b). We then obtain

$$i\hbar\frac{dc_1}{dt}\varphi_1 + i\hbar\frac{dc_2}{dt}\varphi_2 = c_1(W_1 + H^{\mathrm{p}})\varphi_1 + c_2(W_2 + H^{\mathrm{p}})\varphi_2 . \tag{4.4}$$

The resulting equation has a seeming drawback. On the one hand the functions φ_j depend on space and also H^{p} still depends on space. On the other hand the coefficients are supposed to depend only on time and not on space coordinate $\boldsymbol{x}$. Thus we have to get rid of the space dependence of (4.4). In quantum theory it has turned out that this can be easily achieved by multiplying (4.4) by φ_1^* and φ_2^* and then integrating over space. When doing so we use the orthogonality relation (3.23). We further introduce the abbreviations

$$H_{mn}^{\mathrm{p}} = \int \varphi_m^*(\boldsymbol{x})H^{\mathrm{p}}\varphi_n(\boldsymbol{x})\,dV \tag{4.5}$$

or, in bra-ket notation $= \langle m|H^{\mathrm{p}}|n\rangle$. One then obtains

$$i\hbar\frac{dc_1}{dt} = c_1W_1 + c_1H_{11}^{\mathrm{p}} + c_2H_{12}^{\mathrm{p}}$$

$$i\hbar\frac{dc_2}{dt} = c_2W_2 + c_2H_{22}^{\mathrm{p}} + c_1H_{21}^{\mathrm{p}} . \tag{4.6}$$

These are indeed equations independent of space. To discuss the equations (4.6) further we consider the case in which an electric wave acts on the electron. We represent the electric field strength as wave in the form

$$\boldsymbol{E}(\boldsymbol{x},t) = \boldsymbol{E}_0 \cos(\boldsymbol{k}\cdot\boldsymbol{x} - \omega t). \tag{4.7}$$

In cases of practical interest, the extension of atoms is much smaller than the wavelength (of (4.7)) $\lambda = 2\pi/k$. This allows us to replace the coordinate $\boldsymbol{x}$ by an average coordinate $\boldsymbol{x}_0$ which corresponds to the center of gravity of the atom

$$\boldsymbol{E} = \boldsymbol{E}_0 \cos(\boldsymbol{k}\boldsymbol{x}_0 - \omega t). \tag{4.8}$$

By a translation of the origin of the coordinate system we can put $\boldsymbol{x}_0 = 0$. The force $\boldsymbol{F}$ acting on an electron is given by charge times field strength

$$\boldsymbol{F} = (-e)\boldsymbol{E} \tag{4.9}$$

where we have explicitly expressed the negative sign of the electronic charge by the minus sign. For a (spatially) constant force the potential energy is proportional to the distance $\boldsymbol{x}$. Taking into account that $\boldsymbol{E}$ and $\boldsymbol{x}$ are vectors the potential energy is given by

$$V(\boldsymbol{x}) = e\boldsymbol{x}\cdot\boldsymbol{E}_0 \cos \omega t. \tag{4.10}$$

We write it in the form

$$V(\boldsymbol{x}) = \boldsymbol{\vartheta}\cdot\boldsymbol{E} \tag{4.11}$$

which allows for the following interpretation. $\boldsymbol{\vartheta}$ is a dipole moment = charge times distance, whereas $\boldsymbol{E}$ is again the time dependent field strength. Equation (4.11) is a relation well known in electrostatics expressing the energy of a constant dipole moment in a spatially homogeneous electric field. Inserting (4.11) into (4.5) yields for the special case $n = m = 1$

$$H_{11}^{\mathrm{P}} = \int \varphi_1^*(\boldsymbol{x}) e\boldsymbol{x}\varphi_1(\boldsymbol{x})\,\mathrm{d}V\cdot\boldsymbol{E}. \tag{4.12}$$

In many cases of practical interest it turns out that

$$\int \varphi_1^*(\boldsymbol{x})\boldsymbol{x}\varphi_1(\boldsymbol{x})\,\mathrm{d}V = 0 \tag{4.13}$$

vanishes (compare exercise at the end of this section). This fact is expressed by saying that the atom has no static dipole moment. The same is true in many cases for the excited state so that

$$H_{22}^{\mathrm{P}} = \int \varphi_2^* e\boldsymbol{x}\varphi_2\,\mathrm{d}V\cdot\boldsymbol{E} = 0. \tag{4.14}$$

We now turn to matrix elements for which $n \neq m$ and we consider for instance

$$H_{12}^{\mathrm{p}} = \int \varphi_1^* ex\varphi_2 \, \mathrm{d}V \cdot \boldsymbol{E}. \tag{4.15}$$

We assume that this matrix element does not vanish and we abbreviate the integral in (4.15) by $\boldsymbol{\vartheta}_{12}$, so that

$$H_{12}^{\mathrm{p}} = \boldsymbol{\vartheta}_{12} \cdot \boldsymbol{E} \tag{4.16}$$

holds. In the next chapter we shall show how to solve the equations (4.6) under the assumption (4.13), (4.14), (4.16). As is seen, at no instant we made explicit use of the specific form of the wave functions so that the method is quite general. On the other hand we must not forget that we have still to justify why we are allowed to deal only with two levels. We shall come back to this point in later chapters.

Exercise on section 4.1

(1) Let $\varphi(x)$ have the property $\varphi(-x) = \varphi(x)$.

Show $I \equiv \int_{-\infty}^{\infty} \varphi^*(x) x \varphi(x) \, \mathrm{d}x = 0$. Show that the same is true for $\varphi(-x) = -\varphi(x)$.

Hint: The integral remains invariant (unchanged) if $x \to -x$. Show that on the other hand $I \to -I$ under the transformation $x \to -x$, by replacing everywhere x by $-x$.

4.2. Interaction of a two-level system with incoherent light. The Einstein coefficients

We start from equations (4.6) assuming

$$H_{11}^{\mathrm{p}} = H_{22}^{\mathrm{p}} = 0. \tag{4.17}$$

As a first step towards the solution of (4.6) we make the substitutions

$$\begin{aligned} c_1(t) &= d_1 \exp[-\mathrm{i}W_1 t/\hbar] \\ c_2(t) &= d_2 \exp[-\mathrm{i}W_2 t/\hbar] \end{aligned} \tag{4.18}$$

where the coefficients d_1 and d_2 are still unknown time dependent functions. Inserting (4.18) we readily verify that the energies W_1, W_2 drop out from (4.6) and we are left with the equations

$$\mathrm{i}\hbar \frac{\mathrm{d}}{\mathrm{d}t} d_1 = d_2 H_{12}^{\mathrm{p}} \exp[-\mathrm{i}\bar{\omega} t] \tag{4.19}$$

$$\mathrm{i}\hbar \frac{\mathrm{d}}{\mathrm{d}t} d_2 = d_1 H_{21}^{\mathrm{p}} \exp[\mathrm{i}\bar{\omega} t] \tag{4.20}$$

where we have used the abbreviation

$$\bar{\omega} \equiv \omega_{21} = (W_2 - W_1)/\hbar. \tag{4.21}$$

In order to solve (4.19) and (4.20) we apply a procedure which is quite typical for quantum theory. It is called time-dependent perturbation theory. To apply this procedure we assume that the perturbation is comparatively small. We shall see somewhat later what "smallness" means exactly. Since the perturbation is small we may assume that the amplitudes d_1 and d_2 change only slowly in time. Note that without perturbation these amplitudes would be entirely time independent. In a first step we adopt the initial condition

$$d_1(0) = 1, \qquad d_2(0) = 0. \tag{4.22, 23}$$

According to this initial condition at time $t = 0$ the atom is in its lower state. Since the coefficients d_1, d_2 change only slowly with increasing time we may assume that it is a good approximation to insert on the right hand side of (4.19) and (4.20) these initial values (4.22), (4.23). Thus (4.20) now reads

$$\mathrm{i}\hbar \frac{\mathrm{d}}{\mathrm{d}t} d_2 \approx H_{21}^{\mathrm{p}} \exp[\mathrm{i}\bar{\omega}t] \tag{4.24}$$

or after integration over time

$$d_2 = (-\mathrm{i}/\hbar) \int_0^t H_{21}^{\mathrm{p}}(\tau) \exp[\mathrm{i}\bar{\omega}\tau] \,\mathrm{d}\tau. \tag{4.25}$$

We have put the lower limit of the integral equal to 0 so that the coefficient d_2 fulfills the initial condition (4.23). To evaluate the integral in (4.25) further the electric field strength is represented as a superposition of waves whose phases Φ are statistically distributed. We will explain in a minute what is meant by "statistically". We note that this approach is only a model which can be substantiated later by a fully quantum mechanical treatment in which also the electromagnetic field is quantized. Our model shows on the other hand all essential features of the action of an incoherent field on an atom. Our hypothesis thus reads

$$\boldsymbol{E}(t) = \sum_{\omega_\lambda > 0} \boldsymbol{E}_\lambda \exp[\mathrm{i}\Phi_\lambda - \mathrm{i}\omega_\lambda t] + \text{c.c.} \tag{4.26}$$

Inserting it into (4.25) and noting the form of H_{21}^{p} (4.16) we obtain

$$d_2(t) = (-\mathrm{i}/\hbar) \sum_\lambda \boldsymbol{E}_\lambda \cdot \boldsymbol{\vartheta}_{21} \exp[\mathrm{i}\Phi_\lambda] \int_0^t \exp[\mathrm{i}(\bar{\omega} - \omega_\lambda)\tau] \,\mathrm{d}\tau. \tag{4.27}$$

The integral which we abbreviate by S_λ is readily evaluated

$$S_\lambda \equiv \int_0^t \cdots \mathrm{d}\tau = [\mathrm{i}(\bar{\omega} - \omega_\lambda)]^{-1} \{\exp[\mathrm{i}(\bar{\omega} - \omega_\lambda)t] - 1\}. \tag{4.28}$$

We are now interested in the probability with which the level 2 is occupied at a later time t. To this end we form $|d_2|^2 = d_2^* d_2$. Inserting the corresponding expression (4.27) for d_2 and its complex conjugate for d_2^* we obtain

$$|d_2|^2 = \hbar^{-2} \sum_{\lambda,\lambda'} \exp[\mathrm{i}\Phi_\lambda - \mathrm{i}\Phi_{\lambda'}](\boldsymbol{E}_\lambda \cdot \boldsymbol{\vartheta}_{21})(\boldsymbol{E}_{\lambda'}^* \cdot \boldsymbol{\vartheta}_{21}^*) S_\lambda S_{\lambda'}^*. \tag{4.29}$$

We now use our assumption that the phases Φ_λ are statistically distributed. In order to evaluate (4.29) properly we therefore have to supplement it by a statistical average over the phases. We indicate the phase average by brackets. As is shown in the exercises for statistically independent phases we obtain the relation

$$\langle \exp[\mathrm{i}\Phi_\lambda - \mathrm{i}\Phi'_\lambda] \rangle = \delta_{\lambda\lambda'} \tag{4.30}$$

where $\delta_{\lambda\lambda'}$ is the Kronecker symbol $\delta_{\lambda\lambda'} = 1$ for $\lambda = \lambda'$ and $= 0$ otherwise. As a consequence of the Kronecker symbol all terms of the double sum (4.29) vanish unless $\lambda = \lambda'$. This yields

$$\langle |d_2(t)|^2 \rangle = \hbar^{-2} \sum_\lambda |E_\lambda|^2 (\overline{\omega} - \omega_\lambda)^{-2} \cdot 4 \sin^2[(\overline{\omega} - \omega_\lambda)t/2] \cdot |\boldsymbol{e} \cdot \boldsymbol{\vartheta}_{21}|^2 \tag{4.31}$$

where we have used the abbreviation

$$\boldsymbol{E}_\lambda = \boldsymbol{e} E_\lambda. \tag{4.32}$$

$\boldsymbol{e}$ is a unit vector in the direction of the light polarization. Equation (4.32) implies that the incident lightwave is polarized in a single direction. Let us discuss the result (4.31) first by looking at the individual terms. For $\overline{\omega} \neq \omega_\lambda$, i.e. in the absence of resonance the sine function varies periodically, i.e. we don't get any remaining transition for the atomic state 1 into the atomic state 2. On the other hand, in case of resonance $\overline{\omega} = \omega_\lambda$ we notice that $|d_2|^2$ increases proportional to t^2. This is a result which seems to contradict our experience: When we let light impinge on a set of atoms, on the average the number of excited atoms increases linearly with time and not quadratically. For a long time this result was considered as entirely unphysical. Later we will see, however, that this result is very reasonable when we let an atom interact with entirely coherent light.

By postponing this problem to a later chapter let us come back to the case considered here, namely incoherent light. In this case neither the non-resonant nor the resonant case yield the expected result namely a linear increase of occupation number of level 2. This dilemma is resolved, however, when we assume that the incident light field comprises a whole frequency band. This is realized when we deal with light from thermal sources, i.e. incoherent light.

With these ideas in our mind, we continue to treat (4.31). As an intermediate step we differentiate (4.31) with respect to time

$$\frac{\mathrm{d}}{\mathrm{d}t}\langle |d_2(t)|\rangle^2 = \hbar^{-2}\sum_{\lambda} |E_\lambda|^2(\bar{\omega} - \omega_\lambda)^{-1}\cdot 2\sin[(\bar{\omega} - \omega_\lambda)t]|\boldsymbol{e}\cdot\boldsymbol{\vartheta}_{21}|^2. \tag{4.33}$$

The square of the field strength is connected with the intensity. Introducing furthermore an intensity I per frequency interval we put

$$|E_\lambda|^2 = I(\omega_\lambda)\Delta\omega_\lambda. \tag{4.34}$$

By inserting this expression into (4.33) we are left with the evaluation of

$$(1/\hbar^2)\sum_{\lambda} I(\omega_\lambda)\Delta\omega_\lambda \frac{\sin[(\bar{\omega} - \omega_\lambda)t]}{\bar{\omega} - \omega_\lambda}. \tag{4.35}$$

Since usually the frequencies of the incident light wave are continuously distributed we proceed from a summation over individual frequencies ω_λ to an integration replacing $\omega_\lambda - \bar{\omega}$ by ω. Neglecting for the moment being all unimportant factors and assuming furthermore that E_λ does not change close to resonance $\omega_\lambda = \bar{\omega}$ we have to evaluate the expression

$$I(\bar{\omega})\int_{-\infty}^{\infty} \mathrm{d}\omega \sin(\omega t)/\omega \tag{4.36}$$

instead of (4.35). The integral (4.36) can be easily found in tables of integrals. It has the value π. Collecting all factors we obtain the final expression for (4.33) namely

$$\frac{\mathrm{d}}{\mathrm{d}t}\langle |d_2(t)|^2\rangle = \hbar^{-2}2\pi|\boldsymbol{e}\cdot\boldsymbol{\vartheta}_{21}|^2 I(\bar{\omega}). \tag{4.37}$$

We are now in a position to compare this result directly with Einstein's hypothesis on the absorption and stimulated emission of light. The left-hand side of (4.37) is just the temporal change of the occupation number of the upper level per atom. This rate is caused by the incident light under the assumption that initially the atom was in its ground state.

This relation can be easily extended to N atoms. In this case on the left-hand side we then have the temporal change of the numbers of atoms in the excited state. On the right-hand side, we would have the light intensity I, the number of atoms N_1 with initially occupied ground states and a constant coefficient which we may call B_{12}. Replacing I by

$$\rho(\omega) = 2\varepsilon_0 I(\omega) \tag{4.38}$$

where ε_0 is the dielectric constant of vacuum, we are exactly led to

$$(\mathrm{d}N/\mathrm{d}t) = B_{12}N_1\rho \tag{4.38a}$$

which is nothing but Einstein's equation which we got to know in section 2.5. Comparing (4.37) with (4.38a), for the special case $N_1 = 1$ we immediately obtain an explicit expression for the Einstein coefficient

$$B_{12} = \frac{\pi}{\hbar^2 \varepsilon_0} |e \cdot \vartheta_{21}|^2. \tag{4.39}$$

When we repeat the above calculations with the two levels 1 and 2 exchanged we will find exactly the same result. This means that we could have considered the emission process in exactly the same way as the absorption process leading to the same Einstein coefficient. The transitions can take place only when an external field is present. This is why we have to call the emission process, just considered, stimulated emission.

The above outlined approach permits us to derive the Einstein coefficients from a quantum theoretical treatment. There are two unsatisfactory points, however. First of all, we observe that the formalism does not yield any spontaneous emission. Indeed if there is no field then there is no perturbation which can cause any transitions. Within quantum theory this gap can be only surmounted when the electromagnetic field is quantized. We will come back to this point in chapter 7.

Furthermore, the assumption about the statistically distributed phases seems, to some extent, artificial and one should wish that also this result comes out automatically. Indeed we shall see in later chapters that the random phases are again a consequence of a fully quantum mechanical treatment of the light field.

Exercise on section 4.2

(1) We define the phase average by

$$\langle \exp[\mathrm{i}(\Phi_j - \Phi_k)] \rangle = \frac{1}{2\pi} \int_0^{2\pi} \mathrm{d}\Phi_j \frac{1}{2\pi} \int_0^{2\pi} \mathrm{d}\Phi_k \exp[\mathrm{i}(\Phi_j - \Phi_k)].$$

Prove (4.30) by performing the integrations.

4.3. Higher-order perturbation theory

In the preceding paragraph we had assumed that the incident lightwave can be considered as a small perturbation. This made it possible to calculate d_2 or d_1 in a simple manner. This approximation is indeed justified in most cases when light stems from thermal sources. By means of the laser it has become possible, however, to generate light of very high intensity. Therefore in a number of cases our approximation of section 4.2 is no more sufficient. It is necessary to determine higher order terms of the perturbation in a systematic way. As we shall see later, such higher order terms are of fundamental importance in laser theory itself but also in non-linear optics and in non-linear spectroscopy. Since the procedure applies to quite general quantum systems and perturbations, we do not write down the corresponding Hamiltonian explicitly. For visualization the reader may, however, identify the unperturbed Hamiltonian with that of an electron in a potential field, for instance in the hydrogen atom, and the perturbation operator as stemming from an incident lightfield similar to that of section 4.2. We therefore start from the Schrödinger equation

$$H\psi = \mathrm{i}\hbar\frac{\mathrm{d}\psi}{\mathrm{d}t} \tag{4.40}$$

whose Hamiltonian is assumed in the form

$$H = H_0 + H^{\mathrm{P}}. \tag{4.41}$$

We assume that the unperturbed problem has been solved

$$H_0\varphi_n = W_n\varphi_n. \tag{4.42}$$

The corresponding wave functions and energies are distinguished by the quantum number n. This quantum number can be interpreted in various ways, corresponding to the problem under consideration. For instance in the case of the hydrogen atom, the quantum number n stands in reality for the triple n, l, m.

The wave functions φ_n depend on the electron coordinate. Using the bra-ket formalism (compare section 3.2) one can reformulate the whole following procedure so that it applies to any quantum systems, for instance to spin systems. Here, however, to consider a concrete case let us assume the wave functions in the form

$$\varphi_n = \varphi_n(\boldsymbol{x}). \tag{4.43}$$

As is shown in mathematics the wanted solution of eq. (4.40) can be exactly represented as a superposition of the complete set of functions φ_n:

$$\psi(t) = \sum_n d_n(t)\varphi_n. \tag{4.44}$$

In section 4.2 we saw that it is useful to extract from the coefficients c_n an exponential function

$$\exp[-\mathrm{i}W_n t/\hbar].$$

Thus we make the ansatz

$$\psi(t) = \sum_n c_n(t)\exp[-\mathrm{i}W_n t/\hbar]\varphi_n \tag{4.45}$$

instead of (4.44).

The coefficients c_n are still unknown time-dependent functions. Our task is to solve the equation

$$(H_0 + H^{\mathrm{P}})\psi = \mathrm{i}\hbar \frac{\mathrm{d}\psi}{\mathrm{d}t} \tag{4.46}$$

by means of the ansatz (4.45). To this end we insert (4.45) into (4.46) on both sides and use equation (4.42). Carrying out the differentiation with respect to time we find

$$\sum_n c_n(t) \exp[-\mathrm{i}W_n t/\hbar](W_n + H^{\mathrm{P}})\varphi_n$$

$$= \sum_n \exp[-\mathrm{i}W_n t/\hbar]\left(W_n c_n(t) + \mathrm{i}\hbar \frac{\mathrm{d}c_n(t)}{\mathrm{d}t}\right)\varphi_n. \tag{4.47}$$

Clearly, on both sides of (4.47) those terms cancel which contain the factor W_n. In order to get rid of the spatial dependence of the functions in (4.47) we multiply (4.47) from the left with one of the functions φ_m^* and integrate over the whole volume. When doing so we obtain expressions of the form

$$\int \varphi_m^* H^{\mathrm{P}} \varphi_n \,\mathrm{d}V \equiv \langle m | H^{\mathrm{P}} | n \rangle \equiv H_{mn}^{\mathrm{P}} \tag{4.48}$$

where the two expressions on the right-hand sides are only different abbreviations of the integral on the l.h.s. After these manipulations, the system (4.47) reduces to

$$\mathrm{i}\hbar \frac{\mathrm{d}c_m(t)}{\mathrm{d}t} = \sum_n c_n(t) \exp[\mathrm{i}\omega_{mn} t] H_{mn}^{\mathrm{P}} \tag{4.49}$$

where we have used the abbreviation

$$\hbar\omega_{mn} = W_m - W_n. \tag{4.50}$$

To specify the problem we have to fix the initial state at initial time t_0. We assume that a certain state φ_{n_0} was occupied at initial time t_0. This means that the whole expression (4.45) reduces to φ_{n_0} at that time. This implies an initial condition for the coefficients c_n which reads

$$c_n(t_0) = \delta_{n,n_0}. \tag{4.51}$$

We define the "0's" approximation for our iteration procedure. The coefficients of this approximation will be denoted by an upper index 0, i.e. we define

$$c_n^{(0)} \equiv c_n(t_0). \tag{4.52}$$

The basic idea of perturbation theory is as follows. Since the perturbation

H^{P} is relatively small, the coefficients c_n will differ from the 0's approximation only little, at least for sufficiently small times. We therefore use $c_n(t_0)$, given by (4.51), as an approximation for $c_n(t)$ on the r.h.s. of (4.49). This yields

$$\frac{\mathrm{d}}{\mathrm{d}t} c_m^{(1)}(t) = (-\mathrm{i}/\hbar) \exp\left[\mathrm{i}\omega_{mn_0} t\right] H^{\mathrm{P}}_{mn_0}. \tag{4.53}$$

By integration over time we immediately obtain an explicit expression for the c's in first approximation

$$c_m^{(1)}(t) = (-\mathrm{i}/\hbar) \int_{t_0}^{t} \exp\left[\mathrm{i}\omega_{mn_0}\tau\right] H^{\mathrm{P}}_{mn_0} \,\mathrm{d}\tau + \delta_{m,n_0}. \tag{4.54}$$

The upper index (1) now indicates first approximation. The Kronecker symbol on the right hand side of (4.54) takes care of the initial condition (4.51). In the following we have to note that the matrix elements (4.48) may still be time dependent. An explicit example is that of the time-dependent light field of section 4.2. So far we have only recovered the same expression as (4.25) of section 4.2, however now generalized to a system with arbitrarily many levels. To proceed further we may imagine that we now insert an improved expression for $c_n(t)$ on the right-hand side of (4.49), in order to obtain on the left-hand side once again an improved coefficient $c_n^{(2)}$. Continuing this procedure step by step the $l + 1$ step leads us to the relation

$$c_m^{(l+1)}(t) = c_m^{(0)} + (-\mathrm{i}/\hbar) \sum_n \int_{t_0}^{t} c_n^{(l)}(\tau) \exp\left[\mathrm{i}\omega_{mn}\tau\right] H^{\mathrm{P}}_{mn}(\tau) \,\mathrm{d}\tau. \tag{4.55}$$

Using the abbreviation

$$\hat{H}_{mn}(\tau) = \exp\left[\mathrm{i}\omega_{mn}\tau\right] H^{\mathrm{P}}_{mn}(\tau) \tag{4.55a}$$

we can write (4.55) in the form

$$c_m^{(l+1)}(\tau) \equiv c_m^{(0)} + (-\mathrm{i}/\hbar) \sum_n \int_{t_0}^{t} \hat{H}_{mn}(\tau) c_n^{(l)}(\tau) \,\mathrm{d}\tau. \tag{4.55b}$$

The relation (4.55) is a recurrence formula. With its aid we can calculate $c_m^{(l+1)}$ once $c_n^{(l)}$ has been determined. We can learn about the most important aspects when looking at $l = 1$. In this case (4.55) reads

$$c_m^{(2)}(t) = (-\mathrm{i}/\hbar) \sum_{n_1} \int_{t_0}^{t} \hat{H}_{mn_1}(\tau) c_{n_1}^{(1)}(\tau) \,\mathrm{d}\tau + c_m^{(0)}. \tag{4.56}$$

For $l = 0$ we get (4.54) which we write in a slightly generalized form as

$$c_{n_1}^{(1)}(t) = (-\mathrm{i}/\hbar) \sum_{n_2} \int_{t_0}^{t} \hat{H}_{n_1 n_2}(\tau) \,\mathrm{d}\tau c_{n_2}^{(0)} + c_{n_1}^{(0)} \tag{4.57}$$

where the $c_m^{(0)}$ are given constants. It remains our task to express the coefficients in second order, $c_m^{(2)}$, by $c_m^{(0)}$. To this end we insert (4.57) into (4.56) which yields

$$c_m^{(2)}(t) = c_m^{(0)} + (-\mathrm{i}/\hbar)\sum_{n_1}\int_{t_0}^{t}\hat{H}_{mn_1}(\tau)\,\mathrm{d}\tau c_{n_1}^{(0)}$$

$$+ (-\mathrm{i}/\hbar)^2\sum_{n_1,n_2}\int_{t_0}^{t}\hat{H}_{mn_1}(\tau_1)\,\mathrm{d}\tau_1\int_{t_0}^{\tau_1}\hat{H}_{n_1n_2}(\tau_2)\,\mathrm{d}\tau_2 c_{n_2}^{(0)}. \quad (4.58)$$

The expression (4.58) contains quite a number of highly interesting effects which will be discussed in the following sections and, in still much greater detail, in the volume on non-linear optics. Readers who are not too much interested in mere mathematics can end this section here and proceed to the following sections. For the sake of completeness we now demonstrate the general result for $c_n^{(l+1)}$. Eliminating all intermediate c's we obtain

$$c_n^{(l+1)}(t) = c_n^{(0)} + (-i/\hbar)\sum_{n_1}\int_{t_0}^{t}\hat{H}_{nn_1}(\tau)\,\mathrm{d}\tau c_{n_1}^{(0)}$$

$$+ (-i/\hbar)^2\sum_{n_1n_2}\int_{t_0}^{t}\hat{H}_{nn_2}(\tau_1)\,\mathrm{d}\tau_1\int_{t_0}^{\tau_1}\hat{H}_{n_2n_1}(\tau_2)\,\mathrm{d}\tau_2 c_{n_1}^{(0)} + \cdots$$

$$+ (-i/\hbar)^{l+1}\sum_{n_1,n_2\ldots n_{l+1}}\int_{t_0}^{\tau}\hat{H}_{nn_{l+1}}(\tau_1)\,d\tau_1\int_{t_0}^{\tau_1}\hat{H}_{n_{l+1}n_l}(\tau_2)\,d\tau_2\ldots$$

$$\times\int_{t_0}^{\tau_l}\hat{H}_{n_2n_1}(\tau_{l+1})\,d\tau_{l+1}c_{n_1}^{(0)}. \quad (4.59)$$

The coefficients $c_n(t)$ which were originally wanted are obtained by pushing this kind of perturbation theory to infinite order, i.e.

$$c_n(t) = \lim_{l\to\infty} c_n^{(l)}(t). \quad (4.60)$$

The series expansion we have derived here is called a Born series in physics. In section 7.7 a formally different approach to time-dependent perturbation theory will be presented.

4.4. Multi-quantum transitions. Two-photon absorption

In this and the following sections we wish to give some explicit examples which will elucidate some of the physical contents of perturbation theory.

We consider a quantum system, e.g. an atom with three energy levels, 1, 2, 3 (cf. fig. 4.1). We subject this atom to a light field described by its

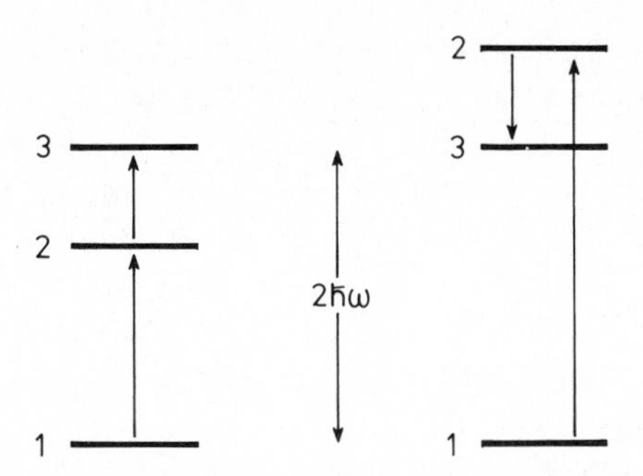

Fig. 4.1. 2-photon absorption illustrated by an example of an atom with 3 levels. Left part: The electron makes a transition from its ground level 1 to its excited level 3 via an intermediate level 2 which lies energetically inbetween the levels 1 and 3. Right part: The electron makes a transition from its groundstate 1 to its excited state 3 via an intermediate level 2 which lies energetically above the level 3. Note that in both cases the energy difference between the intermediate state and the groundstate does not coincide with the energy $\hbar\omega$ of a single light quantum. If the energy difference between the intermediate state and the groundstate is equal to the energy $\hbar\omega$ of the light quantum a physically different process occurs, namely an absorption cascade which consists of a sequence of absorption processes each of which can be treated by first order perturbation theory.

electric field strength. First we assume its frequency ω purely monochromatic such that

$$2\omega = \omega_{31} = (W_3 - W_1)/\hbar. \tag{4.61a}$$

We further assume that

$$\omega \neq \omega_{mn}, \qquad m, n = 1, 2, 3. \tag{4.61b}$$

We now apply the results of section 4.3 to this problem. We assume that the atom is initially in its ground state, 1, and we are interested in the change of the occupation number of level 3. The matrix elements have the general form (compare also (4.15))

$$H^{\mathrm{P}}_{mn} = A_{mn} \exp[-\mathrm{i}\omega t] + \mathrm{c.c.} \tag{4.62}$$

We assume as usual

$$H^{\mathrm{P}}_{mm} = 0, \qquad m = 1, 2, 3 \tag{4.63a}$$

and furthermore

$$H^{\mathrm{P}}_{13} = H^{\mathrm{P}}_{31} = 0 \tag{4.63b}$$

but

$$H^{\mathrm{P}}_{12} \neq 0, \qquad H^{\mathrm{P}}_{21} \neq 0, \qquad H^{\mathrm{P}}_{32} \neq 0, \qquad H^{\mathrm{P}}_{23} \neq 0. \tag{4.63c}$$

Although (4.63b) seems to forbid transitions between levels 1 and 3, we will show, that nevertheless such transitions become possible via a two-step process. To this end we use perturbation theory of second order, i.e. (4.58). We study the change of the occupation number of level 3, i.e. we investigate $c_3^{(2)}(t)$. The individual contributions to $c_3^{(2)}$ on the r.h.s. of (4.58) can be discussed as follows: Due to the initial condition, $c_3^{(0)} = 0$. In the following sum, $c_3^{(0)} = c_2^{(0)} = 0$, and $c_1^{(0)} = 1$. However, $\hat{H}_{31} = 0$ due to our assumption (4.63b). Thus we are left with the study of the last term on the r.h.s. of (4.58). Owing to the initial condition, $c_{n_2}^{(0)} \neq 0$ only for $n_2 = 1$. Furthermore, we have already identified m with 3, $m = 3$. Due to (4.63a, b, c) the only non-vanishing matrix elements $\hat{H}_{3n_1}$, $\hat{H}_{n_1 1}$ are $\hat{H}_{32}$ and $\hat{H}_{21}$, respectively. Consequently, $c_3^{(2)}$ is given by

$$c_3^{(2)}(t) = (-\mathrm{i}/\hbar)^2 \int_{t_0}^{t} \hat{H}_{32}(\tau_1)\,\mathrm{d}\tau_1 \int_{t_0}^{\tau_1} \hat{H}_{21}(\tau_2)\,\mathrm{d}\tau_2 \tag{4.64}$$

so that we have only to evaluate the double integral. As may be shown (compare exercise at the end of this section) the most important contributions to (4.64) stem from the first expression $\propto \exp[-\mathrm{i}\omega t]$ in (4.62). With its use we obtain after integration, choosing $t_0 = 0$

$$c_3^{(2)} = (\mathrm{i}/\hbar) t \frac{A_{32} A_{21}}{W_2 - W_1 - \hbar\omega}. \tag{4.65}$$

From this result it follows that the incident field can cause transitions from the lower level 1 to the upper level 3. This process can be visualized as follows: The electron first goes from level 1 to level 2 by absorption of a light quantum with energy $\hbar\omega$ and then from level 2 to level 3 again by absorption of a quantum $\hbar\omega$. Note that within this process it is by no means necessary that the energy difference $W_2 - W_1$ coincides with $\hbar\omega$. Such a process in which energy is not conserved is called virtual transition. Note, however, that energy conservation is required for the total process

$$W_3 = W_1 + 2\hbar\omega. \tag{4.66}$$

The total process is that of two-photon absorption. To deal with an n-photon absorption process we shall learn to use perturbation theory of nth order.

Since in many cases the incident light is not entirely monochromatic one has to extend the formalism in analogy to section 4.2, taking into account random light phases. One then readily establishes that the occupation number of level 3 increases linearly with time and not quadratically. Since the corresponding formulas are somewhat lengthy and will be treated in the volume on nonlinear optics, we will not present them here (see fig. 4.2).

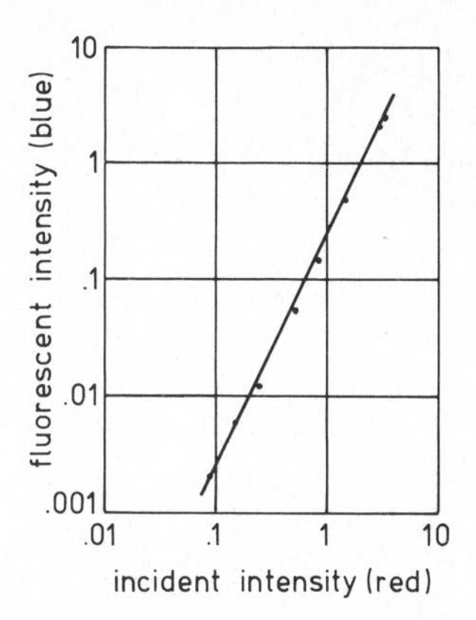

Fig. 4.2. In this experiment done by Kaiser and Garrett (1961) the increase of occupation number of the upper level is measured by observing the fluorescent intensity in the blue (ordinate). The abscissa shows the intensity of the incident red light. The double logarithmic plot shows clearly the increase of the absorption rate being a quadratic function of the intensity of the incident red light.

Exercise on section 4.4

Perform the integrations over time in (4.64) using

$$\hat{H}_{mn}(\tau) = \exp(\mathrm{i}\omega_{mn}\tau)(\exp[-\mathrm{i}\omega t] + \exp[\mathrm{i}\omega t]),$$
$$m, n = 3, 2 \text{ and } 2, 1$$

where $\omega_{mn} \neq \omega$ and $\omega_{32} + \omega_{21} = 2\omega$.
Choose $t_0 = 0$.
Show that for large enough time t, the leading term equals const$\cdot t$.

4.5. Non-resonant perturbations. Forced oscillations of the atomic dipole moment. Frequency mixing

We assume that the frequency of the external perturbation is not in resonance with any of the transition frequencies ω_{mn} of the quantum system and ω_{mn} does not coincide with any multiple of ω. We will see immediately that the coefficients $c_n(t)$ cannot grow in an unlimited way but rather oscillate. We will first investigate the physical meaning of these oscillations. We use wave functions of the form

$$\psi(\boldsymbol{x}, t) = \sum_n c_n(t) \exp[-\mathrm{i}W_n t/\hbar] \varphi_n(\boldsymbol{x}) \tag{4.67}$$

as have been introduced in section (4.3). By means of (4.67), we can calculate the dipole moment

$$\vartheta = \int \psi^*(\boldsymbol{x}, t)(-e\boldsymbol{x})\psi(\boldsymbol{x}, t)\,\mathrm{d}V. \tag{4.68}$$

We have already met a special case of this expression in section (4.1). Furthermore generalizing (4.68) we introduce the abbreviation

$$\vartheta_{mn} = \int \varphi_m^*(\boldsymbol{x}) e\boldsymbol{x} \varphi_n(\boldsymbol{x})\,\mathrm{d}V. \tag{4.69}$$

Inserting (4.67) into (4.68) we obtain

$$\vartheta = -\sum_{m,n} c_m^*(t) c_n(t) \exp[\mathrm{i}\omega_{mn} t] \vartheta_{mn}. \tag{4.70}$$

In (4.67) and thus in (4.70) the coefficients $c_n(t)$ are still unknown. To find them explicitly we evaluate them by means of 1st order time-dependent perturbation theory. We use them in the form

$$c_n(t) \Rightarrow c_n^{(1)}(t) = c_n^{(0)} + a_n \tag{4.71}$$

where the coefficient a_n is proportional to the perturbation

$$a_n \sim H^{\mathrm{p}}. \tag{4.72}$$

We adopt the usual initial condition

$$c_n^{(0)} = \delta_{n,n_0}. \tag{4.73}$$

Furthermore we assume that the atom does not possess constant dipole moments

$$\vartheta_{mm} = 0, \quad \text{for all } m. \tag{4.74}$$

Because the perturbation is assumed small we treat the coefficients a_n as small quantities. We first keep only those terms of (4.70) which are linear in a_n, which is consistent with first order perturbation theory. Thus (4.70) reads

$$\vartheta = -\sum_m \left(a_m^*(t) \exp\left[\mathrm{i}\omega_{mn_0} t\right] \vartheta_{mn_0} + \text{c.c.} \right). \tag{4.75}$$

We now wish to calculate (4.75) explicitly. To this end we use the explicit relation (4.57) of perturbation theory. The corresponding perturbation Hamiltonian is taken in the form

$$H^{\mathrm{p}} = e\boldsymbol{x}\cdot\boldsymbol{E}, \qquad \boldsymbol{E} = 2\boldsymbol{E}_0' \cos\omega t \tag{4.76}$$

so that the matrix elements have the form

$$H_{mn}^{\mathrm{p}}(\tau) = A_{mn}[\exp(-\mathrm{i}\omega\tau) + \text{c.c.}] \tag{4.77}$$

which we just encountered before [cf. (4.15), (4.62)]. As shown below, the dipole moment of the atom oscillates under the influence of the applied periodic field. In this treatment we are not interested in switching-on processes, but rather in the steady state response of atoms to the action of the external field. For this reason we let t_0 go to $-\infty$. In order to suppress the initial oscillations which occur in switching-on processes, we use a mathematical trick: We add a factor $\exp[-\gamma(t-\tau)]$ to the perturbation. It means physically that we slowly switch on the interaction between the atom and the field. After all integrations over time have been performed we let the constant γ go to 0. Using the expressions introduced above we can write the coefficient a_m^* in the form

$$(-\mathrm{i}/\hbar)\int_{-\infty}^{t} \exp[\mathrm{i}\omega_{mn}\tau]H_{mn}^{\mathrm{p}}(\tau)\exp[-\gamma(t-\tau)]\,\mathrm{d}\tau. \tag{4.78}$$

The integral can be done immediately. We readily find

$$a_m^* = (\mathrm{i}/\hbar)A_{mn_0}^*\Big\{\big(-\mathrm{i}\Omega_{mn_0}\big)^{-1}\exp\big[-\mathrm{i}\Omega_{mn_0}t\big] + \big(-\mathrm{i}\Omega_{mn_0}^+\big)^{-1}\exp\big[-\mathrm{i}\Omega_{mn_0}^+t\big]\Big\} \tag{4.79}$$

where we have used the abbreviations

$$\Omega_{mn} = \omega_{mn} - \omega, \qquad \Omega_{mn}^+ = \omega_{mn} + \omega. \tag{4.80}$$

Thus the dipole moment $\boldsymbol{\vartheta}$ acquires the form

$$\boldsymbol{\vartheta} = (1/\hbar)\sum_m A_{mn_0}^*\big\{\Omega_{mn_0}^{-1}\exp[\mathrm{i}\omega t] + \Omega_{mn_0}^{+-1}\exp[-\mathrm{i}\omega t]\big\}\boldsymbol{\vartheta}_{mn_0} + \text{c.c.} \tag{4.81}$$

After some simple algebraic manipulations we find

$$\boldsymbol{\vartheta}(t) = \boldsymbol{E}(t)\sum_m (1/\hbar)|\boldsymbol{\vartheta}_{mn_0}|^2\frac{\omega_{mn_0}}{(\omega_{mn_0})^2-\omega^2} \tag{4.82}$$

where we have used as above the relation

$$\boldsymbol{E} = 2\boldsymbol{E}_0'\cos\omega t. \tag{4.83}$$

From (4.82) it is evident that the atomic dipole moment oscillates exactly in the same manner as the impinging field oscillates. This relation between dipole moment and electric field strength had been treated in classical theory assuming the following model. A harmonically bound particle with mass m and electric charge e is subject to a harmonic electric field. It is found that the particle oscillates in phase with the field. This relation is the basis for the classical theory of dispersion. In (4.82) we have found its quantum mechanical analogue.

We now study whether higher terms of perturbation theory can give similar oscillatory contributions. To this end we consider the second sum in (4.58). To exhibit the new aspects clearly we assume that the incident lightwave is a superposition of two parts with frequencies ω_1 and ω_2 so that the perturbation Hamiltonian has the form

$$H^{\mathrm{P}} = \{\boldsymbol{E}_1 2\cos\omega_1 t + \boldsymbol{E}_2 2\cos\omega_2 t\}e\boldsymbol{x}. \tag{4.84}$$

The corresponding double time integrals in (4.58) can be easily evaluated. We assume as stated in the beginning that

$$\omega_j \neq \omega_{mn}, \qquad 2\omega_j \neq \omega_{mn} \quad (j = 1,2), \qquad \omega_2 \pm \omega_1 \neq \omega_{mn}. \tag{4.85}$$

To demonstrate the general idea we consider a typical term. After performing the integration it reads

$$a_m^{(2)} = -(-1/\hbar)^2 \sum_{n_1} \vartheta_{mn_1}\vartheta_{n_1 n_0} E_1 E_2 (\omega_{mn_1} - \omega_2)^{-1}$$
$$\times (\omega_{mn_0} - \omega_1 - \omega_2)^{-1} \exp\left[\mathrm{i}(\omega_{mn_0} - \omega_1 - \omega_2)t\right]. \tag{4.86}$$

Let us use this expression to evaluate the corresponding dipole moment. We then find

$$\vartheta^{(2)} \propto \sum_{n_1} \vartheta_{mn_0}\vartheta_{mn_1}\vartheta_{n_1 n_0} \exp[-\mathrm{i}(\omega_1 + \omega_2)t]. \tag{4.87}$$

We immediately recognize that the perturbation causes an oscillation of the atomic dipole moment with the sum frequency $\omega_1 + \omega_2$. Taking all the second order terms of (4.58) one may show that in principle all other frequency combinations

$$\begin{aligned} &2\omega_1, \omega_1 + \omega_2, \qquad 2\omega_2, \omega_1 - \omega_2 \\ &0, -\omega_1 + \omega_2, \qquad 0, -\omega_1 - \omega_2 \end{aligned} \tag{4.88}$$

can occur. To obtain this result we have considered only the second order terms of the coefficients $c_m^{(2)}$. For the sake of completeness, we mention that also products stemming from perturbation theory of first order must be taken into account when expressions of the form (4.70) are calculated. We will do this in a systematic manner in Volume 3. In the present context, it has been our main objective to explore the structure and the meaning of the contributions of perturbation theory up to second order. As we know from the theory of electromagnetism, oscillating dipoles can be the source of electromagnetic radiation. In the present case that means that the dipole moments (4.87) can cause electromagnetic waves at frequencies of the form (4.88). This is one of the simplest examples of frequency mixing in which by non-linear interaction two waves of frequencies ω_1 and ω_2 are, so to speak, transformed into new waves with sum and difference frequencies. This example illustrates the origin of many important effects in non-linear optics and such processes will be discussed in greater detail in Volume 3.

4.6. Interaction of a two-level system with resonant coherent light

We now treat eqs. (4.6) under the assumption that the incident light field is purely coherent. Such experiments have become only possible after the advent of the laser and they have led to entirely new phenomena which will also be discussed in Volume 3.

We assume exact resonance between the incident light field and the atomic transition frequency $\omega_0 = (W_2 - W_1)/\hbar$. Again we assume that the atom has no static dipole moment

$$H_{11}^{\mathrm{p}} = H_{22}^{\mathrm{p}} = 0. \tag{4.89}$$

The electric vector is assumed in the form $\boldsymbol{E} = \boldsymbol{E}_0 \cos \omega t$. The perturbation matrix elements (4.15) have the general structure

$$H_{12}^{\mathrm{p}} = H_{21}^{\mathrm{p}*} = a(\exp[\mathrm{i}\omega_0 t] + \exp[-\mathrm{i}\omega_0 t]) \tag{4.90}$$

where

$$a = \tfrac{1}{2}\boldsymbol{E}_0 \cdot \int \varphi_1^*(\boldsymbol{x}) e\boldsymbol{x} \varphi_2(\boldsymbol{x})\, \mathrm{d}V. \tag{4.91}$$

Under this assumption the eqs. (4.6) simplify to

$$\mathrm{i}\hbar \frac{\mathrm{d}}{\mathrm{d}t} c_1 = W_1 c_1 + c_2 a(\exp[\mathrm{i}\omega_0 t] + \exp[-\mathrm{i}\omega_0 t]) \tag{4.92}$$

$$\mathrm{i}\hbar \frac{\mathrm{d}}{\mathrm{d}t} c_2 = W_2 c_2 + c_1 a^*(\exp[\mathrm{i}\omega_0 t] + \exp[-\mathrm{i}\omega_0 t]). \tag{4.93}$$

For further simplification we make again the substitution

$$c_j(t) = \exp[-\mathrm{i}W_j t/\hbar]\, \mathrm{d}_j(t) \tag{4.94}$$

which yields

$$\mathrm{i}\hbar \frac{\mathrm{d}}{\mathrm{d}t} d_1 = d_2 a(1 + \exp[-2\mathrm{i}\omega_0 t]) \tag{4.95}$$

$$\mathrm{i}\hbar \frac{\mathrm{d}}{\mathrm{d}t} d_2 = d_1 a^*(\exp[2\mathrm{i}\omega_0 t] + 1). \tag{4.96}$$

We now assume that d_1 and d_2 change very little over times in which $\exp(2\mathrm{i}\omega_0 t)$ changes. In other words, we assume that $\exp(2\mathrm{i}\omega_0 t)$ oscillates much more quickly than d_1 or d_2. When we integrate eq. (4.95) over a time interval which contains many oscillations of the exponential function but which is small enough so that d_2 has changed little the integral over the exponential function practically vanishes (cf. fig. 4.3). This justifies neglecting the exponential functions $\exp(2i\omega_0 t)$ and $\exp(-2i\omega_0 t)$ compared to unity. This approximation is called the rotating wave approximation. The name may sound somewhat strange, however, we will see the reason for it in section 4.7 when we treat the response of spins to alternating magnetic fields. Under the rotating wave approximation eqs. (4.95), (4.96) acquire the form

$$i\hbar\frac{d}{dt}d_1 = ad_2, \qquad i\hbar\frac{d}{dt}d_2 = a^*d_1. \tag{4.97, 98}$$

To solve these coupled equations we first eliminate d_2. To this end we differentiate the first eq. (4.97) with respect to time and replace the resulting $\mathrm{d}d_2/\mathrm{d}t$ according to the second equation (4.98) by d_1.

Using the abbreviation

$$\Omega = |a|/\hbar \tag{4.99}$$

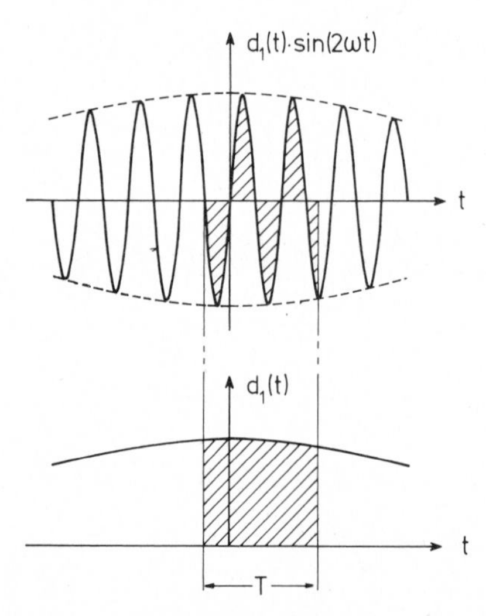

Fig. 4.3. This figure shows how to visualize the approximations explained in the text after eq. (4.96). The upper part shows the function $d_1(t)\sin(2\omega t)$ as a function of time. The dashed curve shows $d_1(t)$ which changes much more slowly than the sin-function. Due to the rapid oscillation of the sine function the total function d_1 sine$(2\omega t)$ changes its sign very rapidly. When we average over a time during which $d_1(t)$ has changed but little, the sine-function has made several oscillations, the positive and negative areas, which we indicate by shaded areas, practically cancel against each other. Lower part: In the absence of rapid oscillations the average over a time interval T leaves $d_1(t)$ unchanged.

we obtain

$$\frac{d^2}{dt^2}d_1 + \Omega^2 d_1 = 0. \tag{4.100}$$

This is the well-known equation of a classical harmonic oscillator. Its general solution has the form

$$d_1(t) = \alpha \cos \Omega t + \beta \sin \Omega t. \tag{4.101}$$

α and β are constants which must be fixed by the initial value of d_1 and d_2. To obtain d_2 we insert (4.101) in (4.97)

$$d_2(t) = (\mathrm{i}\hbar/a)\frac{d}{dt}d_1 = \mathrm{i}(\hbar\Omega/a)[-\alpha \sin \Omega t + \beta \cos \Omega t]. \tag{4.102}$$

We now return to physics and assume that at the initial time $t = 0$ the electron is in quantum state 1. This means that at $t = 0$, $d_1 = 1$ and $d_2 = 0$. This condition can be fulfilled by choosing $\alpha = 1$ and $\beta = 0$. We thus obtain

$$d_1(t) = \cos \Omega t. \tag{4.103}$$

To obtain a particularly simple expression for d_2 we decompose the complex constant, a, into a modulus and phase factor

$$a = |a| \exp[\mathrm{i}\chi]. \tag{4.104}$$

By means of (4.102) we thus obtain

$$d_2(t) = -\mathrm{i} \exp[-\mathrm{i}\chi] \sin \Omega t. \tag{4.105}$$

With the results of (4.103) and (4.105) the coefficients c_j which we originally wanted acquire the form

$$c_1(t) = \exp[-\mathrm{i}W_1 t/\hbar] \cos \Omega t \tag{4.106}$$

$$c_2(t) = -\mathrm{i} \exp[-\mathrm{i}\chi] \exp[-\mathrm{i}W_2 t/\hbar] \sin \Omega t. \tag{4.107}$$

To discuss the meaning of our result we remind the reader that c_1 and c_2 are the coefficients of the unperturbed wave functions occurring in the total wave function (4.3). According to quantum mechanics, $|c_1|^2$ and $|c_2|^2$ tell us the probability of finding the particle in the states 1 or 2, respectively. Plotting the corresponding probability

$$P_1 \equiv |c_1(t)|^2 = \cos^2 \Omega t \tag{4.108}$$

as a function of time we learn how the occupation of that state changes in the course of time. To visualize our result we identify φ_1 with the energetic lower state. At time $t = 0$ this state is occupied whereas the excited state φ_2 is empty. Under the influence of the coherent resonant field the occupation number of the lower state decreases while that of the upper state increases

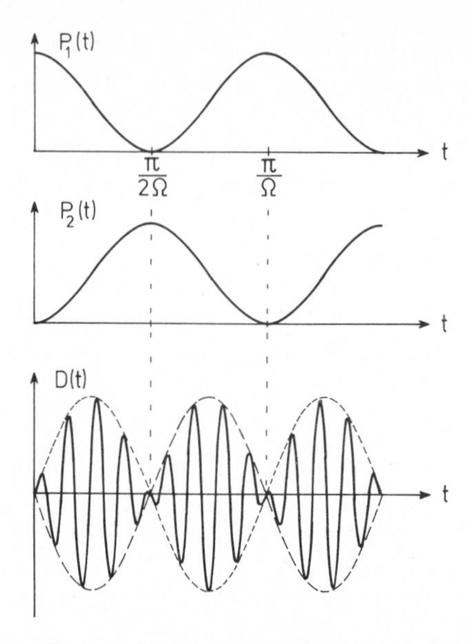

Fig. 4.4. This figure shows the time dependence of various expectation values of an electron within a 2-level atom under the impact of a resonant coherent driving field. The upper part shows the variation of the occupation number of the lower electronic level with time. The middle part shows the variation of the occupation number of the upper level with time. The lower part shows the variation of the dipole moment $D(t)$ (4.111) with time. The dashed envelope is given by $|\vartheta_{12}| \sin 2\Omega t$.

more and more. This means physically that the electron goes over from the lower state to the upper state on account of the absorption process. However, if the interaction between field and atom goes on, the electron does not stay in the upper state but goes down to the lower state with an occupation number described by (cf. fig. 4.4)

$$P_2 \equiv |c_2(t)|^2 = \sin^2 \Omega t. \tag{4.109}$$

Thus under the influence of a coherent resonant field the electron oscillates back and forth between its lower and upper states. It is interesting to study the behavior of the dipole moment

$$D = \int \psi^*(-ex)\psi \, dV \tag{4.110}$$

during this process. By inserting (4.3) with (4.106) and (4.107) in (4.110), we obtain after a short calculation

$$D = \vartheta_{12} \sin \omega_0 t \sin 2\Omega t. \tag{4.111}$$

For simplicity, we have again assumed that the atom does not possess permanent dipole moments and we have assumed ϑ_{12} real. This result can be interpreted as follows: The dipole moment oscillates with the transition frequency ω_0 but its size first increases and then decreases in a way described by $\sin 2\Omega t$. These results were obtained using the rotating wave approximation. The effect of the non-resonant terms on the solution has been discussed by a number of authors. It turns out that the non-resonant terms give rise to a small shift of frequency.

In the case of two-level atoms this shift is called Autler–Townes shift. It is entirely analogous to the socalled Bloch–Siegert shift in spin resonance. We will treat spin resonance in the following chapters.

4.7. The response of a spin to crossed constant and time dependent magnetic fields

Since the spin is a two-level system as is the two-level atom, we may expect interesting analogies between the behaviour of spins and of two-level atoms under the action of external fields. Since these analogies have given rise to important new phenomena we will now discuss spins.

A good deal of important spin resonance experiments are done under the following conditions. We apply a constant magnetic field in z-direction. An additional alternating magnetic field is applied in the x–y-plane (cf. fig. 4.5). We shall see that such experiments lead to interesting spin flip phenomena. Depending on whether these experiments are done by electrons or nuclei the experiments are called electron-spin resonance (ESR) or nuclear magnetic resonance (NMR). Such experiments allow for exact measurements of magnetic moments and are used for structural analysis and especially for the study of relaxation processes in liquids, solids, and biological material. Now let us turn to the corresponding mathematical treatment. We write the magnetic field induction in the form

$$\boldsymbol{B} = \boldsymbol{B}_0 + \boldsymbol{B}^{\mathrm{p}}(t) \tag{4.112}$$

according to the constant part and the alternating part. We choose the vectors of these two fields

$$\boldsymbol{B}_0 = \left(0, 0, B_z^0\right), \qquad \boldsymbol{B}^{\mathrm{p}}(t) = \left(B_x^{\mathrm{p}}(t), B_y^{\mathrm{p}}(t), 0\right). \tag{4.113, 114}$$

Since the Schrödinger equation (3.110) is now time dependent due to (4.112) for its solution we make the ansatz

$$\varphi(t) = c_1(t)\varphi_\uparrow + c_2(t)\varphi_\downarrow \equiv \begin{pmatrix} c_1(t) \\ c_2(t) \end{pmatrix} \tag{4.115}$$

with still unknown, time-dependent coefficients c_1, c_2. To derive equations

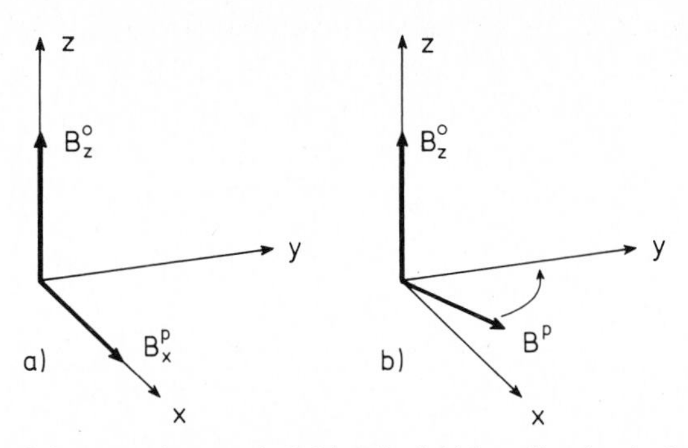

Fig. 4.5. Scheme of the crossed magnetic fields. The field in z-direction is time independent. (a) The time-dependent field oscillates parallel to the x-axis. (b) The time-dependent field rotates with frequency ω in the x–y plane.

for these coefficients we insert (4.115) into (3.110). By using the relations for matrices we readily obtain

$$\left(-\tfrac{1}{2}\hbar\omega_0\right)c_1 - \mu\left(B_x^{\mathrm{p}} - \mathrm{i}B_y^{\mathrm{p}}\right)c_2 = \mathrm{i}\hbar\frac{\mathrm{d}c_1}{\mathrm{d}t} \tag{4.116}$$

$$-\mu\left(B_x^{\mathrm{p}} + \mathrm{i}B_y^{\mathrm{p}}\right)c_1 + \tfrac{1}{2}\hbar\omega_0 c_2 = \mathrm{i}\hbar\frac{\mathrm{d}c_2}{\mathrm{d}t}. \tag{4.117}$$

We have introduced the frequency ω_0 by the relation

$$\hbar\omega_0 = 2\mu B_z^0. \tag{4.118}$$

To simplify the subsequent calculations we assume that the transverse magnetic field rotates with the frequency ω around the z-axis, i.e. we assume it in the form

$$B_x^{\mathrm{p}} = F\cos\omega t, \qquad B_y^{\mathrm{p}} = -F\sin\omega t. \tag{4.119}$$

The specific form of (4.119) allows us to simplify (4.116) and (4.117). To this end we use the relation

$$B_x^{\mathrm{p}} \pm B_y^{\mathrm{p}} = F(\cos\omega t \mp \mathrm{i}\sin\omega t) \equiv F\exp[\mp\mathrm{i}\omega t]. \tag{4.120}$$

We thus obtain instead of (4.116) and (4.117)

$$(-\hbar\omega_0/2)c_1 - \mu F\exp[\mathrm{i}\omega t]c_2 = \mathrm{i}\hbar\frac{\mathrm{d}c_1}{\mathrm{d}t} \tag{4.121}$$

$$-\mu F\exp[-\mathrm{i}\omega t]c_1 + (\hbar\omega_0/2)c_2 = \mathrm{i}\hbar\frac{\mathrm{d}c_2}{\mathrm{d}t}. \tag{4.122}$$

By comparing (4.121) and (4.122) with eqs. (4.92) and (4.93) of section 4.6, a complete analogy is revealed, if the rotating wave approximation is used. Using the following substitutions

$$c_1(t) = d_1(t)\exp[\mathrm{i}\omega_0 t/2], \qquad c_2(t) = d_2(t)\exp[-\mathrm{i}\omega_0 t/2] \tag{4.123}$$

and assuming resonance, i.e. $\omega = \omega_0$, we can immediately write down the solution (again assuming as initial conditions $c_1(0) = 1$, $c_2(0) = 0$).

$$\varphi(t) = \mathrm{i}\sin(\Omega t)\exp[-\mathrm{i}\omega_0 t/2]\varphi_\downarrow + \cos(\Omega t)\exp[\mathrm{i}\omega_0 t/2]\varphi_\uparrow \quad (4.124)$$

where

$$\Omega = \mu F/\hbar. \quad (4.125)$$

To discuss the physical meaning of this result we take expectation values (fig. 4.6). We leave the evaluation of the expectation value as an exercise to the reader. The result reads

$$\langle s_x \rangle = (\hbar/2)\sin(2\Omega t)\sin(\omega_0 t) \quad (4.126)$$

$$\langle s_y \rangle = (\hbar/2)\sin(2\Omega t)\cos(\omega_0 t) \quad (4.127)$$

$$\langle s_z \rangle = (\hbar/2)\cos(2\Omega t). \quad (4.128)$$

According to this result, the spin components in the x,y-plane are composed of a superposition of a rapid precession of the spin with frequency ω_0 and a modulation of frequency 2Ω. Representing the expectation values of s_x, s_y, s_z as a vector we readily obtain fig. 4.7. While the spin precesses around the z-axis it flips from the $+z$-direction into the $-z$-direction and then back so that it oscillates back and forth between these two directions.

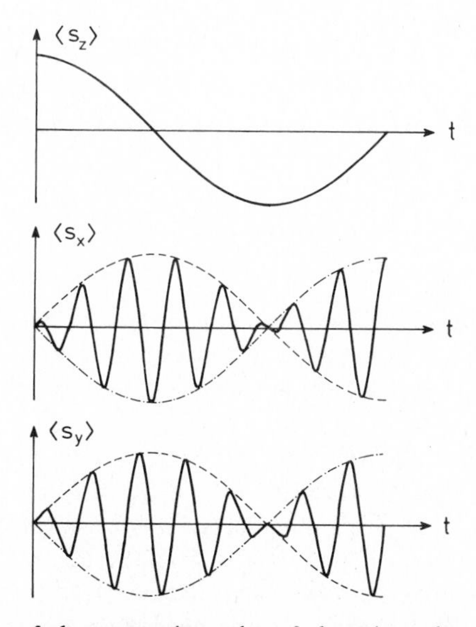

Fig. 4.6. The motion of the expectation value of the spin under the action of crossed magnetic fields according to the arrangement of fig. 4.5. Shown are the individual spin components according to equations (4.126)–(4.128).

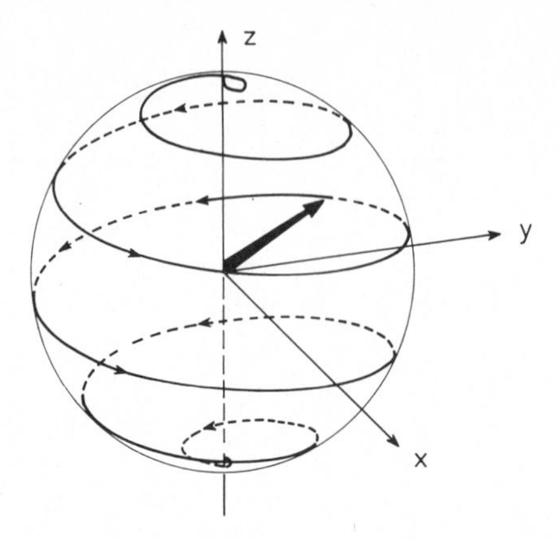

Fig. 4.7. Motion of the expectation value of the spin vector of the same experiment as given in fig. 4.6.

Thus the spin behaves as a spinning top under the impact of external forces.

Let us consider this process once again more quantitatively. At time $t = 0$ we have

$$\langle s_z \rangle = \tfrac{1}{2}\hbar. \tag{4.129}$$

We now wish to determine the time after which the spin has flipped into the horizontal plane, i.e.

$$\langle s_z \rangle = 0. \tag{4.130}$$

This is the case when, according to (4.128) the cosine vanishes, i.e. when

$$2\Omega t = \tfrac{1}{2}\pi \tag{4.131}$$

holds. Thus the flipping time is given by

$$t = \pi/(4\Omega) = \pi\hbar/(4\mu F). \tag{4.132}$$

After this time, the spin has been turned with respect to the vertical axis by an angle $\frac{1}{2}\pi$. A field $\boldsymbol{B}^p$ causing such flipping is called a $\frac{1}{2}\pi$ or 90° pulse. When the field is applied for double the time, the spin is flipped by π or 180°.

These results form the basis of important spin resonance experiments. By application of an external resonant field we can flip the spin from one direction to another. In practical experiments, the magnetic field does not rotate with the spin frequency but has a fixed direction. The then resulting equations have exactly the same form as (4.121), (4.122) except for an oscillatory additional term which is neglected in the rotating wave approximation. [Compare the discussion following (4.95), (4.96).] This explains the notation "rotating wave approximation", because in applying it we pass from a (magnetic field) wave in a constant direction to one rotating with the spin precession.

Exercises on section 4.7

(1) Calculate $\langle s_x \rangle$, $\langle s_y \rangle$, $\langle s_z \rangle$ using (4.124).

(2) Show that $\langle s_x \rangle$, $\langle s_y \rangle$, $\langle s_z \rangle$ which are explicitly given by (4.126), (4.127), (4.128), respectively, obey the equations

$$\frac{d}{dt}\langle s_x \rangle = \mu(\langle s_y \rangle B_z - \langle s_z \rangle B_y) \qquad (*)$$

$$\frac{d}{dt}\langle s_y \rangle = \mu(\langle s_z \rangle B_x - \langle s_x \rangle B_z) \qquad (**)$$

$$\frac{d}{dt}\langle s_z \rangle = \mu(\langle s_x \rangle B_y - \langle s_y \rangle B_x) \qquad (***)$$

or in short

$$\frac{d}{dt}\langle \boldsymbol{s} \rangle = \mu\langle \boldsymbol{s} \rangle \times \boldsymbol{B}.$$

Hint: Use the explicit result of exercise 1, or (4.126)–(4.128).

(3) Calculate numerical values for $\boldsymbol{B}$ to achieve $\pi/2$ or π-pulses. Use (4.131), (4.125) and

$$\hbar = 1.055 \times 10^{-34}\ \text{W s}^2$$

$$\mu = 1.165 \times 10^{-29}\ \text{V s m}$$

$$10^{-6}\ \text{s} \leqslant t \leqslant 1\ \text{s}.$$

Hint: (4.131) and the treatment of section 4.7 are valid only for pulse-duration times t which are much smaller than relaxation times in the system which describe incoherent (phase-destroying) effects, such times may vary from milliseconds up to seconds for real systems.

(4) Let $\boldsymbol{B}^{\mathrm{P}}(t)$ be incoherent. Treat the spin-transition in analogy to section 4.2 repeating the individual steps and replacing $\boldsymbol{E}$ by $\boldsymbol{B}$.

4.8. The analogy between a two-level atom and a spin $\frac{1}{2}$

When we compare the results of section 4.7 to those of section 4.6, far reaching analogies become evident. These analogies have allowed physicists to predict and observe phenomena in the optical region because such phenomena had been found earlier in spin resonance. We now list some analogies.

(i) Analogy between time-independent wave functions and energies. The spin of the electron is exposed to a constant magnetic field in z-direction (compare Section 3.8). The atom is not exposed to any external fields (see table 4.1).

In the preceding sections, we described the two-level atom in terms of the complex expansion coefficients $c_1(t)$, $c_2(t)$. In order to see the full analogy between a spin-$\frac{1}{2}$ and a two-level atom, a description of the two-level atom in terms of the following quantities turns out to be useful

$$\begin{aligned} \tilde{s}_1 &= c_1^* c_2 + c_1 c_2^* \\ \tilde{s}_2 &= i(c_1^* c_2 - c_1 c_2^*) \\ \tilde{s}_3 &= |c_2|^2 - |c_1|^2. \end{aligned} \tag{4.133}$$

These quantities $\tilde{s}_1$, $\tilde{s}_2$, $\tilde{s}_3$ are the counterparts of the expectation values $\langle s_x \rangle$, $\langle s_y \rangle$, $\langle s_z \rangle$ in the spin-$\frac{1}{2}$ case, and we will use them below. $\tilde{s}_1$, $\tilde{s}_2$, $\tilde{s}_3$ can be considered as the components of a vector, which is called the pseudo-spin. We now compare the wave functions in the time-dependent case, i.e. the solutions of the time-dependent Schrödinger equation.

(ii) Wave packets without external alternating fields

Spin	Two-level atom
$\varphi = c_1 \varphi_\downarrow + c_2 \varphi_\uparrow$	$\psi = c_1 \varphi_1 + c_2 \varphi_2$
compare (3.97)	compare section 4.1.

Table 4.1. Analogy between spin and two-level atom

Spin			Two-level atom		
Spin direction	Spin function	Energy		Wave function	Spatial representation
$\downarrow$	$\varphi_\downarrow$	$-\frac{1}{2}\hbar\omega_0$	W_1	φ_1	
$\uparrow$	$\varphi_\uparrow$	$\frac{1}{2}\hbar\omega_0$	W_2	φ_2	

These wave functions are responsible for the following physical phenomena:

Free precession of spin	Free oscillation of dipole moment
$c_1 = a\exp[\mathrm{i}\omega_0 t/2]$	$c_1 = a\exp[-\mathrm{i}W_1 t/\hbar]$
$c_2 = b\exp[-\mathrm{i}\omega_0 t/2]$	$c_2 = b\exp[-\mathrm{i}W_2 t/\hbar]$.

These wave functions allow us to calculate the expectation values of
(a) Magnetic dipole moment of the spin $(e/m)\langle \boldsymbol{s}\rangle$, (3.105).
(b) The electric dipole moment of the electron of the two-level atom

$$\vartheta = \int (c_1^*\varphi_1^* + c_2^*\varphi_2^*)(-e\boldsymbol{x})(c_1\varphi_1 + c_2\varphi_2)\,\mathrm{d}V. \tag{4.134}$$

Under the assumption that $\vartheta_{11} = \vartheta_{22} = 0$, where

$$\vartheta_{jk} = \int \varphi_j^*(-e\boldsymbol{x})\varphi_k\,\mathrm{d}V \tag{4.135}$$

we have

$$\vartheta = c_1^* c_2 \vartheta_{12} + c_1 c_2^* \vartheta_{21}. \tag{4.136}$$

We decompose $\boldsymbol{\vartheta}_{12} = \boldsymbol{\vartheta}_{21}^*$ into its real and imaginary part

$$\boldsymbol{\vartheta}_{12} = \vartheta'_{12} + i\vartheta''_{12}. \tag{4.137}$$

By using (4.137), (4.136), and (4.134) we obtain

$$\vartheta = 2\vartheta'_{12}\tilde{s}_1 + 2\vartheta''_{12}\tilde{s}_2. \tag{4.138}$$

Furthermore, the difference of the expectation values of the occupation numbers of levels 2 and 1 can be written as

$$|c_2|^2 - |c_1|^2 = \tilde{s}_3. \tag{4.139}$$

With these expressions in mind, we may establish the following analogies:

spin	"pseudo"-spin
$\langle s_x\rangle = \hbar\|a\|\|b\|\cos\omega_0(t-t_0)$	$\tilde{s}_1 = 2\|a\|\|b\|\cos\omega_0(t-t_0)$
$\langle s_y\rangle = \hbar\|a\|\|b\|\sin\omega_0(t-t_0)$	$\tilde{s}_2 = 2\|a\|\|b\|\sin\omega_0(t-t_0)$
$\langle s_z\rangle = \frac{\hbar}{2}(\|b\|^2 - \|a\|^2)$	$\tilde{s}_3 = (\|b\|^2 - \|a\|^2)$

where

$$ab^* = |a|\,|b|\exp[\mathrm{i}\omega_0 t_0], \qquad ab^* = |a|\,|b|\exp[\mathrm{i}\omega_0 t_0].$$

(Compare exercise (1) on section 3.8.) Note that this analogy is formal so that, e.g. $\tilde{s}_3$ must not be interpreted as a quantity proportional to the z-component of the electric dipole moment. On the other hand, such analogies will prove useful below when we discuss the response of a two-level atom to external fields.

(iii) Response of a spin or a two-level atom to a coherent resonant field

spin	pseudo-spin of two-level atom
$\langle s_x \rangle = (\hbar/2)\sin 2\Omega t \sin \omega_0 t,$	$\tilde{s}_1 = \sin 2\Omega t \sin \omega t$
$\langle s_y \rangle = (\hbar/2)\sin 2\Omega t \cos \omega_0 t,$	$\tilde{s}_2 = \sin 2\Omega t \cos \omega t$
$\langle s_z \rangle = (\hbar/2)\cos 2\Omega t,$	$\tilde{s}_3 = -\cos 2\Omega t$
compare (4.126)–(4.128)	compare (4.108),(4.109).

(iv) Equations of motion of expectation values.
The equations of motion for both the spin-$\frac{1}{2}$ and the two-level atom subject to an external field may be written in the following form:

spin	pseudo-spin of two-level atom
$\frac{\mathrm{d}}{\mathrm{d}t}\langle \boldsymbol{s} \rangle = \mu \langle \boldsymbol{s} \rangle \times \boldsymbol{B},$	$\frac{\mathrm{d}}{\mathrm{d}t}\langle \tilde{\boldsymbol{s}} \rangle = \Omega \times \langle \tilde{\boldsymbol{s}} \rangle$
with	
$\boldsymbol{B} = (B_x, B_y, B_z),$	$\Omega = (2\vartheta \boldsymbol{E}, 0, \omega)$

The transition of the electron from its upper level to its lower level and back (compare section 4.6) can be put in parallel to the up and down motion of a spinning top via the spin analogy. Therefore, the corresponding phenomenon shown by electrons is often called optical nutation. The reader should be warned, however, that the expression "optical nutation" is sometimes used by some other authors in a somewhat different sense.

(v) Free induction decay
Let us consider an ensemble of spins in a sample in a constant magnetic field. In their lowest states the spins will point downwards and their individual spin functions are $\varphi_{\downarrow}$. These states have no dipole moment in the x–y-plane. Now let us apply a resonant alternating magnetic field with a $\frac{1}{2}\pi$-pulse. Then all spins are brought to the x–y-plane in which they start to rotate with frequency ω_0. In many practical cases, the magnetic field $\boldsymbol{B}_0$ is inhomogeneous (for various reasons) so that each individual spin senses its individual field $\boldsymbol{B}_{0,j}$, where j distinguishes the different spins. In section 3.8 we saw that the magnitude of the constant magnetic field determines the precession frequencies of the spins. Thus, for different magnetic fields different precession frequencies will result. To the spin of each electron or proton a magnetic moment corresponds which oscillates in the same way

as the spin does. According to the theory of electromagnetism, an oscillating magnetic dipole moment emits electromagnetic radiation. Thus the spin precession leads to electromagnetic radiation. However, because the spins precess with different frequencies, their emitted fields get out of phase according to the spread of dipole moments (cf. fig. 4.8c). As a result the intensity of the emitted radiation decreases. The whole process is called free induction decay.

Owing to the analogy between spins and two-level systems, the same process may occur, e.g. for electrons of atoms with two energy levels. First at time $t = 0$ all electrons of the sample are in their lower states. Then a short light pulse of the type of a $\frac{1}{2}\pi$-pulse is applied which brings the electrons into a mixed state composed of the upper and lower level (compare section 4.6). The further motion of each individual electron is described by the wave packet

$$\psi = 2^{-1/2}\{\exp[-iW_1 t/\hbar]\varphi_1 - i\exp[-iW_2 t/\hbar]\varphi_2\}$$

which is connected with an oscillating electric dipole moment, as shown above.

Such experiments have been done, for instance, with atoms embedded in solids. Due to different surroundings the individual atoms possess somewhat different electronic energy levels so that the frequency of the electronic oscillation, described by the wave packet is somewhat different for each atom. Since each electron acts as an oscillating electric dipole, it emits electromagnetic radiation. Due to the spread of oscillation frequencies, after a certain time the oscillating electric dipoles get out of phase. As a consequence, the emitted light pulse decays and we are dealing with the free induction decay.

(vi) Spin- and photon-echo

We first give a qualitative description of this effect, which will then be treated quantitatively. Let us consider an ensemble of spins or electrons in a sample where initially all spins or electrons are in their ground states. We first apply a $\frac{1}{2}\pi$-pulse to the spins (or electrons) by which we prepare their wave functions in a mixed state and the spins (or electrons) start their free induction decay. In the echo-experiment, some time after the $\frac{1}{2}\pi$-pulse had been applied, a π-pulse is applied. It has the effect of rotating the spins by 180° around the 1-axis which is shown in figs. 4.8. After the π-pulse, the spins precess further with the original speed and, as is evident from that figure, are in phase again after a time which approximately equals the time between the $\frac{1}{2}\pi$- and π-pulse. Thus they can emit radiation as before. The restoration of the original state of the spins (all spins in phase) can be achieved as long as we can neglect irreversible processes causing an

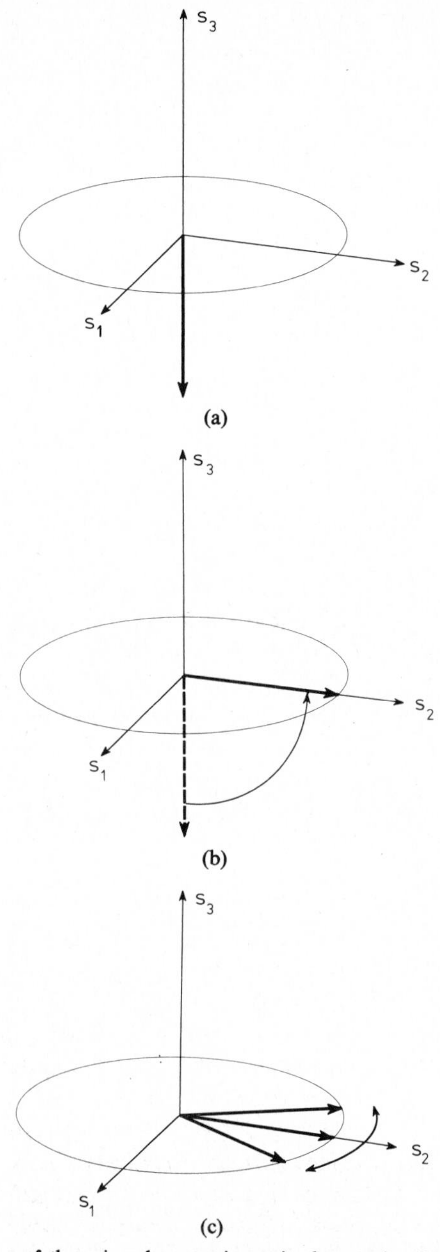

Figs. 4.8(a)–(f). Scheme of the spin-echo experiment in the rotating frame. The figures and their legends apply both to the spin and the pseudo-spin. ($s_1 \leftrightarrow s_x$, $s_2 \leftrightarrow s_y$, $s_3 \leftrightarrow s_z$). (a) At time $t = 0$ all spins are in their ground states. (b) By applying a $\frac{1}{2}\pi$-pulse, the spins are flipped into the s_2-direction. (The final direction of the spin-vector in the horizontal plane

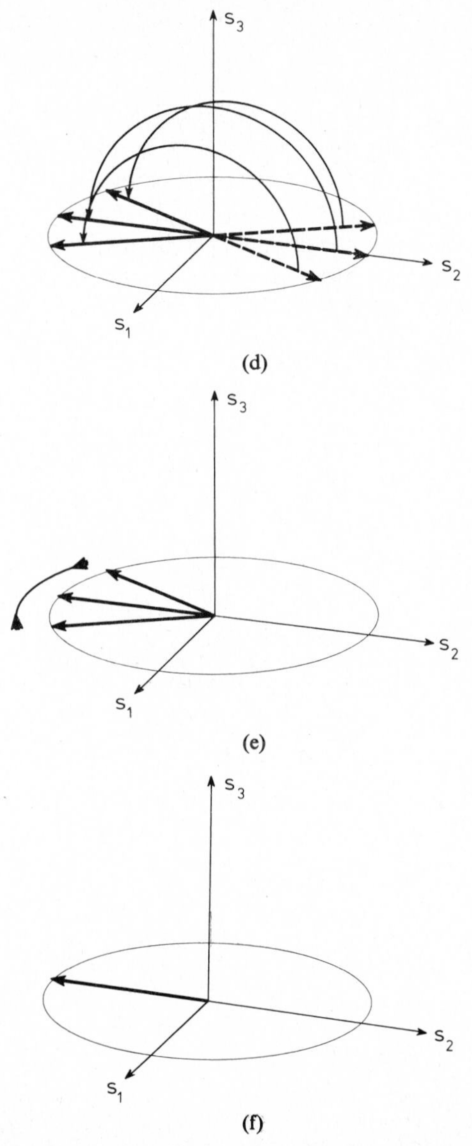

depends on the initial phase angle between field and dipole matrix element). (c) Without external driving field, the spins start to precess around the s_3-axis. As each spin has its own precession frequency, i.e. the sample shows inhomogeneous broadening (compare text), in the rotating frame the spins get out of phase, i.e. they show free induction decay. (d) A π-pulse flips the spins by an angle of π around the s_1-axis as indicated. (e) After the π-pulse, the spins precess further with their own frequency. (f) The phase lags have been cancelled, i.e. all spins are in phase again. Thus the ensemble of (in phase) oscillating spins emits an observable signal, i.e. the "echo" (of the initially exciting $\frac{1}{2}\pi$-pulse).

irreversible dephasing of spins. Quite a similar experiment can be done with electrons and we leave it as an exercise to the reader to discuss the effect of a $\frac{1}{2}\pi$-pulse and a subsequent π-pulse.

We now treat the whole process quantitatively (see also fig. 4.9). We consider an ensemble of spins or of two-level atoms with different transition frequencies $\omega = (W_2 - W_1)/\hbar$. Since the spread of transition frequencies (or, equivalently, energies W_1, W_2) is usually caused by a superposition of many effects (for instance inhomogeneities in solids, if the atoms considered here are embedded in a solid), one may often assume a Gaussian distribution $f(\omega)$ of ω around a center frequency $\bar{\omega}$:

$$f(\omega) = \frac{N}{\sqrt{\pi}\,\Delta\omega} \exp\left[-\frac{(\omega - \bar{\omega})^2}{\Delta\omega^2} \right] \tag{4.140}$$

where

$$\int_{-\infty}^{\infty} f(\omega)\,\mathrm{d}\omega = N \tag{4.141}$$

is the total number of electrons in our sample. As above, we write the wave function of a single spin (or electron) in the form

$$\varphi = c_1\varphi_1 + c_2\varphi_2 \quad (\text{or} = c_1\varphi_\downarrow + c_2\varphi_\uparrow). \tag{4.142}$$

We shall simplify our analysis by considering only a single spin (or electron) assumed to be in resonance with the external field. The effect of detuning is taken into account by averaging the resulting dipole moments

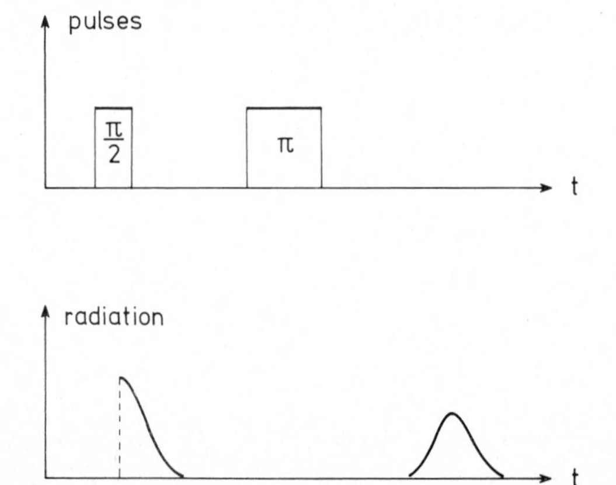

Fig. 4.9. Typical time sequence in an echo-experiment. The spins or atoms are excited with a $\frac{1}{2}\pi$-pulse, thus starting free induction decay. After some time T a π-pulse is applied, thus after approximately $2T$ the spins or dipoles are in phase again which produces an observable signal.

over the frequency distribution $f(\omega)$. We assume that in the beginning all spins (or electrons) are in their ground states

$$c_1(0) = 1, \qquad c_2(0) = 0. \tag{4.143}$$

First we apply a $\frac{1}{2}\pi$-pulse ($\boldsymbol{E} = 2\boldsymbol{E}_0 \cos \omega t$) and end up with

$$c_1(t) = \frac{1}{\sqrt{2}} \exp[-\mathrm{i}\omega_1 \tau_1] \exp[-\mathrm{i}\omega_1 t],$$

$$c_2(t) = \frac{-\mathrm{i}}{\sqrt{2}} \exp[-\mathrm{i}\omega_2 \tau_1] \exp[-\mathrm{i}\omega_2 t], \tag{4.144}$$

where

$$\omega_j = W_j/\hbar, \qquad \boldsymbol{\vartheta}_{12} \cdot \boldsymbol{E}_0' = \hbar\Omega, \qquad \Omega\tau_1 = \tfrac{1}{4}\pi \tag{4.145}$$

where $t = 0$ now corresponds to the end of the $\frac{1}{2}\pi$-pulse. By calculating the dipole moment gives

$$\boldsymbol{P}_\omega \equiv \langle \boldsymbol{P} \rangle = -\boldsymbol{\vartheta}_{12} \sin \omega(t + \tau_1). \tag{4.146}$$

Experimentally, one observes the averaged macroscopic dipole moment

$$\bar{\boldsymbol{P}} = \int_0^\infty f(\omega) \boldsymbol{P}_\omega \, \mathrm{d}\omega \tag{4.147}$$

which may be easily evaluated

$$\bar{\boldsymbol{P}} = -N\boldsymbol{\vartheta}_{12} \sin \bar{\omega}(t + \tau_1) \exp\left[-\tfrac{1}{4}(\Delta\omega(t + \tau_1))^2\right] \tag{4.148}$$

where we used the integral

$$\frac{1}{\sqrt{\pi}\,\Delta\omega} \int_{-\infty}^{\infty} \exp\left[-\frac{(\omega - \bar{\omega})^2}{\Delta\omega^2}\right] \sin \omega t \, \mathrm{d}\omega = \sin \bar{\omega} t \exp\left[-\tfrac{1}{4}(\Delta\omega t)^2\right] \tag{4.149}$$

and the fact that we can approximate $\int_0^\infty \mathrm{d}\omega$ by $\int_{-\infty}^\infty \mathrm{d}\omega$ provided $\Delta\omega \ll \bar{\omega}$.

Remember, that in (4.148) t starts from zero (end of the $\frac{1}{2}\pi$-pulse). Thus, the observed polarization decays rapidly due to the exponential, i.e. the sample shows free induction decay. Equations (4.144) describe the free motion of our spins (or electrons) after the excitation. Let us apply, some time T after the first pulse, a π-pulse, i.e. we specialize eqs. (4.144) for $t = T$ and then use the resulting quantities c_1, c_2 as initial conditions in (4.157). This gives us, together with (4.156), which is also derived in the exercise 3 below

$$c_1(t) = -\frac{1}{\sqrt{2}} \exp\left[-\mathrm{i}(\omega_1\tau_2 + \omega_2(\tau_1 + T) + \omega_1 t)\right]$$

$$c_2(t) = -\frac{\mathrm{i}}{\sqrt{2}} \exp\left[-\mathrm{i}(\omega_2\tau_2 + \omega_1(\tau_1 + T) + \omega_2 t)\right] \tag{4.150}$$

where, evidently, the time t now runs from the end of the π-pulse. A single dipole moment would read

$$\boldsymbol{P}_\omega = \boldsymbol{\vartheta}_{12} \sin\omega(t - t_0) \tag{4.151}$$

with

$$t_0 = T + \tau_1 - \tau_2 \tag{4.152}$$

as is easily seen by inserting (4.150) into (4.134). Finally, let us consider the macroscopic (averaged) dipole moment $\bar{\boldsymbol{P}}$. By using (4.151) and (4.149), $\bar{\boldsymbol{P}}$ is given by

$$\bar{\boldsymbol{P}} = N\boldsymbol{\vartheta}_{12} \sin\bar{\omega}(t - t_0)\exp\left[-\tfrac{1}{4}(\Delta\omega(t - t_0))^2\right] \tag{4.153}$$

i.e. the "echo" shows up with a delay of $t_0 \approx 2T$ after the first $\frac{1}{2}\pi$-pulse where the π-pulse was applied at time $t = T$ after the excitation.

Exercises on section 4.8

(1) Free induction decay:
Calculate the decay time of freely precessing spins in an inhomogeneous time-independent magnetic field B_z with spread ΔB_z.
Hint: Calculate

$$\frac{1}{\Delta\omega}\int_{\omega_0-\Delta\omega/2}^{\omega_0+\Delta\omega/2} \langle s_x\rangle \,\mathrm{d}\omega, \quad \text{where} \quad \omega \propto B_z.$$

(2) Discuss what $\frac{1}{2}\pi$- and π-pulse means for the wave functions of electrons.
(3) A two-level atom driven by a coherent external field may be described by the wave function

$$\psi = c_1(t)\varphi_1 + c_2(t)\varphi_2 \tag{4.154}$$

where the coefficients c_j obey the equations

$$\mathrm{i}\frac{\mathrm{d}}{\mathrm{d}t}c_1 = \omega_1 c_1 + \boldsymbol{\vartheta}_{12}\cdot\boldsymbol{E}(t)c_2, \quad \mathrm{i}\frac{\mathrm{d}}{\mathrm{d}t}c_2 = \omega_2 c_2 + \boldsymbol{\vartheta}_{12}\cdot\boldsymbol{E}(t)c_1. \tag{4.155}$$

(Compare sections 4.6 and 4.7.) Solve (4.155) for arbitrary initial conditions $c_1(0)$, $c_2(0)$. Show that the solutions read

$$c_1(t) = c_1(0)\exp[-i\omega_1 t], \qquad c_2(t) = c_2(0)\exp[-i\omega_2 t] \tag{4.156}$$

in the case of free motion, i.e. $\boldsymbol{E} = 0$, whereas in the case of an external field the solutions read

$$c_1(t) = [c_1(0)\cos\Omega t - ic_2(0)\sin\Omega t]\exp[-i\omega_1 t]$$

$$c_2(t) = [c_2(0)\cos\Omega t - ic_1(0)\sin\Omega t]\exp[-i\omega_2 t] \tag{4.157}$$

where

$$\Omega = \boldsymbol{\vartheta}_{12}\cdot\boldsymbol{E}_0/\hbar, \qquad \boldsymbol{E}(t) = 2\boldsymbol{E}_0\cos\omega t, \qquad \omega_2 - \omega_1 = \omega. \tag{4.158}$$

4.9. Coherent and incoherent processes

An important comment on our previous discussions must be made with respect to the impact of incoherent processes. In all the chapters thus far, we have treated the quantum system of electrons or spins neglecting any damping effects. This is clearly visible from the fact that the dipole moments of spins or electrons oscillate at a certain frequency ad infinitum. On the other hand, we have distinguished between coherent and incoherent electromagnetic fields and we saw that such fields cause quite different behavior with respect to electronic transitions. In the case of incoherent fields the electron goes from its lower state to its upper state with a certain transition probability per second, where any phase relations of the lower and upper states are ignored.

On the other hand, under the impact of a coherent field an electron wave packet with well defined phases of the lower and upper state could be established (more precisely speaking, the relative phase between lower and upper state is well defined all the time). In reality, the motion of an electron or a spin is subject to various kinds of perturbations which steadily cause fluctuations of the relative phase between lower and upper states. Thus in reality the oscillation of the dipole moment decays in a rather short time depending on the individual system. Therefore, all the experiments which we have discussed so far using coherent excitation, are supposed to be done in such a short time that the internal dephasing of electronic or spin dipole moments can be neglected. We will come back to the question how to take care of such dephasing effects (more precisely speaking to the question of damping and fluctuations) in chapter 9.

Exercise on section 4.9

(1) In exercise (2), section 4.7 we got to know the equations of motion for the expectation values of the spin operators $\langle s_x \rangle$, $\langle s_y \rangle$, $\langle s_z \rangle$ when magnetic fields are applied. It is known experimentally that the interaction of the spin with its surrounding causes a damping of the phase of the spin and a relaxation of $\langle s_z \rangle$ towards an average value s_0. To take these effects phenomenologically into account, Bloch introduced into the equations (*), (* *), (* * *) of exercise 2 on section 4.7 the following additional terms:

$$\frac{\mathrm{d}}{\mathrm{d}t}\langle s_x \rangle = \cdots - \frac{1}{T_2}\langle s_x \rangle$$

$$\frac{\mathrm{d}}{\mathrm{d}t}\langle s_y \rangle = \cdots - \frac{1}{T_2}\langle s_y \rangle$$

$$\frac{\mathrm{d}}{\mathrm{d}t}\langle s_z \rangle = \cdots + \frac{s_0 - \langle s_z \rangle}{T_1}.$$

T_1 and T_2 are called the longitudinal and transverse relaxation times, respectively.

Exercise: Write down the full Bloch equations replacing the dots above by the expressions (*), (* *), (* * *). Solve the Bloch equations for $\boldsymbol{B} = (0, 0, B_z)$, where (a) B_z is time independent, $= B_{z,0}$, and (b) $B_z = B_{z,0} + B_{z,1} \sin \omega' t$.

5. Quantization of the light field

5.1. Example: A single mode. Maxwell's equations

In this chapter we will deal with the electromagnetic field in classical physics which is described by Maxwell's equations. We therefore remind the reader briefly of those equations. (For their illustration cf. fig. 5.1)

(i) The induction equation. According to it a temporal change of magnetic induction causes a curl of the electric field.

$$\text{curl}\, \boldsymbol{E} = -\frac{\partial \boldsymbol{B}}{\partial t}. \tag{5.1}$$

(ii) An electric current or/and a temporal change of dielectric displacement causes a curl of the magnetic field strength $\boldsymbol{H}$.

$$\text{curl}\, \boldsymbol{H} = \boldsymbol{j} + \frac{\partial \boldsymbol{D}}{\partial t}. \tag{5.2}$$

(iii) The magnetic induction $\boldsymbol{B}$ has no sources.

$$\text{div}\, \boldsymbol{B} = 0. \tag{5.3}$$

(iv) The source of the dielectric displacement is a charge density.

$$\text{div}\, \boldsymbol{D} = \rho. \tag{5.4}$$

$\boldsymbol{D}$ and $\boldsymbol{E}$ as well as $\boldsymbol{B}$ and $\boldsymbol{H}$ are connected by phenomenological equations

$$\boldsymbol{D} = \varepsilon\varepsilon_0 \boldsymbol{E}, \qquad \boldsymbol{B} = \mu\mu_0 \boldsymbol{H} \tag{5.5, 6}$$

where ε is the dielectric constant, ε_0 the dielectric constant in vacuum, μ the magnetic susceptibility and μ_0 the magnetic susceptibility in vacuum. ε and μ can be determined by experiments. It is a goal of modern theory, especially quantum mechanics, to derive ε and μ by means of a microscopic theory. The relations (5.5) and (5.6) seem to indicate a linear relation between $\boldsymbol{D}$ and $\boldsymbol{E}$ on the one hand and $\boldsymbol{B}$ and $\boldsymbol{H}$ on the other. It

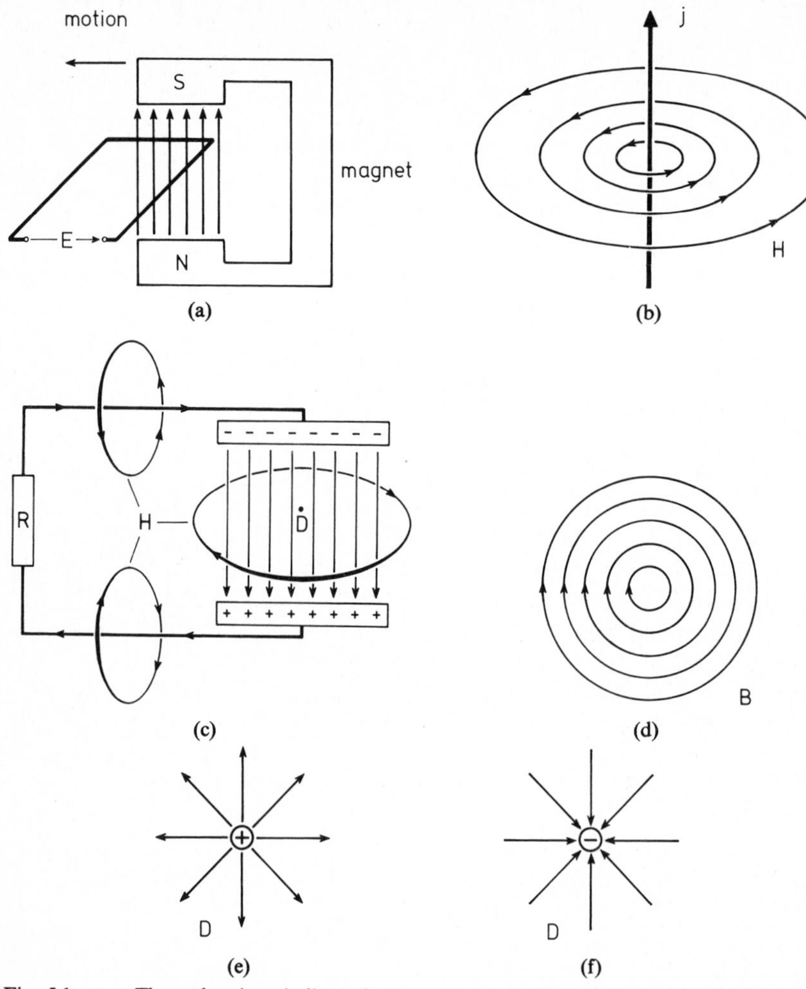

Fig. 5.1. a–e. These drawings indicate how one may visualize the meaning of Maxwell's equations. (a) According to the induction equation, a temporal change of the magnetic flux within the conducting loop causes a curl of the electric field strength which causes a flux of electric current in the closed loop (wire). (b) According to eq. (5.2), but without the term $\partial \boldsymbol{D}/\partial t$, an electric current $\boldsymbol{j}$ causes a circular magnetic field. (c) This figure explains the meaning of (5.2) if the last term $\partial D/\partial t$ is kept. A condenser charged with positive and negative charges is discharged. This causes a change of the dielectric displacement $\boldsymbol{D}$ between the condenser plates. The temporal change of $\boldsymbol{D}$ causes a curl of the magnetic field strength. The individual loops of the magnetic field strength along the whole circuitry go over into each other continuously and are never created or vanish. Thus eq. (5.2) secures that even in spatial regions where there is no material current $\boldsymbol{j}$ there holds $\boldsymbol{H} \neq 0$, or more precisely speaking, the second term on the right hand side of (5.2) takes care of the fact that div.curl $\boldsymbol{H} = 0$. (d) Eq. (5.3) tells us that the lines of magnetic induction have neither sinks nor sources. Thus they must be closed. (e) According to eq. (5.4) the dielectric displacement $\boldsymbol{D}$ has its sources or sinks at positive or negative charges, respectively.

follows from section 4.5, however, that this is only true for weak fields, whereas for higher fields nonlinear relations result. Such relations and their physical results will be the central topic of Volumes 2 and 3.

It is well known that Maxwell's equations describe the light field as electromagnetic waves. We know, however, that light can manifest itself as particles – photons. This leads us to the question how to quantize the electromagnetic field. To this end we first specialize the above equations to vacuum, i.e. $\varepsilon = \mu = 1$, and to the case that no charges or currents are present. Thus the above equations reduce to

$$\operatorname{curl} \boldsymbol{E} = -\frac{\partial \boldsymbol{B}}{\partial t}, \qquad \operatorname{curl} \boldsymbol{B} = \varepsilon_0 \mu_0 \frac{\partial \boldsymbol{E}}{\partial t} \tag{5.7, 8}$$

$$\operatorname{div} \boldsymbol{E} = 0, \qquad \operatorname{div} \boldsymbol{B} = 0. \tag{5.9, 10}$$

It can be shown that $\mu_0 \varepsilon_0$ has the dimension of the inverse of the square of a velocity

$$\mu_0 \varepsilon_0 = 1/c^2. \tag{5.11}$$

It will turn out (below) that this velocity is identical with the light velocity in vacuum. We shall henceforth use the relation (5.11). Now let us consider a special case namely a standing electric wave with wave vector k and with its electric vector in z-direction (cf. fig. 5.2)

$$\boldsymbol{E} = (0, 0, E_z) \tag{5.12}$$

where

$$E_z = p(t)\mathcal{N} \sin kx \tag{5.13}$$

and where $p(t)$ is a still unknown function of time. Equation (5.13) defines a mode between two infinitely extended mirrors. To derive the corresponding magnetic induction we insert (5.13) into (5.7). One can convince oneself readily that only the y-component of this equation is non-vanishing.

$$-\frac{\partial}{\partial x} E_z = -\frac{\partial B_y}{\partial t}. \tag{5.14}$$

Since the left-hand side of this equation is proportional to $\cos(kx)$ it suggests that we put B_y proportional to $\cos(kx)$. This leads us to the ansatz

$$B_y = q(t)(\mathcal{N}/c) \cos kx \tag{5.15}$$

where we have included the factor $1/c$ for later convenience. This factor gives p and q the same physical dimension. Inserting (5.15) into (5.14) yields

$$\frac{\mathrm{d}q}{\mathrm{d}t} = \omega p \tag{5.16}$$

where we have used the abbreviation

$$\omega = ck. \tag{5.17}$$

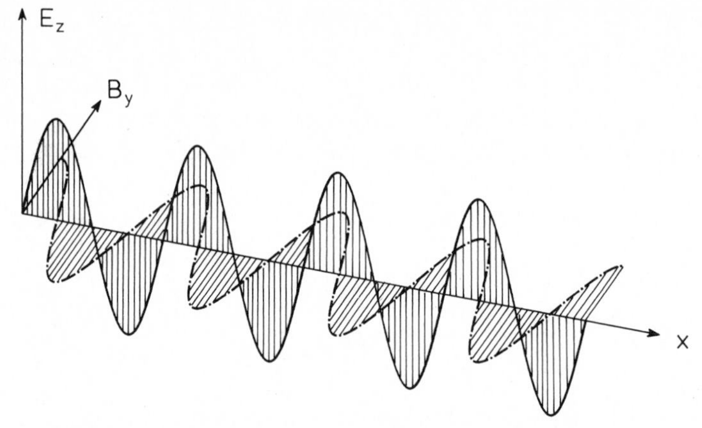

Fig. 5.2. An electromagnetic wave propagating in x-direction.

Since k is a wave number and c a velocity, ω in (5.17) is a circular frequency. Inserting E (5.13) and B (5.15) into (5.8) yields

$$\frac{\mathrm{d}p}{\mathrm{d}t} = -\omega q. \tag{5.17a}$$

Now let us consider equations (5.16) and (5.17a) in more detail. First of all we can differentiate eq. (5.16) again with respect to time and eliminate p from it by means of (5.17a). This yields

$$\frac{\mathrm{d}^2 q}{\mathrm{d}t^2} + \omega^2 q = 0. \tag{5.18}$$

This equation is the well-known equation of an harmonic oscillator with a circular frequency ω. Equations (5.16) and (5.17a) can be written in a very elegant form introducing the Hamiltonian

$$H = \tfrac{1}{2}\omega(p^2 + q^2). \tag{5.19}$$

With its aid we can write (5.16) and (5.17a) in the form

$$\frac{\mathrm{d}q}{\mathrm{d}t} = \frac{\partial H}{\partial p}, \qquad \frac{\mathrm{d}p}{\mathrm{d}t} = -\frac{\partial H}{\partial q}. \tag{5.20, 21}$$

We have encountered such equations in section 3.3 when we dealt with the classical harmonic oscillator. Comparing (5.20) and (5.21) with equations (3.49) and (3.50) of chapter 3 we recognize that here again we are dealing with the Hamiltonian equations of an harmonic oscillator. This then allows us to definitely identify p with the momentum and q with the coordinate of an harmonic oscillator. With this identification we have the key in our hands to quantize the electromagnetic field. This is done by a purely formal analogy. In section 3.3 we saw how to quantize the motion of the harmonic oscillator. Here we want to do exactly the same. To put this analogy between the harmonic oscillator and the electromagnetic field on firm ground we show that H (5.19) is identical with the energy of the electromagnetic field mode. According to electrodynamics, the energy density is defined by

$$U(\boldsymbol{x}) = \tfrac{1}{2}(\boldsymbol{E}\cdot\boldsymbol{D} + \boldsymbol{B}\cdot\boldsymbol{H}). \tag{5.22}$$

Specializing this expression to vacuum we insert

$$\boldsymbol{D} = \varepsilon_0 \boldsymbol{E}, \qquad \boldsymbol{B} = \mu_0 \boldsymbol{H} \tag{5.23}$$

and thus obtain

$$U(\boldsymbol{x}) = \tfrac{1}{2}\left(\varepsilon_0 \boldsymbol{E}^2 + \frac{1}{\mu_0}\boldsymbol{B}^2\right). \tag{5.24}$$

We obtain the total energy in a given volume V by integrating (5.22) over that volume

$$\overline{U} = \int U(\boldsymbol{x})\,\mathrm{d}^3\boldsymbol{x}. \tag{5.25}$$

In the case of our one-dimensional example it suffices to integrate over the spatial direction x

$$\overline{U} = \int_0^L U(x)\,\mathrm{d}x. \tag{5.26}$$

By inserting (5.13) and (5.15) into the energy expression (5.26) we obtain

$$\overline{U} = \tfrac{1}{2}\mathcal{N}^2\varepsilon_0\left\{\int_0^L (p^2\sin^2 kx + q^2\cos^2 kx)\,\mathrm{d}x\right\}. \tag{5.27}$$

The integration over x can easily be performed using

$$\int_0^L \sin^2 kx\,\mathrm{d}x = \tfrac{1}{2}L, \qquad \int_0^L \cos^2 kx\,\mathrm{d}x = \tfrac{1}{2}L \tag{5.28, 29}$$

so that we are left with

$$\overline{U} = \tfrac{1}{4}L\mathcal{N}^2\varepsilon_0(p^2 + q^2). \tag{5.30}$$

We find exactly the same function of p and q as occurring in (5.19). However, this identification now allows us to determine the still unknown normalization factor $\mathcal{N}$. Comparing (5.30) with (5.19) yields

$$\mathcal{N} = \sqrt{\frac{\omega}{\varepsilon_0}}\sqrt{\frac{2}{L}}\,. \tag{5.31}$$

Now let us return to the quantization problem. We wish to utilize the analogy between the Hamiltonian (5.19) with that of the harmonic oscillator. It is convenient to use its Hamiltonian in the form (3.59). The equivalence of (5.19) with (3.59) is achieved by putting

$$p = \sqrt{\hbar}\,\pi, \qquad q = \sqrt{\hbar}\,\xi \tag{5.32, 33}$$

so that the Hamiltonian (5.19) acquires exactly the same form

$$H = \frac{\hbar\omega}{2}(\pi^2 + \xi^2). \tag{5.34}$$

Here, however, we know what the quantum version looks like. We have to replace π by the operator $\partial/i\partial\xi$ exactly in analogy to section 3.3. By exploiting that analogy further we introduce creation and annihilation operators by

$$\frac{1}{\sqrt{2}}\left(-\frac{\partial}{\partial\xi} + \xi\right) = b^+, \qquad \frac{1}{\sqrt{2}}\left(\frac{\partial}{\partial\xi} + \xi\right) = b \tag{5.35, 36}$$

or, solving for p and q

$$p = \mathrm{i}\sqrt{\frac{\hbar}{2}}\,(b^+ - b), \qquad q = \sqrt{\frac{\hbar}{2}}\,(b^+ + b). \tag{5.37, 38}$$

The creation and annihilation operators b^+ and b obey the commutation relation [cf. (3.69)]

$$bb^+ - b^+b = 1. \tag{5.39}$$

By using (5.37) and (5.38), we can express the free fields $\boldsymbol{E}$ and $\boldsymbol{B}$ by means of these operators in the form

$$E_z = \mathrm{i}(b^+ - b)\sqrt{\frac{\hbar}{2}}\,\mathcal{N}\sin kx \tag{5.40}$$

$$B_y = (b^+ + b)\sqrt{\frac{\hbar}{2}}\,\frac{\mathcal{N}}{L}\cos kx \tag{5.40a}$$

or

$$E_z = i(b^+ - b)\sqrt{\frac{\hbar\omega}{2\varepsilon_0}}\sqrt{\frac{2}{L}}\,\sin kx \tag{5.41}$$

$$B_y = (b^+ + b)\sqrt{\frac{\hbar\omega\mu_0}{2}}\sqrt{\frac{2}{L}}\,\cos kx. \tag{5.41a}$$

The normalization factor is given by

$$\mathcal{N} = \sqrt{\frac{\omega}{\varepsilon_0}}\sqrt{\frac{2}{L}}, \qquad \varepsilon_0\mu_0 = 1/c^2. \tag{5.42}$$

With the transformations (5.35) and (5.36), the Hamiltonian (5.34) can be expressed by the creation and annihilation operators exactly as in section 3.3 and yields

$$H = \hbar\omega\left(b^+ b + \tfrac{1}{2}\right). \tag{5.43}$$

We leave it as an exercise to the reader to convince yourself that this Hamiltonian could be also derived by inserting (5.41) and (5.41a) into (5.26).

For a number of problems dealing with the interaction between electrons and the electromagnetic field it will turn out that we need a third quantity aside from $\boldsymbol{E}$ and $\boldsymbol{B}$, namely the vector potential $\boldsymbol{A}$. $\boldsymbol{A}$ is connected with the magnetic induction by

$$\boldsymbol{B} = \operatorname{curl} \boldsymbol{A}. \tag{5.44}$$

A further relation holds between the electric field strength and the vector potential namely

$$\boldsymbol{E} = -\frac{\partial \boldsymbol{A}}{\partial t} - \operatorname{grad} V \tag{5.45}$$

where V is the scalar potential. As is shown in electrodynamics, for given $\boldsymbol{E}$ and $\boldsymbol{B}$, $\boldsymbol{A}$ and V are not uniquely determined. Here all we need to know is that $\boldsymbol{A}$ can be made unique by an additional requirement. In our book we choose the "Coulomb gauge"

$$\operatorname{div} \boldsymbol{A} = 0. \tag{5.46}$$

Choosing $\boldsymbol{B}$ in the form (5.41a) one readily convinces oneself that the relations (5.44) and (5.46) are fulfilled by

$$A_z = -\left(b^+ + b\right)\sqrt{\frac{\hbar\omega\mu_0}{2}}\sqrt{\frac{2}{L}}\,\frac{1}{k}\sin kx. \tag{5.47}$$

Therefore, the vector potential belonging to the mode (5.12), (5.13) or (5.41a) is given by

$$\boldsymbol{A} = (0, 0, A_z) \tag{5.48}$$

with (5.47).

Let us summarize the above results. When we quantize the electromagnetic field, the electric field strength, the magnetic induction, and the vector potential become operators that can be expressed by the familiar creation and annihilation operators b^+, b of a harmonic oscillator. The total energy of the field also becomes an operator of the form $\hbar\omega(b^+b + \frac{1}{2})$. To complete the formalism we will do the following. First, we establish the Schrödinger equation of a single mode and discuss its solution. Since the creation and annihilation operators b, b^+ play an eminent role in quantum optics, we will discuss a number of their properties. Eventually (cf. section 5.8) we shall show how we can quantize not only a single mode, but the complete light field composed of many modes.

5.2. Schrödinger equation for a single mode

The Schrödinger equation belonging to the Hamiltonian (5.43) reads

$$\hbar\omega b^+ b\Phi = W\Phi \tag{5.49}$$

where we have shifted the origin of the energy scale so that $\frac{1}{2}\hbar\omega$ is absorbed in W. The wave functions will be denoted from now on by Φ. As we have seen in section 3.3 the energy levels are given by

$$W_n = n\hbar\omega. \tag{5.50}$$

This expression permits us to say that the field mode $\propto \sin(kx)$ is occupied with n photons each of energy $\hbar\omega$. The wave function reads

$$\Phi_n = \frac{1}{\sqrt{n!}}(b^+)^n\Phi_0 \tag{5.51}$$

where the state with no photon present, or in other words, the vacuum state is defined by

$$b\Phi_0 = 0. \tag{5.52}$$

We now have to apply the general scheme of quantum mechanics to the electromagnetic field, i.e. we must establish a table corresponding to table 3.1 in section 3.2. The observables are the electric field strength and the magnetic induction, and, in a way, the vector potential $\boldsymbol{A}$. Also the energy is an observable. In quantum theory all these observables become operators. Measured values must be now compared with expectation values. We quote as an explicit example the following expectation value

$$\langle\Phi_n|E_z|\Phi_n\rangle = \mathrm{i}\langle\Phi_n|(b^+ - b)|\Phi_n\rangle\sqrt{\frac{\hbar\omega}{2\varepsilon_0}}\sqrt{\frac{2}{L}}\ \sin kx. \tag{5.53}$$

Since b^+, b have nothing to do with the spatial coordinate x we have been able to extract $\sin(kx)$ and all other constants out of the quantum mechanical expectation value. From the exercises of section 3.3, we know that the first bracket vanishes, i.e. we obtain

$$\langle\Phi_n|E_z|\Phi_n\rangle = 0. \tag{5.54}$$

Similarly we find

$$\langle\Phi_n|B_y|\Phi_n\rangle = 0. \tag{5.55}$$

This result seems surprising because we know from (5.50) that the field mode is occupied with a certain number of photons so that the energy is non-vanishing and we might expect a non-vanishing amplitude. This puzzle can be resolved as follows. In classical theory the expectation value (5.53) has an analogue when we imagine that the phases of $\boldsymbol{e}$ and $\boldsymbol{b}$ are unknown and an average is made over them. Then we should also find (5.54) and (5.55). Indeed we will see below that a fixed photon number entirely destroys the knowledge of phases. However, in eqs. (5.54) and (5.55) we have used just those wave functions with the fixed photon number.

Exercises on section 5.2

(1) Establish a table corresponding to table 3.1 on page 72 for the electric field strength $\boldsymbol{E}$, the magnetic induction $\boldsymbol{B}$, energy density and total energy (each time for a single mode).
Hint: (example)

Observable	Operator	Expectation value
electric field strength		
$E_z(x,t)$	$E_z(x,t) = \mathrm{i}(b^+ - b)\sqrt{\dfrac{\hbar\omega}{2\varepsilon_0}}\sqrt{\dfrac{2}{L}}\ \sin kx$	$\langle\Phi\vert E_z(x,t)\vert\Phi\rangle$

(2) Evaluate the expectation values for $\boldsymbol{e}^2(x,t)$, $\boldsymbol{b}^2(x,t)$ and the energy density for the solutions Φ_n of (5.49).
(3) Calculate the following commutators:

$$[\boldsymbol{E}, b^+], \quad [\boldsymbol{B}, b], \quad [b^+b, b], \quad [H, \boldsymbol{E}],$$
$$[\boldsymbol{E}, b], \quad [\boldsymbol{B}, b^+], \quad [b^+b, b^+], \quad [H, \boldsymbol{B}],$$

where $\boldsymbol{e}$ is the electric field strength $(0, 0, E_z)$, $\boldsymbol{B}$ the magnetic induction $(0, B_y, 0)$, and H the Hamiltonian (5.43).
Note in particular that non-commuting of the operators H and $\boldsymbol{E}$ (or H and $\boldsymbol{B}$) implies, that the energy of the field mode and its electric field strength (or magnetic induction) cannot be simultaneously precisely measured (compare section 3.2).

5.3. Some useful relations between creation and annihilation operators

In this section we derive some relations which we will use later on quite frequently in this, and the following books. All these relations are based on the commutation relation

$$bb^+ - b^+ b = 1. \tag{5.56}$$

To understand the following property, the reader should recall the meaning of commutation relations: We always have to imagine that both sides have to be applied to an (arbitrary) wave-function. As we know, it is possible to apply b^+ (or b) several times to a wave-function Φ, i.e. to form $(b^+)^n\Phi$. This leads us to the question of studying the properties of $(b^+)^n$ within a commutation relation. The following relation can be derived from (5.56):

$$b(b^+)^n - (b^+)^n b = n(b^+)^{n-1} \tag{5.57}$$

where n is an integer $n = 1, 2, \ldots$ To indicate how this relation can be proved we choose $n = 2$. We then rewrite the left hand side in the following way

$$\begin{aligned} b(b^+)^2 - (b^+)^2 b &= bb^+b^+ - b^+b^+b + b^+bb^+ - b^+bb^+ \\ &= (bb^+ - b^+b)b^+ + b^+(bb^+ - b^+b). \end{aligned}$$

Making use of the commutation relation (5.56) yields

$$\underbrace{(bb^+ - b^+b)}_{1}b^+ + b^+\underbrace{(bb^+ - b^+b)}_{1} = 2b^+. \tag{5.58}$$

Relation (5.57) can then be proven by complete induction (compare exercise at the end of this section). In an analogous way we may prove the relation

$$b^+b^n - b^nb^+ = -nb^{n-1}. \tag{5.59}$$

In sections (5.4) and (5.5) we will encounter wave functions of the form

$$\Phi = \sum_{n=0}^{\infty} c_n(b^+)^n\Phi_0. \tag{5.60}$$

By writing this in the form

$$\Phi = \left\{\sum_{n=0}^{\infty} c_n(b^+)^n\right\}\Phi_0 \tag{5.61}$$

we are led to study the properties of such a sum over operators:

$$f(b^+) \equiv \sum_{n=0}^{\infty} c_n(b^+)^n. \tag{5.62}$$

For such functions the relation

$$bf(b^+) - f(b^+)b = \frac{\partial f(b^+)}{\partial b^+} \tag{5.63}$$

can be derived. Similarly the relation

$$b^{+}f(b) - f(b)b^{+} = -\frac{\partial f(b)}{\partial b} \tag{5.64}$$

holds (compare exercise). When we choose f as exponential function $\exp(\alpha b^{+})$, (5.63) reads

$$b\exp[\alpha b^{+}] - \exp[\alpha b^{+}]b = \alpha\exp[\alpha b^{+}]. \tag{5.65}$$

Similarly (5.64) is replaced by

$$b^{+}\exp[\alpha b] - \exp[\alpha b]b^{+} = -\alpha\exp[\alpha b]. \tag{5.66}$$

We now want to study the expression

$$f(\alpha) = \exp[\alpha b^{+}b]b\exp[-\alpha b^{+}b]. \tag{5.67}$$

To this end we differentiate it with respect to α which yields

$$\frac{\partial f(\alpha)}{\partial \alpha} = \exp[\alpha b^{+}b](b^{+}bb - bb^{+}b)\exp[-\alpha b^{+}b] \tag{5.68}$$

where we have strictly preserved the sequence of operators. (5.68) can be rearranged in the form

$$\frac{\partial f(\alpha)}{\partial \alpha} = \exp[\alpha b^{+}b]\left\{\underbrace{(b^{+}b - bb^{+})}_{=-1}b\right\}\exp[-\alpha b^{+}b] \tag{5.69}$$

which allows us to make use of the commutation relation (5.56). Using again the definition (5.67) on the right-hand side of (5.69) we obtain the differential equation

$$\partial f/\partial\alpha = -f. \tag{5.70}$$

It is solved by

$$f(\alpha) = e^{-\alpha}f(0) \tag{5.71}$$

where we deduce from (5.67) that

$$f(0) = b. \tag{5.72}$$

Therefore the final result reads

$$f(\alpha) = e^{-\alpha}b. \tag{5.73}$$

In putting (5.67) equal to (5.73), the fundamental relation yielded is

$$\exp[\alpha b^{+}b]b\exp[-\alpha b^{+}b] = e^{-\alpha}b. \tag{5.74}$$

In a similar way, we obtain the relation

$$\exp[\alpha b^{+}b]b^{+}\exp[-\alpha b^{+}b] = e^{\alpha}b^{+}. \tag{5.75}$$

Readers interested in more details of the b^{+}, b calculus are referred to my book: Quantum Field Theory of Solids (North-Holland, Amsterdam, 1976).

Exercises on section 5.3

These exercises are somewhat formal and more intended for the mathematically interested readers.

(1) Prove (5.57) by complete induction.
Hint: (5.57) is correct for $n = 1$. Assume that it has been proven up to $n = n_0$. Show, that then (5.57) is also correct for $n = n_0 + 1$.
Further hint: Rewrite

$$b(b^+)^{n_0+1} - (b^+)^{n_0+1}b$$

as

$$\left[b(b^+)^{n_0} - (b^+)^{n_0}b\right]b^+ + (b^+)^{n_0}(bb^+ - b^+b).$$

(2) Prove (5.63), (5.64).
Hint: Insert on both sides (5.62) and use (5.57) or (5.59).

(3) Derive the Baker–Hausdorff theorem

$$\exp[\alpha b + \beta b^+] = \exp[\alpha b]\exp[\beta b^+]\exp\left[-\tfrac{1}{2}\alpha\beta\right] \qquad (*)$$

$$\exp[\alpha b + \beta b^+] = \exp[\beta b^+]\exp[\alpha b]\exp\left[\tfrac{1}{2}\alpha\beta\right]. \qquad (**)$$

α, β are complex numbers.
Hint: To prove $(*)$, consider the operator $\Omega(t)$ defined by

$$\exp[t(\alpha b + \beta b^+)] = \exp[t\alpha b]\Omega(t), \qquad \Omega(0) = 1. \qquad (***)$$

Differentiate $(***)$ with respect to t and use the definition of $\Omega(t)$ to simplify the resulting expressions. In this way a differential equation for $\Omega(t)$ results. Multiply this equation by $\exp[-t\alpha b]$ from the left and use (5.63), (5.64) to eliminate the exponentials. Solve this equation for $\Omega(t)$ under the appropriate initial condition.
To prove $(**)$, use the definition

$$\exp[t(\alpha b + \beta b^+)] = \exp[t\beta b^+]\Omega(t), \qquad \Omega(0) = 1$$

and proceed as before.

5.4. Solution of the time dependent Schrödinger equation for a single field mode. Wave packets.

The time dependent Schrödinger equation corresponding to the time independent equation (5.49) reads

$$H_{0,1}\Phi = \mathrm{i}\hbar \frac{\mathrm{d}\Phi}{\mathrm{d}t}. \tag{5.76}$$

A simple example of a solution is given by the superposition of a 0-photon and a single-photon function

$$\Phi = c_0\Phi_0 + c_1\Phi_1 \exp[-\mathrm{i}\omega t]. \tag{5.77}$$

By inserting (5.77) into (5.76) we find that the corresponding terms cancel on the right- and left-hand side. The most general solution is given by a superposition of functions with all possible photon numbers

$$\Phi = c_0\Phi_0 + c_1\Phi_1 \exp[-\mathrm{i}\omega t] + \cdots + c_n\Phi_n \exp[-\mathrm{i}n\omega t] + \cdots . \tag{5.78}$$

For what follows it is interesting to study expectation values of b and b^+ with respect to the solutions (5.77) or (5.78). We introduce the abbreviation

$$\langle b\rangle = \langle\Phi|b|\Phi\rangle \tag{5.79}$$

and evaluate this expectation value for the explicit example (5.77). Also inserting (5.77) into (5.79) yields

$$\begin{aligned}\langle b\rangle = |c_0|^2\langle\Phi_0|b|\Phi_0\rangle + c_0^*c_1\langle\Phi_0|b|\Phi_1\rangle \exp[-\mathrm{i}\omega t]\\ + c_1^*c_0\langle\Phi_1|b|\Phi_0\rangle \exp[\mathrm{i}\omega t] + |c_1|^2\langle\Phi_1|b|\Phi_1\rangle.\end{aligned} \tag{5.80}$$

When we use the relations

$$\langle\Phi_0|b|\Phi_0\rangle = \langle\Phi_1|b|\Phi_1\rangle = \langle\Phi_1|b|\Phi_0\rangle = 0 \tag{5.81}$$

and $\langle\Phi_0|b|\Phi_1\rangle = 1$ (compare exercise 6 of section 3.3) we can reduce this expression to

$$\langle b\rangle = c_0^*c_1 \exp[-\mathrm{i}\omega t]. \tag{5.82}$$

Differentiating this expression with respect to time we obtain the equation

$$\frac{\mathrm{d}}{\mathrm{d}t}\langle b\rangle = -\mathrm{i}\omega\langle b\rangle. \tag{5.83}$$

We leave it as an exercise to the reader to demonstrate that this relation holds even if we use the most general solution of the Schrödinger equation (5.76). We also leave it as an exercise to the reader to convince himself that

$$\frac{\mathrm{d}}{\mathrm{d}t}\langle b^+\rangle = \mathrm{i}\omega\langle b^+\rangle \tag{5.84}$$

holds.

Exercises on section 5.4

(1) Prove that (5.77) satisfies (5.76).
Hint: Insert (5.77) into (5.76) and compare the individual terms Φ_n.

(2) Show that Φ, (5.77) or generally (5.78) is normalized,

$$\langle \Phi | \Phi \rangle = 1, \tag{$*$}$$

provided

$$\sum_{n=0}^{\infty} |c_n|^2 = 1.$$

Hint: Insert (5.78) into (*), multiply term by term and use

$$\langle \Phi_n | \Phi_m \rangle = \delta_{nm}.$$

(3) Calculate the expectation value for the electric field strength (single mode) using (5.77). Compare the result with (5.54).

(4) Show that (5.83) holds for

$$\langle b \rangle \equiv \langle \Phi | b | \Phi \rangle \tag{$*\,*$}$$

where Φ is given by (5.78).
Hint: Insert (5.58) into (* *), multiply the sums term by term and use (compare exercise 6 of section 3.3)

$$\langle \Phi_n | b | \Phi_m \rangle = \sqrt{m}\, \delta_{n,m-1}.$$

(5) Derive (5.84).
Hint: Same as before, but

$$\langle \Phi_n | b^+ | \Phi_m \rangle = \sqrt{m+1}\, \delta_{n,m+1}.$$

(6) Repeat exercise (3), but now with Φ = (5.78) (a repetition of former exercise).

5.5. Coherent states

In the preceding section we have written down the most general solution of the time-dependent Schrödinger equation of a single mode. Choosing time $t = 0$ this solution acquires the form

$$\Phi = c_0\Phi_0 + c_1\Phi_1 + \cdots + c_n\Phi_n + \cdots \tag{5.85}$$

where the functions Φ_n can be written more explicitly by means of the photon creation operator in the form

$$\Phi_n = \left(1/\sqrt{n!}\right)(b^+)^n\Phi_0. \tag{5.86}$$

In laser theory as well as in non-linear optics a special form of (5.86) plays a fundamental role. In it the coefficients have the form

$$c_n = \left(\alpha^n/\sqrt{n!}\right)\exp\left[-|\alpha|^2/2\right]. \tag{5.87}$$

As is known from quantum mechanics the absolute square of c_n gives us the probability of finding an n-photon state when the photon number (or, equivalently, the field energy) of the single mode is measured. This probability reads explicitly

$$|c_n|^2 = \left(|\alpha|^{2n}/n!\right)\exp\left[-|\alpha|^2\right]. \tag{5.88}$$

This expression is well known in probability theory because it is identical with the Poisson distribution (fig. 5.3).

Inserting (5.87) and (5.86) into (5.85) we obtain

$$\Phi = \exp\left[-|\alpha|^2/2\right]\sum_{n=0}^{\infty}(\alpha^n/n!)(b^+)^n\Phi_0. \tag{5.89}$$

The sum over n in front of Φ_0 is strongly reminiscent of the usual exponential function. The only difference lies in the fact that b^+ is an operator still acting on Φ_0 whereas the exponential function is originally defined for numbers. However, in quantum mechanics one writes that sum containing powers of b^+ in a formal way again as an exponential function (compare section 5.3). We therefore write

$$\Phi = \exp\left[-|\alpha|^2/2\right]\exp\left[\alpha b^+\right]\Phi_0. \tag{5.90}$$

When doing so we have to keep in mind that b^+ is still an operator acting

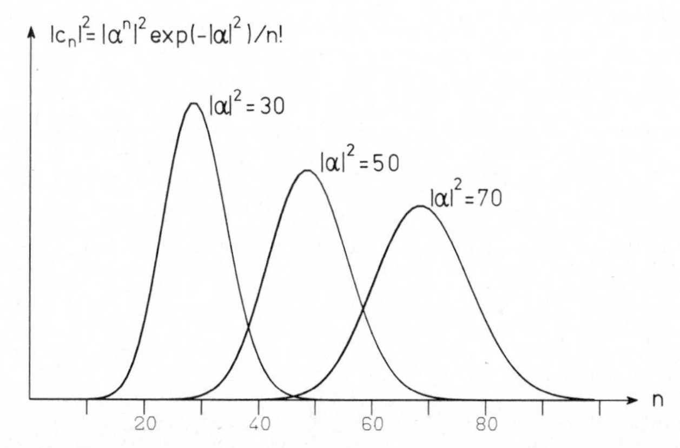

Fig. 5.3. The Poisson distribution as a function of n shown for different parameter values α.

on the subsequent function Φ_0. We now want to show that (5.90) has quite specific properties. To this end we apply the annihilation operator b on Φ. We make use of the commutation relation (5.65) we have derived earlier in chapter (5.3) namely

$$b\exp[\alpha b^+] - \exp[\alpha b^+]b = \alpha\exp[\alpha b^+] \tag{5.91}$$

which we now write a little bit differently namely

$$b\exp[\alpha b^+] = \alpha\exp[\alpha b^+] + \exp[\alpha b^+]b.$$

When we apply both sides of this relation on Φ_0 and multiply it by $\exp(-|\alpha|^2/2)$ we obtain

$$b\Phi = \alpha\Phi. \tag{5.92}$$

Thus we see that the application of b on Φ is equivalent to multiplying Φ by α. In the sense of quantum mechanics Φ is an eigenfunction to the annihilation operator b. This has important consequences. When we have to calculate any expectation value by means of such a Φ we have simply to replace the operator b by α, and correspondingly the creation operator b^+ by α^* provided all annihilation operators b stand on the right-hand side and all creation operators b^+ on the left-hand side of the total operation Ω.

We illustrate this statement by a few examples and leave their proof to the reader as an exercise

$$\langle\Phi|b|\Phi\rangle = \alpha, \qquad \langle\Phi|b^+b|\Phi\rangle = |\alpha|^2 \tag{5.93, 94}$$

$$\langle\Phi|E_z|\Phi\rangle = \mathrm{i}\mathcal{N}(\alpha^* - \alpha)\sin kx. \tag{5.95}$$

The eigenfunctions Φ_α belonging to different α's are not orthogonal to each other. We rather find

$$\langle\Phi_\alpha|\Phi_\beta\rangle = \exp\left(-\tfrac{1}{2}|\alpha|^2 - \tfrac{1}{2}|\beta|^2 + \alpha^*\beta\right). \tag{5.96}$$

(For a proof and further discussion see exercise 4 hereafter.)

The function (5.90) can be also obtained as solution of a Schrödinger equation. To this end we consider the Schrödinger equation

$$(\hbar\omega b^+ b + \gamma^* b^+ + \gamma b)\Phi = W\Phi. \tag{5.97}$$

This equation which describes the "displaced harmonic oscillator" appears quite frequently in quantum physics (compare exercise 3 hereafter).

We claim that (5.90) with appropriately chosen α is a solution of (5.97). We insert (5.90) into (5.97) and use the abbreviation

$$\mathcal{N} = \exp[-|\alpha|^2/2]. \tag{5.98}$$

We make further use of (5.92). This yields

$$\mathcal{N}\exp(\alpha b^+)\{\hbar\omega\alpha b^+ + \gamma^* b^+ + \gamma\alpha\}\Phi_0 = W\mathcal{N}\exp(\alpha b^+)\Phi_0. \tag{5.99}$$

We multiply both sides from the left side by $\mathcal{N}^{-1}\exp(-\alpha b^+)$. Since the functions $b^+ \Phi_0$ and Φ_0 are linearly independent, the resulting equation can be only fulfilled if the coefficients of $b^+ \Phi_0$ and of Φ_0 respectively, vanish. This yields

$$\hbar\omega\alpha + \gamma^* = 0 \tag{5.100}$$

and

$$\gamma\alpha = W. \tag{5.100a}$$

From (5.100) we derive

$$\alpha = -\gamma^*/(\hbar\omega). \tag{5.101}$$

From (5.100) and (5.101) the eigenvalue

$$W = -|\gamma|^2/(\hbar\omega) \tag{5.102}$$

results. In conclusion we quote an important maximum property of a coherent state. One may show quite generally, i.e. for arbitrary quantum states, that

$$\gamma_{11} = \frac{\langle\Phi|b|\Phi\rangle\langle\Phi|b^+|\Phi\rangle}{\langle\Phi|b^+b|\Phi\rangle} \tag{5.103}$$

is $\leqslant 1$. However, the states for which $\gamma_{11} = 1$ are precisely the coherent states.

Exercises on section 5.5

(1) Prove (a) (5.93), (b) (5.94), (c) (5.95).
Hints:
(a) Multiply (5.92) with $\langle\Phi|$ from the left and form $\langle\Phi|b|\Phi\rangle$. Use $\langle\Phi|\Phi\rangle = 1$.
(b) Use

$$\langle\varphi|b\chi\rangle = \langle b^+\varphi|\chi\rangle \tag{$*$}$$

(exercise 4 of section 3.3) and (5.92).
(c) Use (5.93).

(2) Prove $\langle\Phi|(b^+)^m b^n|\Phi\rangle = (\alpha^*)^m\alpha^n$, where Φ is a coherent state.
Hint: Use $(*)$ of exercise (1b) m times and use (5.93) n times.

(3) Equation (5.97) occurs at several occasions in quantum mechanics and quantum optics. To see this, treat the following exercises:
(a) Quantum mechanics: the displaced harmonic oscillator. Subject the harmonic oscillator of section 3.3 to a constant force K_0, so that $\dot{p} = -fq + K_0$. (We call this oscillator displaced, because the equilibrium position $q_0 = 0$ for $K_0 = 0$ is shifted to $q_0' = K_0/f$).

Derive the additional term to the Hamiltonian (3.46), stemming from K_0. Transform p and q into b^+, b. Then (5.97) results, where

$$\gamma, \gamma^* \propto K_0.$$

(b) Quantum mechanics: the forced harmonic oscillator. Repeat the same steps as in (a), for a time dependent K_0.
(c) Quantum optics (electronics). Start from Maxwell's equations (5.1)–(5.4), put for the electric current $\boldsymbol{j} = (0, 0, j_z)$, $j_z = j_0 \cos kx$, $\rho = 0$, $\boldsymbol{D} = \varepsilon_0 \boldsymbol{E}$, $\boldsymbol{B} = \mu_0 \boldsymbol{H}$.
Go through the same procedure as in section 5.1 and show that in H, (5.19), an additional term containing j_0 arises. Perform the same quantization procedure following (5.32). Note that j_0 can be time dependent.

(4) Prove the "unorthogonality" relation (5.95).
Hint: Insert (5.89) for α and β in $\langle \Phi_\alpha | \Phi_\beta \rangle$, multiply term by term, use $\langle \Phi_m | \Phi_n \rangle = \delta_{m,n}$ and sum up again.

5.6. Time-dependent operators. The Heisenberg picture

In all our considerations we have treated b and b^+ as given time independent operators which act on wave functions which in the general case may be time dependent. The wave functions had to obey the time-dependent Schrödinger equation. This representation is called the Schrödinger representation or the Schrödinger picture. In quantum optics and laser theory as well as in other fields of quantum mechanics another representation is frequently used. To this end remember what we have seen in section 5.4. It turned out there that the expectation values of b and b^+ obey the equations (5.83), (5.84), quite independently of the original wave functions. This leads us quite naturally to the question of whether we can find operators which are time-dependent and which automatically obey the equations (5.83) and (5.84) i.e.

$$\frac{d}{dt} b^+ = i\omega b^+, \qquad \frac{d}{dt} b = -i\omega b. \tag{5.104, 105}$$

In this case, the total time dependence of the physical process is contained in the operators whereas the wave functions are now supposed to be entirely time independent. This representation is called the Heisenberg representation. We now show in a systematic way how this Heisenberg picture can be derived. To make our procedure as transparent as possible we treat the explicit example of a single mode. Later on we shall see, however, that the Heisenberg picture has a quite universal meaning in quantum theory.

We consider as an explicit example the expectation value

$$\langle \Phi(t)|b|\Phi(t)\rangle \tag{5.106}$$

where $\Phi(t)$ is given by a general solution of the time-dependent Schrödinger equation

$$\mathrm{i}\hbar\frac{\mathrm{d}\Phi(t)}{\mathrm{d}t} = \hbar\omega b^+ b\Phi(t). \tag{5.107}$$

Our goal will it be to cast (5.106) into a form where b is a time-dependent operator and $\Phi(t)$ is replaced by a time-independent function. Since this problem can be solved not only for the Schrödinger equation (5.107) but quite generally, we start with the time-dependent Schrödinger equation

$$\mathrm{i}\hbar\frac{\mathrm{d}\Phi}{\mathrm{d}t} = H\Phi \tag{5.108}$$

where H is a general Hamilton operator. We seek a formal solution of (5.108). If H were not an operator but a number, (5.108) is an ordinary first order differential equation whose solution reads

$$\Phi(t) = \exp(-\mathrm{i}Ht/\hbar)\Phi(0). \tag{5.109}$$

However, this relation remains valid in the case that H is an operator and $\Phi(0)$ a wave function given at time $t = 0$. We have seen above that we can define the exponential function of an operator (formerly, e.g. b or b^+) by its power series expansion

$$\exp(-\mathrm{i}Ht/\hbar) = \sum_{n=0}^{\infty} \frac{1}{n!}(-\mathrm{i}H/\hbar)^n t^n. \tag{5.110}$$

By inserting (5.109) with (5.110) into (5.108) and comparing both sides term by term we readily verify that (5.108) is solved by (5.109). In the special case of $H = \hbar\omega b^+ b$, $\Phi(t)$ reads

$$\Phi(t) = \exp(-i\omega b^+ bt)\Phi(0), \tag{5.109a}$$

$\Phi(0)$ is an arbitrary wave function, which can be represented as a superposition of the solutions of the time independent Schrödinger equation

$$b^+ b\Phi_n = n\Phi_n$$

in the form

$$\Phi(0) = \sum_{n=0}^{\infty} c_n\Phi_n.$$

Now let us insert (5.109) into (5.106). Hereby we have to take into account that $\Phi(t)$ on the left-hand side is to be replaced by $\Phi(0)\exp(\mathrm{i}Ht/\hbar)$

according to the rules derived in exercise 2 below. Thus we obtain

$$\langle\Phi(t)|b|\Phi(t)\rangle = \langle\Phi(0)|\underbrace{\exp[\mathrm{i}Ht/\hbar]\,b\exp[-\mathrm{i}Ht/\hbar]}_{b(t)}|\Phi(0)\rangle. \tag{5.111}$$

When we introduce a new operator $b(t)$ as indicated by the bracket our goal to rewrite the original expression (5.106) has been reached. We now turn to the question how to determine $b(t)$. This is best answered by asking which equation is obeyed by $b(t)$. To this end we differentiate $b(t)$ with respect to t which yields

$$\begin{aligned}\frac{\mathrm{d}b(t)}{\mathrm{d}t} &= \frac{\mathrm{d}}{\mathrm{d}t}\exp[\mathrm{i}Ht/\hbar]\,b\exp[-\mathrm{i}Ht/\hbar]\\ &= \exp[\mathrm{i}Ht/\hbar]\frac{\mathrm{i}}{\hbar}\underbrace{(Hb-bH)}_{-\hbar\omega b}\exp[-\mathrm{i}Ht/\hbar].\end{aligned} \tag{5.112}$$

Since the commutator between H and b yields $-\hbar\omega b$, (5.112) transforms into

$$\frac{\mathrm{d}b(t)}{\mathrm{d}t} = -\mathrm{i}\omega\exp[\mathrm{i}Ht/\hbar]\,b\exp[-\mathrm{i}Ht/\hbar]. \tag{5.113}$$

The right-hand side is identical with $b(t)$ so that the desired operator equation reads

$$\frac{\mathrm{d}}{\mathrm{d}t}b(t) = -\mathrm{i}\omega b(t). \tag{5.114}$$

This is exactly what we had expected from our previous section 5.4 [compare (5.83) and (5.84)]. The first-order differential equation (5.114) can be readily solved and yields

$$b(t) = \mathrm{e}^{-\mathrm{i}\omega t}b(0). \tag{5.115}$$

Since at time $t=0$ the commutation relation $bb^+ - b^+b = 1$ is fulfilled it follows from (5.115) that $b(t)$ fulfills that relation for all times

$$b(t)b^+(t) - b^+(t)b(t) = 1 \tag{5.116}$$

as can be seen by inserting (5.115) into (5.116).

While our above procedure was quite useful in the special case of the operator b we now formulate a procedure in a way which lends itself immediately to general applications. To this end we rewrite the second row of (5.112) in the form

$$\begin{aligned}\frac{\mathrm{d}b(t)}{\mathrm{d}t} = \frac{\mathrm{i}}{\hbar}\{&H\exp(\mathrm{i}Ht/\hbar)b\exp(-\mathrm{i}Ht/\hbar)\\ &-\exp(\mathrm{i}Ht/\hbar)b\exp(-\mathrm{i}Ht/\hbar)H\}\end{aligned} \tag{5.117}$$

which can be written in a shorter way

$$\frac{\mathrm{d}b(t)}{\mathrm{d}t} = \frac{\mathrm{i}}{\hbar}(Hb(t) - b(t)H). \tag{5.118}$$

Or introducing the definition of commutators we eventually find

$$\frac{\mathrm{d}b(t)}{\mathrm{d}t} = \frac{\mathrm{i}}{\hbar}[H, b(t)]. \tag{5.119}$$

This is the fundamental relation which is obeyed by operators in the Heisenberg picture. This relation is valid for any operator provided it is time independent in the Schrödinger picture.

There may be cases in which the operator has an additional intrinsic time dependence. For example

$$\Omega(t) = b(t)\sin\omega t. \tag{5.120}$$

When we wish to derive an equation of motion for Ω we have to take into account this additional time dependence by a partial derivative, so that (5.119) must be replaced by

$$\frac{\mathrm{d}\Omega(t)}{\mathrm{d}t} = \frac{\mathrm{i}}{\hbar}[H, \Omega(t)] + \frac{\partial\Omega(t)}{\partial t}. \tag{5.121}$$

In our present example the partial derivative refers to the time dependence of the sine function. Formulas (5.119)–(5.121) have been derived by means of an example where H is time-independent. We note that (5.119) and (5.121) remain valid even if H depends on time explicitly.

Exercises on section 5.6

(1) Heisenberg picture and spin.
As stated above, the Heisenberg picture does not only apply to the harmonic oscillator, but to arbitrary quantum systems. A nice example is provided by spin operators. Solve the following problem:
Derive the Heisenberg equations of motion for s_x, s_y, s_z using the Hamiltonian $H = -\frac{e}{m}\boldsymbol{B}\cdot\boldsymbol{s}$ (cf. (3.110)), where $\boldsymbol{s} = (s_x, s_y, s_z)$. Compare the resulting equations with the equations for $\langle s_x\rangle$, $\langle s_y\rangle$, $\langle s_z\rangle$, exercise 2 of section 4.7.

(2) Show

$$\langle \exp(-\mathrm{i}Ht/\hbar)\Phi|\chi\rangle = \langle\Phi|\exp(\mathrm{i}Ht/\hbar)\chi\rangle$$

where H is the Hamiltonian.
Hints: Use the definition of bra and ket. Expand $\exp(\pm\mathrm{i}Ht/\hbar)$ into a power series and use that H is a Hermitian operator.

5.7. The forced harmonic oscillator in the Heisenberg picture

The great advantage of the Heisenberg picture lies in the fact that the equations of motion for operators are often strongly reminiscent of classical equations of motion. In particular, this is so for the field operators b and b^+. As an example we consider the equations of motion of a forced harmonic oscillator (compare exercise 3b of section 5.5). The Schrödinger equation of the forced harmonic oscillator can be formulated by means of b and b^+ and reads:

$$H\Phi \equiv (\hbar\omega b^+ b + \gamma^* b^+ + \gamma b)\Phi = \mathrm{i}\hbar \frac{\mathrm{d}\Phi}{\mathrm{d}t}. \tag{5.122}$$

We have encountered the corresponding time-independent equation in (5.97). Now we admit, that γ and γ^* may be time-dependent. While it is a rather formidable task to solve this Schrödinger equation with time-dependent $\gamma(t)$'s the solution of the corresponding problem in the Heisenberg picture is rather simple. To this end we start from the Heisenberg equation

$$\frac{\mathrm{d}}{\mathrm{d}t} b(t) = \frac{\mathrm{i}}{\hbar}[H, b(t)] \tag{5.123}$$

where H is given by the Hamiltonian of eq. (5.122). We leave it as an exercise to the reader to verify by means of the commutation relation (5.116) that the Heisenberg equation of motion now reads

$$\frac{\mathrm{d}}{\mathrm{d}t} b(t) = -\mathrm{i}\omega b(t) - \mathrm{i}\gamma^*(t)/\hbar \tag{5.124}$$

$$\frac{\mathrm{d}}{\mathrm{d}t} b^+(t) = \mathrm{i}\omega b^+(t) + \mathrm{i}\gamma(t)/\hbar. \tag{5.125}$$

The solution of (5.124) can be found in complete analogy to that of ordinary differential equations and reads

$$b(t) = -\frac{\mathrm{i}}{\hbar}\int_{t_0}^{t} \exp[-\mathrm{i}\omega(t-\tau)]\gamma^*(\tau)\,\mathrm{d}\tau + b(t_0)\exp[-\mathrm{i}\omega(t-t_0)]. \tag{5.126}$$

$b(t_0)$ is the operator b in the Schrödinger picture.

Later we will see, in section 8.2, that the driven (or forced) oscillator is the prototype of a coherent field arising from a classical source.

Exercises on section 5.7

(1) Derive (5.124) and (5.125) from (5.123).
Hint: Use $H = \hbar\omega b^+(t)b(t)$ and (5.116).

(2) Calculate

$$\langle\Phi(t)|b|\Phi(t)\rangle$$

$$\langle\Phi(t)|b^+|\Phi(t)\rangle$$

$$\langle\Phi(t)|b^+b|\Phi(t)\rangle$$

where Φ obeys equation (5.122) and (a) $\Phi(0) = \Phi_n$ [solution of (5.49)] (b) $\Phi(0) = \Phi_\alpha$ [coherent state, (5.90)].
Hint: Go over to the Heisenberg representation and use the solution (5.126). (Readers who want to check their results are referred to section 8.2.)

(3) Convince yourself that K_0 or j_z of exercise 3 of section 5.5 may be time dependent giving rise to the same Hamiltonian as before (i.e. to the l.h.s. of eq. (5.122)).

5.8. Quantization of lightfield: The general multimode case

In this section we resume our considerations of section 5.1 where we showed how to quantize a single mode of the electromagnetic field. The reader is advised to briefly recapitulate that section.

We describe the electromagnetic field by the vector of the electric field strength $\boldsymbol{E}$ and the vector of the magnetic induction $\boldsymbol{B}$. Without charges and currents Maxwell's equations in vacuum read

$$\text{curl}\,\boldsymbol{E} = -\frac{\partial \boldsymbol{B}}{\partial t}, \qquad \text{curl}\,\boldsymbol{B} = \varepsilon_0\mu_0\frac{\partial \boldsymbol{E}}{\partial \boldsymbol{T}}, \tag{5.127, 128}$$

$$\text{div}\,\boldsymbol{E} = 0, \qquad \text{div}\,\boldsymbol{B} = 0. \tag{5.129, 130}$$

Instead of (5.12) and (5.13) we now consider the general case in which $\boldsymbol{E}$ is described as a superposition of modes $\boldsymbol{u}_\lambda(\boldsymbol{x})$. $\boldsymbol{u}$ may describe standing or running waves but it might describe also spherical waves or still more complicated configurations depending on the physical problem. Since $\boldsymbol{E}$ is a vector, $\boldsymbol{u}$ must describe by its vector character the direction of polarization. We distinguish the different modes by an index λ. In this way we generalize (5.13)

$$\boldsymbol{E}(\boldsymbol{x},t) = -\sum_\lambda p_\lambda(t)\mathcal{N}_\lambda\boldsymbol{u}_\lambda(\boldsymbol{x}). \tag{5.131}$$

The minus-sign is exhibited explicitly for sake of later convenience. In a corresponding manner we generalize (5.15) to the hypothesis

$$\boldsymbol{B}(\boldsymbol{x},t) = \sum_\lambda q_\lambda\frac{\mathcal{N}_\lambda}{c}\boldsymbol{v}_\lambda(\boldsymbol{x}) \tag{5.132}$$

where we shall determine the connection between the functions $\boldsymbol{v}_\lambda$ with $\boldsymbol{u}_\lambda$ in a minute. Inserting (5.131) and (5.132) into (5.127) leads us to

$$-\sum_\lambda p_\lambda(t)\mathcal{N}_\lambda \operatorname{curl} \boldsymbol{u}_\lambda(\boldsymbol{x}) = -\sum_\lambda \frac{\partial q_\lambda}{\partial t}\frac{\mathcal{N}_\lambda}{c}\boldsymbol{v}_\lambda(\boldsymbol{x}). \tag{5.133}$$

To solve this equation we require that it be fulfilled for each index λ individually. It can be shown that this requirement follows rigorously from the fact that $\boldsymbol{u}$ and $\boldsymbol{v}$ are linearly independent functions. The resulting equations can be further decomposed into an equation which must be fulfilled by the time-dependent functions q_λ and p_λ

$$\frac{\partial}{\partial t} q_\lambda = \omega_\lambda p_\lambda \tag{5.134}$$

and by the space dependent functions $\boldsymbol{u}_\lambda$ and $\boldsymbol{v}_\lambda$

$$\operatorname{curl} \boldsymbol{u}_\lambda(\boldsymbol{x}) = \frac{\omega_\lambda}{c}\boldsymbol{v}_\lambda(\boldsymbol{x}). \tag{5.135}$$

As can be shown rigorously these relations follow uniquely except for an arbitrary constant which we called ω_λ. It will turn out soon that ω_λ is the mode frequency. When we insert (5.131) and (5.132) into (5.128) and make the corresponding steps we obtain

$$\frac{\partial}{\partial t} p_\lambda = -\omega_\lambda q_\lambda \tag{5.136}$$

and

$$\operatorname{curl} \boldsymbol{v}_\lambda(\boldsymbol{x}) = \frac{\omega_\lambda}{c}\boldsymbol{u}_\lambda(\boldsymbol{x}). \tag{5.137}$$

From (5.129) and (5.130) we readily deduce

$$\operatorname{div} \boldsymbol{u}_\lambda(\boldsymbol{x}) = 0, \qquad \operatorname{div} \boldsymbol{v}_\lambda(\boldsymbol{x}) = 0. \tag{5.138, 139}$$

Thus a solution of Maxwell's equations is reduced to two kinds of problems, namely to the solution of the equations for the spatial field modes (5.135), (5.137), (5.138), (5.139) and to those for the time dependent amplitudes q_λ, p_λ, namely (5.134) and (5.136). As is shown in electrodynamics the equations of the field modes can be solved uniquely provided adequate boundary conditions are given. We will not discuss this problem here but refer the reader to corresponding textbooks.

In analogy to chapter 5.1 we expect that the quantization procedure refers to q_λ and p_λ. To this end we repeat the mean steps of section 5.1. The total energy of the electromagnetic field in the volume V under consideration is given by

$$U = \int \mathrm{d}V \tfrac{1}{2}\left(\varepsilon_0 \boldsymbol{E}^2 + \frac{1}{\mu_0}\boldsymbol{B}^2\right). \tag{5.140}$$

We assume that the field modes are orthogonal and normalized according to

$$\int \boldsymbol{u}_\lambda(\boldsymbol{x}) \boldsymbol{u}_{\lambda'}(\boldsymbol{x})\, \mathrm{d}V = \delta_{\lambda,\lambda'} \tag{5.141}$$

$$\int \boldsymbol{v}_\lambda(\boldsymbol{x}) \boldsymbol{v}_{\lambda'}(\boldsymbol{x})\, \mathrm{d}V = \delta_{\lambda,\lambda'}. \tag{5.142}$$

Again we must refer the reader to the theory of electromagnetism in which it is shown that these relations can be fulfilled indeed. Inserting (5.131) and (5.132) into (5.140) we readily obtain

$$U = \tfrac{1}{2}\varepsilon_0 \sum_\lambda (p_\lambda^2 + q_\lambda^2)\mathcal{N}_\lambda^2. \tag{5.143}$$

(5.143) can be considered as a sum

$$U = \sum_\lambda H_\lambda \tag{5.144}$$

over Hamiltonians

$$H_\lambda = \tfrac{1}{2}\hbar\omega_\lambda (p_\lambda^2 + q_\lambda^2). \tag{5.145}$$

Each of these Hamiltonians has exactly the same form as the one considered in section 5.1. Furthermore the p's and q's obey equations namely (5.134) and (5.136) which are completely analogous to the corresponding ones (5.16) and (5.17a) in section 5.1. This allows us to perform again exactly the same steps as in that section which leads us to the following relations

$$\mathcal{N}_\lambda = \sqrt{\frac{\hbar\omega_\lambda}{\varepsilon_0}} \tag{5.146}$$

(note, that the spatial modes are now normalized differently)

$$q_\lambda = \frac{1}{\sqrt{2}}(b_\lambda^+ + b_\lambda) \tag{5.147}$$

$$p_\lambda = \frac{\mathrm{i}}{\sqrt{2}}(b_\lambda^+ - b_\lambda). \tag{5.148}$$

Inserting these quantities into (5.131) and (5.132) we obtain our final result

$$\boldsymbol{E}(\boldsymbol{x},t) = \sum_\lambda \mathrm{i}(b_\lambda - b_\lambda^+)\sqrt{\hbar\omega_\lambda/(2\varepsilon_0)}\, \boldsymbol{u}_\lambda(\boldsymbol{x}), \tag{5.149}$$

$$\boldsymbol{B}(\boldsymbol{x},t) = \sum_\lambda (b_\lambda + b_\lambda^+)\sqrt{\hbar\omega_\lambda\mu_0/2}\, \boldsymbol{v}_\lambda(\boldsymbol{x}). \tag{5.150}$$

For later purposes we present the explicit decomposition of $\boldsymbol{E}$ into its positive and negative frequency parts:

$$E^{(+)}(x,t) = \sum_{\lambda} i b_{\lambda} \sqrt{\hbar\omega_{\lambda} / (2\varepsilon_0)}\, u_{\lambda}(x) \tag{5.149a}$$

$$E^{(-)}(x,t) = -\sum_{\lambda} i b_{\lambda}^{+} \sqrt{\hbar\omega_{\lambda} / (2\varepsilon_0)}\, u_{\lambda}(x). \tag{5.149b}$$

$\boldsymbol{B}$ can be decomposed correspondingly.

Now let us anticipate that $b_{\lambda}^{+}, b_{\lambda}$ have become operators (see below). Generalizing section 5.1 to the multimode case we obtain the Hamiltonian in the form

$$H = \sum_{\lambda} \hbar\omega_{\lambda}\left(b_{\lambda}^{+} b_{\lambda} + \tfrac{1}{2}\right) \tag{5.151}$$

where we shall drop in the following the zero-point energy

$$\tfrac{1}{2}\sum_{\lambda} \hbar\omega_{\lambda}. \tag{5.152}$$

As we know, quantization is achieved when we subject the quantities $b_{\lambda}, b_{\lambda}^{+}$ to commutation relations. Generalizing our former results (5.39) we are readily led to require

$$b_{\lambda} b_{\lambda}^{+} - b_{\lambda}^{+} b_{\lambda} = 1. \tag{5.153}$$

However, the question is open what commutation relation we shall require when the b_{λ}'s belong to different indices λ, λ'.

Since the different modes can be interpreted as describing different quantum systems which are not dynamically coupled, it is suggestive to require that the operators $b_{\lambda}, b_{\lambda'}^{+}$ commute for different λ's. Thus we generalize (5.153) to

$$b_{\lambda} b_{\lambda'}^{+} - b_{\lambda'}^{+} b_{\lambda} = \delta_{\lambda\lambda'} \tag{5.154}$$

and require in addition

$$b_{\lambda} b_{\lambda'} - b_{\lambda'} b_{\lambda} = 0, \qquad b_{\lambda}^{+} b_{\lambda'}^{+} - b_{\lambda'}^{+} b_{\lambda}^{+} = 0. \tag{5.155, 156}$$

Of course, our requirement with respect to the commutation relation is somewhat heuristic but we know from numerous calculations and their comparison with experiments that these commutation relations are the correct ones. In section 5.1 we considered in addition to $\boldsymbol{E}$ and $\boldsymbol{B}$ the vector potential $\boldsymbol{A}$. In the absence of a scalar potential V, $\boldsymbol{E}$ and $\boldsymbol{A}$ are connected by the equations

$$\boldsymbol{B} = \operatorname{curl} \boldsymbol{A}, \qquad \boldsymbol{E} = -\frac{\partial}{\partial t}\boldsymbol{A}. \tag{5.157}$$

To find a suitable form for $\boldsymbol{A}$ we decompose it into still unknown field modes according to

$$A(x,t) = \sum_\lambda a_\lambda(t) w_\lambda(x). \tag{5.158}$$

Inserting this expression as well as (5.131) into (5.157) and comparing the individual terms we are led to the equations

$$-p_\lambda \mathcal{N}_\lambda u_\lambda(x) = -\frac{\partial a_\lambda}{\partial t} w_\lambda(x). \tag{5.159}$$

This relation can be decomposed into

$$w_\lambda(x) = -\mathcal{N}_\lambda u_\lambda(x) \tag{5.160}$$

and

$$p_\lambda = -\frac{\partial a_\lambda}{\partial t}. \tag{5.161}$$

By comparing (5.161) with the relation (5.134) leads us to the choice

$$a_\lambda = -\frac{1}{\omega_\lambda} q_\lambda. \tag{5.162}$$

Putting all terms together we obtain for the vector potential A

$$A(x,t) = \sum_\lambda (b_\lambda + b_\lambda^+) \sqrt{\frac{\hbar}{2\omega_\lambda \varepsilon_0}}\, u_\lambda(x). \tag{5.163}$$

Since b_λ and b_λ^+ become operators, so do the electric field strength $E(x)$, the magnetic induction $B(x)$, and the vector potential $A(x)$. When we use the Schrödinger picture, these operators are time independent; when we use the Heisenberg picture, they will be time dependent.

In conclusion of this section we can now do the last step of our quantization procedure, namely we can write down the Schrödinger equation. The time-dependent Schrödinger equation reads

$$H\Phi = \mathrm{i}\hbar \frac{\mathrm{d}\Phi}{\mathrm{d}t} \tag{5.164}$$

and its time-independent version

$$H\Phi = W\Phi. \tag{5.165}$$

The Hamiltonian is given by (5.151). Identifying the index λ with numbers $1, 2, \ldots$ we may very simply describe the solutions of (5.165). Using the abbreviation

$$\{n\} = \{n_1, n_2, n_3 \ldots\} \tag{5.166}$$

we may write the solution as

$$\Phi_{\{n\}} = \frac{1}{\sqrt{n_1!n_2!n_3!\dots n_\lambda!\dots}}(b_1^+)^{n_1}(b_2^+)^{n_2}(b_3^+)^{n_3}\dots(b_\lambda^+)^{n_\lambda}\dots\Phi_0. \tag{5.167}$$

Φ_0 is the vacuum state which has the property

$$b_\lambda\Phi_0 = 0. \tag{5.168}$$

The energy W of the field belonging to the function (5.167) is given by

$$W = \sum_\lambda \hbar\omega_\lambda\left(n_\lambda + \tfrac{1}{2}\right). \tag{5.169}$$

Finally we mention that the Φ's obey an orthonormality relation namely

$$\langle\Phi_{\{n\}}|\Phi_{\{m\}}\rangle = \delta_{n_1m_1}\delta_{n_2m_2}\delta_{n_3m_3}\dots\delta_{n_Nm_N}. \tag{5.170}$$

We refer the reader to the exercises where it is indicated how to prove that (5.167) solves (5.165).

5.8.1. Quantization of the electromagnetic field by means of its decomposition into running plane waves

In section 5.1 and in the present section we have treated the following problem. We considered the electromagnetic field enclosed in a region, or, more physically speaking, within a cavity. As a consequence the field has to obey certain boundary conditions. These conditions are essential to fix the expansion functions $\boldsymbol{u}_\lambda(\boldsymbol{x}), \boldsymbol{v}_\lambda(\boldsymbol{x})$. As is evident from (5.13) we treated standing waves. We shall make ample use of this type of approach in laser physics (compare Volume 2). Of course, there are experimental arrangements other than that of a given cavity possible. For instance we can make a measurement of the field using a different experimental setup. Then we have to expand the field into the corresponding new types of modes.

There are other experiments for which standing waves are not the adequate description, but rather running plane waves. Important examples will be provided by nonlinear optics where we shall study the propagation of such waves in "nonlinear" crystals. Another example, which will be treated in section 7.6, is spontaneous or stimulated emission. In this case no boundaries are present and it is "natural" to decompose the field into running plane waves. In classical physics, such a decomposition reads

$$\begin{aligned}\boldsymbol{E}(\boldsymbol{x},t) = \sum_\lambda \boldsymbol{e}_\lambda\mathfrak{N}_\lambda\{&B_\lambda\exp(-\mathrm{i}\omega_\lambda t)\exp(\mathrm{i}\boldsymbol{k}_\lambda\boldsymbol{x})\\ &+B_\lambda^*\exp(\mathrm{i}\omega_\lambda t)\exp(-\mathrm{i}\boldsymbol{k}_\lambda\boldsymbol{x})\}\end{aligned} \tag{5.171}$$

where the individual expressions have the following meaning:

λ	index numerating the individual waves
$\boldsymbol{e}_\lambda$	vector of polarization of wave λ
B_λ	complex amplitude of wave λ
ω_λ	circular frequency
$\mathfrak{N}_\lambda$	a normalization factor which we shall specify below
$\boldsymbol{k}_\lambda$	wave vector of wave λ.

Since the normalization of waves in infinite space provides some formal difficulties (which one may overcome, however), we shall use a well-known trick. We subject the wave-functions

$$\exp(\mathrm{i}\boldsymbol{k}_\lambda \boldsymbol{x}) \tag{5.172}$$

to periodic boundary conditions (compare exercise 3 below).

We may quantize the fields $\boldsymbol{E}$ and $\boldsymbol{B}$ in a way quite analogous to the procedure we have outlined above in this chapter. We therefore quote only the final result. Again the amplitudes $B_\lambda \exp(-\mathrm{i}\omega_\lambda t)$ and $B_\lambda^* \exp(\mathrm{i}\omega_\lambda t)$ become operators b_λ and b_λ^+, respectively. The expansions for $\boldsymbol{E}$ and $\boldsymbol{B}$ read

$$\boldsymbol{E}(\boldsymbol{x}) = \sum_\lambda \boldsymbol{e}_\lambda \sqrt{\frac{\hbar\omega_\lambda}{2\varepsilon_0 V}} \left\{\gamma^* b_\lambda \exp[\mathrm{i}\boldsymbol{k}_\lambda \boldsymbol{x}] + \gamma b_\lambda^+ \exp[-\mathrm{i}\boldsymbol{k}_\lambda \boldsymbol{x}]\right\} \tag{5.173}$$

$$\boldsymbol{B}(\boldsymbol{x}) = \sum_\lambda \hat{\boldsymbol{k}} \times \boldsymbol{e}_\lambda \sqrt{\frac{\hbar\omega_\lambda \mu_0}{2V}} \left\{\gamma^* b_\lambda \exp[\mathrm{i}\boldsymbol{k}_\lambda \boldsymbol{x}] + \gamma b_\lambda^+ \exp[-\mathrm{i}\boldsymbol{k}_\lambda \boldsymbol{x}]\right\} \tag{5.174}$$

where $\hat{\boldsymbol{k}} = \boldsymbol{k}/|\boldsymbol{k}|$. γ is a complex constant with

$$|\gamma| = 1. \tag{5.175}$$

A choice often made is $\gamma = -i$. The vector potential $\boldsymbol{A}$ reads

$$\boldsymbol{A} = -\mathrm{i}\sum_\lambda \boldsymbol{e}_\lambda \sqrt{\frac{\hbar}{2\omega_\lambda \varepsilon_0 V}} \left\{\gamma^* b_\lambda \exp[\mathrm{i}\boldsymbol{k}_\lambda \boldsymbol{x}] - \gamma b_\lambda^+ \exp[-\mathrm{i}\boldsymbol{k}_\lambda \boldsymbol{x}]\right\}. \tag{5.176}$$

As it turns out, the operators b_λ, b_λ^+ again obey the commutation relations (5.153), (5.154), (5.155), (5.156). The Hamilton operator, the Schrödinger equation, etc. are identical with our former results.

In conclusion, let us once again stress that the kind of expansion we choose depends on the physical problem. In free space (without boundaries), we can use also spherical waves instead of plane waves. A decomposition of $\boldsymbol{E}, \boldsymbol{B}, \boldsymbol{A}$ into these waves means that we want to treat experiments (at least "Gedankenexperiments") in which photons belonging to such waves are measured.

Exercises on section 5.8

The following exercises are arranged as follows. By (1), (2), and (3) the reader will learn what typical solutions ("modes") of the classical Maxwell's equations look like explicitly and what their orthogonality properties are.

Exercises (4)–(6) deal with important properties of quantized lightfields: (4) deals with the solutions of the time independent Schrödinger equation, and by (5) the reader will learn about a basic difference between the expectation values of the fields described by two different types of wave functions.

(1) Standing waves of the electromagnetic field in a rectangular cavity with perfectly conducting walls (see fig. 5.4). Its side lengths are L_1, L_2, L_3 in the x, y, z-directions, respectively. $\boldsymbol{E}$ and $\boldsymbol{B}$ obey Maxwell's equations and the boundary conditions

$$E_{\text{tangential}} = 0 \tag{$*$}$$

$$B_{\text{normal}} = 0 \quad \text{at the walls.} \tag{$**$}$$

Decompose $\boldsymbol{E}$ and $\boldsymbol{B}$ into modes $\boldsymbol{u}_\lambda$ and $\boldsymbol{v}_\lambda$ and show that $\boldsymbol{u}_\lambda, \boldsymbol{v}_\lambda$ fulfill the equations (5.135), (5.137), (5.138), (5.139) and the boundary conditions

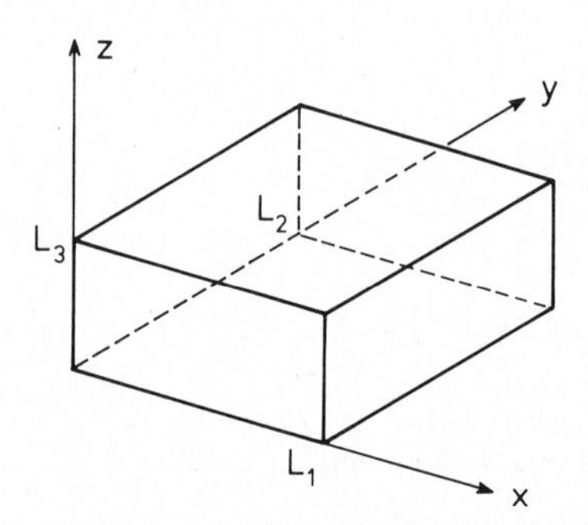

Fig. 5.4. This figure belongs to exercise 1 on section 5.8. Arrangement of the perfectly conducting walls.

$(*)$, $(**)$ provided

$$u_{\lambda,x} = e_{\lambda,x}\sqrt{\frac{2}{V}}\ \cos k_x x \sin k_y y \sin k_z z,$$

$$u_{\lambda,y} = e_{\lambda,y}\sqrt{\frac{2}{V}}\ \sin k_x x \cos k_y y \sin k_z z,$$

$$u_{\lambda,z} = e_{\lambda,z}\sqrt{\frac{2}{V}}\ \sin k_x x \sin k_y y \cos k_z z, \qquad (***)$$

$$v_{\lambda,x} = \tilde{e}_{\lambda,x}\sqrt{\frac{2}{V}}\ \sin k_x x \cos k_y y \cos k_z z,$$

$$v_{\lambda,y} = \tilde{e}_{\lambda,y}\sqrt{\frac{2}{V}}\ \cos k_x x \sin k_y y \cos k_z z,$$

$$v_{\lambda,z} = \tilde{e}_{\lambda,z}\sqrt{\frac{2}{V}}\ \cos k_x x \cos k_y y \sin k_z z.$$

$\boldsymbol{e}_\lambda$ and $\tilde{\boldsymbol{e}}_\lambda$ are constant unit vectors with components $(e_{\lambda,x}, e_{\lambda,y}, e_{\lambda,z})$ and $(\tilde{e}_{\lambda,x}, \tilde{e}_{\lambda,y}, \tilde{e}_{\lambda,z})$, respectively:

$$\boldsymbol{e}_\lambda \cdot \boldsymbol{k} = 0, \qquad \tilde{\boldsymbol{e}}_\lambda \cdot \boldsymbol{k} = 0, \qquad \tilde{\boldsymbol{e}}_\lambda \cdot \boldsymbol{e}_\lambda = 0$$

$$\boldsymbol{k} = \left(\frac{\pi}{L_1}n, \frac{\pi}{L_2}m, \frac{\pi}{L_3}l\right), \qquad n, m, l \text{ integer}, \qquad V = L_1 L_2 L_3.$$

Hint: Insert $\boldsymbol{u}_\lambda$ and $\boldsymbol{v}_\lambda$ into eqs. (5.135), (5.137)–(5.139). Convince yourself that $\boldsymbol{u}_\lambda$ and $\boldsymbol{v}_\lambda$ fulfill the boundary conditions $(*)$, $(**)$ at $x = 0, L_1$, $y = 0, L_2$, $z = 0, L_3$.

(2) Prove (5.141), (5.142) for the solutions given in $(***)$.
Hint: Insert $(***)$ into (5.141), (5.142) and use

$$\int_0^L \sin\frac{n\pi}{L}x \sin\frac{m\pi}{L}x\,dx = \frac{L}{2}\delta_{n,m}$$

$$\int_0^L \cos\frac{n\pi}{L}x \cos\frac{m\pi}{L}x\,dx = \frac{L}{2}\delta_{n,m}.$$

(3) Plane wave solutions:
We require the $\boldsymbol{E}$ and $\boldsymbol{B}$ are periodic within the cube with side lengths L:

$$\boldsymbol{E}(x + L, y, z) = \boldsymbol{E}(x, y, z)$$

$$\boldsymbol{E}(x, y + L, z) = \boldsymbol{E}(x, y, z)$$

$$\boldsymbol{E}(x, y, z + L) = \boldsymbol{E}(x, y, z)$$

and correspondingly for $\boldsymbol{B}$.

Show that the modes $\boldsymbol{u}_\lambda$ and $\boldsymbol{v}_\lambda$ fulfill their corresponding equations (5.135), (5.137)–(5.139) provided

$$\boldsymbol{u}_\lambda(\boldsymbol{x}) = \boldsymbol{e}_\lambda L^{-3/2} \exp[\mathrm{i}\boldsymbol{k}_\lambda \cdot \boldsymbol{x}]$$
$$\boldsymbol{v}_\lambda(\boldsymbol{x}) = \tilde{\boldsymbol{e}}_\lambda L^{-3/2} \exp[\mathrm{i}\boldsymbol{k}_\lambda \cdot \boldsymbol{x}]$$

where

$$\boldsymbol{e}_\lambda \perp \tilde{\boldsymbol{e}}_\lambda, \qquad \boldsymbol{k}_\lambda = \left(\frac{2\pi n_1}{L}, \frac{2\pi n_2}{L}, \frac{2\pi n_3}{L}\right), \qquad n_i \text{ integer}.$$

Precisely speaking, in this case the index λ can be identified with (n_1, n_2, n_3).

(4) Show that (5.167) solves (5.165) with W given by (5.169).
Hint: Use the decomposition (5.151) dropping $\frac{1}{2}\Sigma_\lambda \hbar\omega_\lambda b_\lambda^+ b_\lambda$. Consider $b_{\lambda_0}^+ b_{\lambda_0} \Phi_{\{\boldsymbol{n}\}}$. Use that $b_{\lambda_0}^+, b_{\lambda_0}$ commute with all other b_λ^+, b_λ, $\lambda \neq \lambda_0$. Now consider $b_{\lambda_0}^+ b_{\lambda_0} (b_{\lambda_0}^+)^{n_{\lambda_0}} \Phi_0$. Consult section 3.3, 5.3.

(5) Calculate

$$\langle \Phi | \boldsymbol{E}(\boldsymbol{x}, t) | \Phi \rangle \qquad (*)$$

for (a)

$$\Phi = \Phi_{\{\boldsymbol{n}\}},$$

(5.167), and (b)

$$\Phi_\alpha = \exp\left[-\tfrac{1}{2}\sum_\lambda |\alpha_\lambda|^2\right] \exp\left[\sum_\lambda \alpha_\lambda b_\lambda^+\right] \Phi_0 .$$

Hint: Decompose $\boldsymbol{E}$ according to (5.149) and write Φ_α in the form of a product

$$\Phi = \left[\prod_\lambda \exp\left[-\tfrac{1}{2}|\alpha_\lambda|^2 + \alpha_\lambda b_\lambda^+\right]\right] \Phi_0$$

While in the case (a), $(*)$ vanishes, in case (b) we obtain

$$(*) = \sum_\lambda \mathrm{i}(\alpha_\lambda - \alpha_\lambda^*) \sqrt{\hbar\omega_\lambda / 2\varepsilon_0}\, \boldsymbol{u}_\lambda(\boldsymbol{x}),$$

i.e. $\boldsymbol{E}$ takes a form entirely corresponding to the classical expression.

(6) Show that (5.154)–(5.156) are valid also in the Heisenberg picture.
Hint: Multiply (5.154)–(5.156) from the left by $U^{-1} = \exp(\mathrm{i}Ht/\hbar)$ and from the right by $U = \exp(-\mathrm{i}Ht/\hbar)$. Write, for instance,

$$U^{-1} b_\lambda b_{\lambda'}^+ U = U^{-1} b_\lambda U U^{-1} b_{\lambda'}^+ U$$

and use

$$b_\lambda(t) = U^{-1} b_\lambda U$$

etc.

5.9. Uncertainty relations and limits of measurability*

In quantum mechanics, operators Ω_1, Ω_2 which do not commute cannot be measured simultaneously with absolute exactness (cf. section 3.2). There is always an uncertainty left, the degree of which is expressed by uncertainty relations. A measure for the uncertainty in the measurement of the observable with the operator Ω is the root mean square deviation:

$$\Delta\Omega = \left[\left\langle(\Omega - \langle\Omega\rangle)^2\right\rangle\right]^{1/2} \tag{5.177}$$

where the average refers to a single wave function:

$$\langle\Omega\rangle = \langle\Phi|\Omega|\Phi\rangle. \tag{5.178}$$

For hermitian operators one can show quite generally

$$\Delta\Omega_1 \cdot \Delta\Omega_2 \geqslant \tfrac{1}{2}|\langle[\Omega_1, \Omega_2]\rangle| \tag{5.179}$$

where

$$[\Omega_1, \Omega_2] = \Omega_1\Omega_2 - \Omega_2\Omega_1.$$

A famous example is in quantum mechanics $\Omega_1 = p$ (momentum), $\Omega_2 = q$ (position). Because $[p, q] = -i\hbar$ one finds:

$$\Delta p \cdot \Delta q \geqslant \tfrac{1}{2}\hbar. \tag{5.180}$$

which is the Heisenberg uncertainty relation (cf. section 1.9). We are here primarily concerned with the electromagnetic fields, the photon number, and the phase of light.

Field and photon number

The question arises how accurately can one measure the field strength and the number of photons in a certain mode λ_0 simultaneously? Since the b_λ^+'s, b_λ's commute for different λ's, we need only consider the same mode λ in the expansion of $\boldsymbol{E}$ or $\boldsymbol{B}$ which can be obtained from (5.149) and (5.150). The time-dependent mode amplitude is represented by the operators $b_{\lambda_0}^+$ and b_{λ_0}, so that we have merely to determine the commutation relation between $(b^+ + b)$ and $i(b - b^+)$ on the one hand and $n = b^+b$ on the other. (From now on we drop the index λ or λ_0). On account of (5.179), and

$$[b^+, b^+b] = -b^+, \qquad [b, b^+b] = b \tag{5.181, 182}$$

we find

$$\Delta(b^+ + b)\Delta n \geqslant \tfrac{1}{2}|\langle -b^+ + b\rangle| \tag{5.183}$$

$$\Delta i(b - b^+)\Delta n \geqslant \tfrac{1}{2}|\langle b^+ + b\rangle|. \tag{5.184}$$

*This section is rather formal and can be skipped, because no explicit use of its results will be later made in this book.

If the averages on the right hand sides are evaluated for states, for which $\langle \pm b^+ + b \rangle$ does not vanish, it is evident, that a small uncertainty of n necessitates a big one of $\Delta(b^+ \pm b)$ and vice versa. Because the electric (or magnetic) field strength is proportional to b^+ and b, a precise measurement of the photon number is connected with an uncertainty of the field amplitudes (This is the case even if no light quanta at all are present).

If we use states for which the right hand sides of (5.183) and (5.184) vanish, at a first glance it would seem possible to determine $(b^+ \pm b)$ and n exactly in a simultaneous measurement. We want to show that this conclusion is misleading, because in these cases one factor on the left hand side of these equations vanishes:

1) If we use eigenstates of the number operator $b^+ b$, both sides vanish,
2) The same is true, if we use certain coherent states.

Thus in these two cases the relations (5.183), (5.184) fail to give us information about the uncertainty of n, when a coherent state is measured and vice versa. Therefore one is forced for these two cases, at least, to calculate Δn or $\Delta(b^+ + b)$ directly:

5.9.1. *A coherent state Φ_α (5.90) is given.*

Inserting Φ_α into $(\Delta n)^2$ we obtain, according to the definition (5.177)

$$\begin{aligned}(\Delta n)^2 &= \langle \Phi_\alpha | b^+ b b^+ b | \Phi_\alpha \rangle - \langle \Phi_\alpha | b^+ b | \Phi_\alpha \rangle^2 \\ &= \langle \Phi_\alpha | b^+ b^+ b b | \Phi_\alpha \rangle + \langle \Phi_\alpha | b^+ b | \Phi_\alpha \rangle - \langle \Phi_\alpha | b^+ b | \Phi_\alpha \rangle^2 \\ &= \langle \Phi_\alpha | b^+ b | \Phi_\alpha \rangle = |\alpha|^2 = \bar{n}. \end{aligned} \tag{5.185}$$

The mean square deviation of n, determined for a coherent state, equals the average number of photons, $\bar{n}$, in that state. The relative fluctuations decrease, however, with $\bar{n}$: $\Delta n / \bar{n} = 1/\sqrt{\bar{n}}$.

5.9.2. *An eigenstate of $b^+ b$ is given: Φ_n (5.86).*

We obtain

$$\begin{aligned}(\Delta(b^+ + b))^2 = \langle \Phi_n | b^{+2} + (b^+ b + b b^+) + b^2 | \Phi_n \rangle \\ - \langle \Phi_n | b^+ + b | \Phi_n \rangle^2. \end{aligned} \tag{5.186}$$

Due to (5.54), (5.53) the second expectation value vanishes while the first reduces to

$$(5.186) = 2n + 1. \tag{5.187}$$

The same is found for

$$(\Delta \mathrm{i}(b - b^+))^2.$$

The mean square deviation of $(b + b^+)$ and $\mathrm{i}(b - b^+)$ determined for a state with a definite photon number, goes like the photon number.

Phase and photon number

Heuristic considerations. Since the classical analogue of b, b^+ is a complex amplitude, one may try to decompose these operators into a real amplitude and a phase factor

$$b^+ = \sqrt{n}\exp[-\mathrm{i}\varphi] \tag{5.188}$$

and

$$b = \sqrt{n}\exp[\mathrm{i}\varphi] \tag{5.189}$$

where φ and $\sqrt{n}$ are operators. We assume for the time being that such a representation does exist. Then it follows from the commutation relation (5.56) that

$$\exp[\mathrm{i}\varphi]n - n\exp[\mathrm{i}\varphi] = \exp[\mathrm{i}\varphi]. \tag{5.190}$$

This relation is satisfied if φ and n satisfy the commutation relation

$$\varphi n - n\varphi = -\mathrm{i} \tag{5.191}$$

and thus on account of (5.179) the uncertainty relation $\Delta n \, \Delta\varphi \geqslant \frac{1}{2}$ follows.

There are, however, two serious difficulties:

(1) The decompositions (5.188), (5.189) do not exist.

(2) It is necessary to derive (5.191) from (5.190) and not vice versa.

We demonstrate in a qualitative manner that the decompositions (5.188), (5.189) cannot exist. It is obviously required that

$$\exp[-\mathrm{i}\varphi]\exp[\mathrm{i}\varphi] = 1. \tag{5.192}$$

In expressing $\exp[-\mathrm{i}\varphi]$ by $b^+/\sqrt{n}$ and $\exp[\mathrm{i}\varphi]$ by $b/\sqrt{n}$, we immediately run into difficulties: If we calculate these exponentials in the n-representation, $\sqrt{n}$ vanishes in those matrix-elements which contain the vacuum. This difficulty can easily be overcome, however, by replacing $\sqrt{n}$ by $\sqrt{n+1}$. One is thus led to a new decomposition of b, b^+ which allows us to replace the uncertainty relation (5.191) by one which can be rigorously derived. We now turn to an:

Exact treatment. We write according to Glowgower and Susskind

$$b^+ = E_+(b^+b + 1)^{1/2}, \qquad b = (b^+b + 1)^{1/2}E_-. \tag{5.193*}$$

*The operator introduced in (5.193) is not to be mixed up with the electric field.

Since $(b^+b+1)^{1/2}$ possesses an inverse, E_+ and E_- are uniquely defined by

$$E_+ = b^+(b^+b+1)^{-1/2}, \qquad E_- = (b^+b+1)^{-1/2}b \tag{5.194}$$

where $E_\pm$ are obviously the substitutes for the former $\exp[\mp i\varphi]$. The E's have the property of creating normalized eigenfunctions in the n-representation

$$\Phi_{n\pm1} = E_\pm \Phi_n. \tag{5.195}$$

Neither b^+, b nor E_-, E_+ are Hermitian operators, however, for which an uncertainty relation of the form (5.179) can be deduced. We therefore define

$$S = \frac{1}{2i}(E_- - E_+), \qquad (\sim \sin\varphi\,!) \tag{5.196}$$

and

$$C = \tfrac{1}{2}(E_+ + E_-), \qquad (\sim \cos\varphi\,!). \tag{5.197}$$

The operators S and C have a continuous eigenvalue spectrum in the interval from -1 to $+1$. Using (5.179) one derives the following uncertainty relations:

$$\begin{aligned} \Delta n \cdot \Delta S &\geqslant \tfrac{1}{2}|\langle C\rangle| \\ \Delta n \cdot \Delta C &\geqslant \tfrac{1}{2}|\langle S\rangle|. \end{aligned} \tag{5.198}$$

From (5.198) the relation

$$U \equiv (\Delta n)^2 \frac{(\Delta S)^2 + (\Delta C)^2}{\langle S\rangle^2 + \langle C\rangle^2} \geqslant \tfrac{1}{4} \tag{5.199}$$

follows. In particular one can show for coherent states that $\frac{1}{2} \geqslant U \geqslant \frac{1}{4}$. The eqs. (5.198) are of no use, if eigenstates of n are involved, and again one has to evaluate $\Delta C, \Delta S$ by the given wave function

$$(\Delta C)^2 = \tfrac{1}{4}\langle \Phi_n | E_-^2 + E_-E_+ + E_+E_- + E_+^2 | \Phi_n\rangle = \tfrac{1}{2} \tag{5.200}$$

and

$$(\Delta S)^2 = \tfrac{1}{2}. \tag{5.201}$$

Note for comparison, that

$$\frac{1}{2\pi}\int_0^{2\pi} \sin^2\varphi \, d\varphi = \tfrac{1}{2}.$$

For a definite value of n, C and S can take any value between -1 and $+1$, so that the "phase" is completely undetermined.

We mention some useful results for expectation values taken with respect to the coherent states (5.90)

$$\langle S^2\rangle_\alpha = \tfrac{1}{2} - \tfrac{1}{4}\exp[-\bar{n}] - \tfrac{1}{2}\exp[-\bar{n}]\bar{n}(1-2\xi)\sum_{\nu=0}^{\infty}\frac{\bar{n}^\nu}{\nu![(\nu+1)(\nu+2)]^{1/2}} \tag{5.202}$$

where

$$\xi = \frac{(\mathrm{Im}\alpha)^2}{|\alpha|^2}, \qquad \bar{n} = |\alpha|^2 \tag{5.203}$$

$$\langle S\rangle_\alpha = \exp[-\bar{n}](\mathrm{Im}\alpha)\sum_{\nu=0}^{\infty}\frac{\bar{n}^\nu}{\nu!\,(\nu+1)^{1/2}} \tag{5.204}$$

$$\langle C^2 + S^2\rangle_\alpha = 1 - \tfrac{1}{2}\exp[-\bar{n}] \tag{5.205}$$

$$(\langle C\rangle_\alpha)^2 + (\langle S\rangle_\alpha)^2 = \bar{n}\exp[-2\bar{n}]\left(\sum_{\nu=0}^{\infty}\frac{\bar{n}^\nu}{\nu!\,(\nu+1)^{1/2}}\right)^2 \approx 1 - \frac{1}{4\bar{n}}. \tag{5.206}$$

Here $\bar{n}$ always means the mean photon number.

$$(\Delta S)_\alpha(\Delta C)_\alpha \geqslant \tfrac{1}{4}\exp[-\bar{n}]. \tag{5.207}$$

From these considerations it is evident that for photon numbers $\bar{n} \gg 1$ (say 10) the old "heuristic" considerations hold, whereas for photon numbers of order 1 at least quantitative differences occur. In the maser or laser process a great number of photons is involved, so that one may safely use the more elementary treatment. On the other hand, when the measurement of phases of single light quanta is discussed, for instance in counting experiments, the exact treatment must be used.

6. Quantization of electronwave field

6.1. Motivation

In the previous chapter we have learned how to quantize the light field. In it we started from field equations (Maxwell's equations). The fields $\boldsymbol{E}$ and $\boldsymbol{B}$ were expanded into modes, for instance the electric field strength was represented in the form

$$\boldsymbol{E}(\boldsymbol{x}) = \sum_{\lambda} (b_{\lambda}^{+} + b_{\lambda}) \mathcal{N}_{\lambda} \boldsymbol{u}_{\lambda}(\boldsymbol{x}). \tag{6.1}$$

In it the functions $\boldsymbol{u}_{\lambda}(\boldsymbol{x})$ obeyed classical field equations. On the other hand, the amplitudes b, b^{+} became operators subject to certain commutation relations. We then obtained the Schrödinger equation by starting from an expression for the energy density or, more precisely speaking, for the total field energy. It has turned out that it is possible and even necessary to do exactly the same for electrons and other elementary particles. The reason why we did not come across this problem earlier is mainly historical and partly pedagogical. From the historical development, we know that electrons manifested themselves first as particles and only later as waves. This is due to the fact that the wavelength of electrons is so short that electron diffraction, etc. could become accessible in experiments performed only rather late. However, in principle, we could equally well start first from experiments showing the wave character of electrons and then proceed from there to particles, much the same as we did with the light field. In our present case, we know the equation describing the wave character of electrons. It is nothing but the Schrödinger equation

$$\left(-\frac{\hbar^2}{2m}\Delta + V\right)\psi = i\hbar\frac{\mathrm{d}\psi}{\mathrm{d}t}. \tag{6.2}$$

When we take the analogy between the light field and electron field seriously we just have to repeat the steps which led us to the quantization of the light field for the electron wave field. We will do this in the next section 6.2.

6.2. Quantization procedure

We note that a somewhat more rigorous way of doing the quantization is based on Lagrangians. To avoid unnecessary complications we use a more straightforward method, however, and refer the interested reader to books dealing with the quantization of the electron wave field, e.g. my own book Quantum Field Theory of Solids (North-Holland 1976). Now, let us rather exploit the analogy with the light field. The individual field modes of the electron obey the time-independent Schrödinger equation

$$\left(-\frac{\hbar^2}{2m}\Delta + V\right)\varphi_j(\boldsymbol{x}) = W_j\varphi_j(\boldsymbol{x}). \tag{6.3}$$

We expand a general wave function $\psi(\boldsymbol{x})$ (which is an analogue to the electric field strength eq. (6.1) into a superposition of eigenfunctions ("modes")

$$\psi(\boldsymbol{x}) = \sum_j a_j\varphi_j(\boldsymbol{x}). \tag{6.4}$$

Now, however, we have to note a difference between the Schrödinger wave field and the electromagnetic field. While the electromagnetic field is directly measurable and therefore a real quantity, the electron wave field is intrinsically a complex quantity. Therefore we have to consider the complex conjugate of eq. (6.4) as well. Having the analogy with the quantization of the light field in mind, we expect that the functions $\varphi_j(\boldsymbol{x})$ remain ordinary functions in the quantization procedure, whereas the coefficients a_j become operators. For this reason we denote the complex conjugate of $\varphi_j(\boldsymbol{x})$ as usual by $\varphi_j^*(\boldsymbol{x})$, whereas operators a_j have as conjugate a_j^+, which are called the hermitian conjugate of a_j. Since $\psi(\boldsymbol{x})$ also becomes an operator, its hermitian conjugate is denoted by $\psi^+(x)$. We thus obtain in addition to eq. (6.4)

$$\psi^+(\boldsymbol{x}) = \sum_j a_j^+\varphi_j^*(\boldsymbol{x}). \tag{6.5}$$

For what follows, we will assume as usual that the functions φ_j obey the orthogonality relations

$$\int \varphi_j^*(\boldsymbol{x})\varphi_{j'}(\boldsymbol{x})\,\mathrm{d}^3\boldsymbol{x} = \delta_{jj'}. \tag{6.6}$$

To derive the appropriate new type of Schrödinger equation which describes the quantized electron wave field we start from an expression for the total energy. An expression which plays a role analogous to the field energy U is the expectation value of the Hamiltonian H occurring in eq. (6.2)

$$\hat{H} = \int \psi^+ H \psi \, d^3x. \tag{6.7}$$

To express eq. (6.7) by means of the amplitudes a_j^+, a_j we insert eqs. (6.4) and (6.5) into (6.7). Making use of eq. (6.3) and the orthogonality relation (6.6), we obtain

$$\hat{H} = \sum_{jj'} a_j^+ a_{j'} \int \varphi_j^*(x) \underbrace{\left\{ -\frac{\hbar^2}{2m}\Delta + V(x) \right\} \varphi_{j'}(x)}_{W_{j'}\varphi_{j'}(x)} d^3x, \tag{6.8}$$

and eventually

$$\hat{H} = \sum_j W_j a_j^+ a_j. \tag{6.9}$$

Comparing eq. (6.9) with the expression (5.151) of section 5.8 reveals a striking analogy. It makes us think that a_j^+ and a_j must now become operators describing the creation and annihilation of electrons in the state j. It is not our purpose to describe all details of this quantization. We will just give a short motivation and then a description of the properties of the operators a_j^+, a_j. Readers interested in more details are referred to my above-mentioned book. Again we introduce a vacuum state Φ_0 by the definition

$$a_j \Phi_0 = 0 \tag{6.10}$$

for all js. A state in which a single electron occupies the state j is then described by

$$a_j^+ \Phi_0. \tag{6.11}$$

However, an important difference between the operators of photons and of electrons occurs. As we have mentioned in section 3.5, no two electrons may occupy exactly the same state (when the electron spin is taken into account). This is the well-known Pauli exclusion principle. However, we can always apply twice the creation operator a_j^+ on the vacuum state

$$a_j^+ a_j^+ \Phi_0, \tag{6.12}$$

at least in a formal way.

Since this two-electron state must not exist we require that (6.12) vanishes. However, we could apply a_j^+ twice to any other state as well. Since in all these cases the resulting state must not exist we require that

$$a_j^+ a_j^+ \Phi \tag{6.13}$$

vanishes for any state Φ. In quantum mechanics this is expressed in operator form by the requirement

$$(a_j^+)^2 = 0 \tag{6.14}$$

The relation (6.14) is, of course, different from commutation relations known from photons. It is nevertheless possible to build up a consistent formalism when the following commutation relations are postulated

$$\begin{aligned} a_j^+ a_k + a_k a_j^+ &= \delta_{jk}, \\ a_j^+ a_k^+ + a_k^+ a_j^+ &= 0, \\ a_j a_k + a_k a_j &= 0. \end{aligned} \tag{6.15}$$

In the following we will not make use of the full potentialities of this formalism but restrict ourselves to applications in quantum optics. In particular, we will mainly be concerned with single-electron states only. We just mention that the formalism allows one to cope with many-body problems in a very elegant fashion. A total state with n electrons occupying the quantum states $j_1 \cdots j_n$ is described by

$$\Phi_{\{j\}} = a_{j_1}^+ a_{j_2}^+ \ldots a_{j_n}^+ \Phi_0. \tag{6.16}$$

It is eigenfunction to the Schrödinger equation

$$\hat{H}\Phi = W_{\text{tot}}\Phi. \tag{6.17}$$

With a total energy given by the sum of the individual energies W_j:

$$W_{\text{tot}} = W_{j_1} + W_{j_2} + \cdots + W_{j_n}. \tag{6.18}$$

The functions of eq. (6.16) are orthonormal, i.e.

$$\langle \Phi_{\{j\}} | \Phi_{\{k\}} \rangle = \delta_{j_1 k_1} \delta_{j_2 k_2} \cdots \delta_{j_n k_n}. \tag{6.19}$$

In particular, we have for the vacuum state

$$\langle \Phi_0 | \Phi_0 \rangle = 1. \tag{6.20}$$

Exercises on section 6.2:

(1) Show that

$$\Phi = a_k^+ \Phi_0 \tag{$*$}$$

solves (6.17) with $W_{\text{tot}} = W_k$.
Hint: Insert ($*$) into eq. (6.17) and use eq. (6.15).

(2) Show that

$$\Phi = a_{k_1}^+ a_{k_2}^+ \Phi_0 \qquad (**)$$

solves (6.17) with $W_{\text{tot}} = W_{k_1} + W_{k_2}$.
Hint: Same as for exercise (1).

(3) Prove eq. (6.19)
Hint: Use the auxiliary relations

$$\langle \Phi^{(1)} | a_k^+ \Phi^{(2)} \rangle = \langle a_k \Phi^{(1)} | \Phi^{(2)} \rangle$$

and

$$\langle \Phi^{(1)} | a_k \Phi^{(2)} \rangle = \langle a_k^+ \Phi^{(1)} | \Phi^{(2)} \rangle .$$

These relations are introduced by definition in analogy to corresponding relations for b^+, b. Use, in addition eqs. (6.15) and (6.10).
(4) It is necessary and possible to formulate expectation values in the new formalism. Here are a few examples:
When we consider the usual Schrödinger equation (cf. section 3.1) as a wave equation, we are led to consider $|\psi(\boldsymbol{x})|^2$ as an intensity, or, since we are dealing with matter, as a density of matter, and correspondingly, to interpret $e|\psi(\boldsymbol{x})|^2$ as charge density. Although we know that this interpretation is too naive, it is helpful in the present context. Namely, it allows us to apply the general scheme of quantum mechanics table 1, page 72, to our present case:

Table 1

Observable	Operator	Expectation value
Charge density $e\|\psi(\boldsymbol{x})\|^2$	$e\psi^+(\boldsymbol{x})\psi(\boldsymbol{x}) = e\Sigma_{jk} a_j^+ a_k \varphi_j^*(\boldsymbol{x})\varphi_k(\boldsymbol{x})$.	$\langle \Phi \| e\psi^+(\boldsymbol{x})\psi(\boldsymbol{x}) \| \Phi \rangle$

Supplement this scheme for potential energy $\int \psi^*(\boldsymbol{x}) V(\boldsymbol{x}) \psi(\boldsymbol{x}) \mathrm{d}^3\boldsymbol{x}$ and kinetic energy

$$\frac{1}{2m} \int \psi^*(\boldsymbol{x}) \left(\frac{\hbar}{i} \nabla \right)^2 \psi(\boldsymbol{x}) \mathrm{d}^3\boldsymbol{x}!$$

(5) In continuation of exercise (4), calculate

$$\langle \Phi | e\psi^+(\boldsymbol{x})\psi(\boldsymbol{x}) | \Phi \rangle, \qquad \langle \Phi | \int \psi^+(\boldsymbol{x}) V(\boldsymbol{x}) \psi(\boldsymbol{x}) \mathrm{d}^3\boldsymbol{x} | \Phi \rangle$$

for the wave functions $(*)$, $(**)$ of exercises (1) and (2).
Hint: Use first eqs. (6.4), (6.5) and then exercise (4).

7. The interaction between light field and matter

7.1. Introduction: Different levels of description

When we wish to treat the interaction between light field and matter we have to distinguish between different levels of description. These levels are partly caused by historical development but also by the kind of problems we have to treat. These levels of description are:

(1) Fully classical treatment. The light field is described by Maxwell's equations, while the constituents of matter such as nucleons and electrons obey Newton's equations. Typically three kinds of problems occur:

(a) The electromagnetic field is given and we study the motion of classical charges caused by the field.

(b) Classical charges or currents are given and the fields created by them have to be determined.

(c) In a third class of problems, the collective motion of particles and fields is studied, the dynamics of one system causing the dynamics of the other system and vice versa. Important examples are magneto-hydrodynamics and plasma physics. Some explicit examples will be provided by the exercises at the end of this section.

The next step is usually called:

(2) The semiclassical approach. In it the field is treated classically by Maxwell's equations whereas matter is treated by the Schrödinger equation. Typical problems are:

(a) The space- and time-dependent, electromagnetic field is given and we study the temporal change of wave functions of particles under the influence of the field. We encountered a number of typical cases in chapter 4.

(b) Space- and time-dependent wave functions are given. To utilize them in the classical Maxwell's equation, one first forms expectation values of charge distributions and currents, which then act as source terms in Maxwell's equations. We will treat this case in the exercise below and to a great extent in Volumes 2 and 3.

(c) In a number of cases we have to study phenomena in which the field determines the wavefunctions but the latter in turn determine the field. The most important example of this type of approach in quantum optics is the laser (compare Volume 2), but many phenomena of non-linear optics are treated by this approach, too.
(3) The semiquantum theoretical approach. The light field is treated quantum mechanically whereas charges and currents are given classical quantities. This kind of problem leads to the forced harmonic oscillator of a field mode which we treated in ch. 5, section 5.7. This example shows that a coherent driving force can cause a coherent field.
(4) Fully quantum mechanical approach. Both the light field and motion of particles are quantized. This treatment will be the main object of the following part of our book because here the full realm of quantum optics becomes accessible.

We close this introductory section with a remark concerning coherence. From the above said it appears that to obtain coherent fields we need a coherent motion of charges. On the other hand to obtain a coherent motion of charges, e.g. of electrons in atoms, we need coherent driving fields. So we are in a way confronted with a vicious circle. We will see in the second volume that the coupled system field-matter is capable of producing such coherent motions in a self-organized way so that the vicious circle can be broken.

Exercises on section 7.1

The following exercises will elucidate our above statements:
(1a) Fully classical treatment: solve the equation of motion of a particle with coordinate $\boldsymbol{q}$, mass m and charge e

$$m\frac{\mathrm{d}^2}{\mathrm{d}t^2}\boldsymbol{q} = e\boldsymbol{E}, \qquad (*)$$

(α) for a constant field $\boldsymbol{E}$, (β) for the field $\boldsymbol{E} = \boldsymbol{E}_0 \cos \omega t$.
Hints: integrate ($*$) twice, adding integration constants. Another example for (1a) (compare section 1.4). A charged particle is elastically bound to the nucleus: Solve (in one dimension)

$$m\frac{\mathrm{d}^2 q}{\mathrm{d}t^2} + \gamma\frac{\mathrm{d}q}{\mathrm{d}t} + fq = eE_0 \cos \omega t. \qquad (**)$$

Hint: Write $\cos \omega t = \frac{1}{2}[\exp(i\omega t) + \exp(-i\omega t)]$ and try
$q(t) = q_1 \exp(i\omega t) + q_2 \exp(-i\omega t)$ ("forced oscillation").
(b) Charge densities ρ and current densities $\boldsymbol{j}$ are given as functions of space and time. In the case of oscillatory charges or currents, the solutions

of Maxwell's equations are electromagnetic waves. The derivation of these solutions is beyond the scope of these exercises. However, the following model is illuminating.

Take a "one-dimensional" model, in which $E_z(x,t) = a(t)\sin kx$, $B_y(x,t) = b(t)\cos kx$,

$$j_z(x,t) = \underbrace{j_0 \sin \omega t \sin kx}_{\text{driving term}} + \underbrace{\sigma E_z(x,t)}_{\text{damping}},$$

Solve Maxwell's equations and determine $a(t)$ and $b(t)$! Discuss how $\boldsymbol{E}$ and $\boldsymbol{B}$ depend on ω and σ!

(c) Classical model of plasma oscillations. Consider charged particles moving in an electric field:

$$m\frac{\mathrm{d}}{\mathrm{d}t}\boldsymbol{v}_i = e\boldsymbol{E}(\boldsymbol{x}_i), \qquad \boldsymbol{v}_i = \frac{\mathrm{d}}{\mathrm{d}t}\boldsymbol{x}_i \tag{$*$}$$

where the index i distinguishes the different particles. Show that the averaged charge density ρ may oscillate with frequency

$$\Omega_p^2 = \frac{e^2 n}{m\varepsilon_0}$$

where n is the particle density.

Hint: "Multiply" ($*$) by $(e/\Delta V)\Sigma_{i\in\Delta V}$ and perform the average on both sides

$$\sum_{i\in\Delta V} \boldsymbol{v}_i \to \Delta N \boldsymbol{v}, \qquad \sum_{i\in\Delta V} \boldsymbol{E}(\boldsymbol{x}_i) \to \Delta N \boldsymbol{E}.$$

$\boldsymbol{E}(\boldsymbol{x}_i)$ is assumed to vary slowly over the small volume ΔV. Then express the l.h.s. of the resulting equation using the current density $\boldsymbol{j}$, take the divergence on both sides of that equation and use eq. (5.4) and the continuity equation

$$\frac{\partial \rho}{\partial t} + \operatorname{div} \boldsymbol{j} = 0.$$

Identify the frequency in the resulting oscillator equation for the charge density ρ.

This exercise may serve as an example of a semiclassical dispersion theory (theory of the dielectric constant). Consider a set of atoms located at sites $\boldsymbol{x}_j$ within a sample. A classical, oscillating electric field $\boldsymbol{E}$ can cause oscillating dipole moments of the atoms. In a classical model, the atoms are described by the model of exercise (1), ($**$), their dipole moments read $\boldsymbol{\vartheta} = e\boldsymbol{q}(t)$.

In the semiclassical theory, this problem has been dealt with in section 4.5, its solution for weak fields is given by eq. (4.82). We denote the dipole moment of the atom located at $\boldsymbol{x}_j$ by $\boldsymbol{\vartheta}_j$. As shown in classical electrodynamics the dipole moments give rise to a macroscopic polarization $\boldsymbol{P}(\boldsymbol{x},t) = \Sigma_j \delta(\boldsymbol{x} - \boldsymbol{x}_j)\boldsymbol{\vartheta}_j$, where δ is Dirac's δ-function. The dielectric displacement $\boldsymbol{D}$ is connected with $\boldsymbol{E}$ and $\boldsymbol{P}$ by

$$\boldsymbol{D} = \varepsilon_0 \boldsymbol{E} + \boldsymbol{P}.$$

We can now formulate our exercise: With aid of exercise (1) (* *) or with eq. (4.81) calculate the dielectric constant, which is defined by

$$\boldsymbol{D} = \varepsilon\varepsilon_0 \boldsymbol{E}.$$

Hint: Replace $\boldsymbol{P}(\boldsymbol{x},t)$ by a spatial average which contains many atoms, but still the same expressions of $\boldsymbol{\vartheta}_j$, i.e. write $\boldsymbol{P}(\boldsymbol{x},t)$ as $\boldsymbol{P}(\boldsymbol{x},t) \approx (N/V)\boldsymbol{\vartheta}$ (N: number of atoms in volume V). Use $\boldsymbol{\vartheta}$ as resulting from (1) (classical theory) or $\boldsymbol{\vartheta}$ as given by eq. (4.82) (semiclassical theory).
Discuss the dependence of ε on ω. Why is this approach only valid for weak fields?
(*Hint:* consult section 4.5).

7.2. Interaction field–matter: Classical Hamiltonian, Hamiltonian operator, Schrödinger equation

In classical physics, the motion of a particle with mass m and charge e is described by the Lorenz equation

$$m(\mathrm{d}\boldsymbol{v}/\mathrm{d}t) = e\boldsymbol{E} + e\boldsymbol{v} \times \boldsymbol{B}. \tag{7.1}$$

In it $\boldsymbol{v}$ is the velocity of the particle, $\boldsymbol{E}$ the electric field strength and $\boldsymbol{b}$ the magnetic induction. We will denote the coordinate of the particle by

$$\boldsymbol{x} = (x_1, x_2, x_3) \equiv (x, y, z). \tag{7.2}$$

It is known from electrodynamics that the electric field strength $\boldsymbol{E}$ as well as the magnetic induction $\boldsymbol{B}$ can be derived from a scalar potential $\tilde{V}(\boldsymbol{x},t)$ and a vector potential $\boldsymbol{A}(\boldsymbol{x},t)$ by the following relations

$$\boldsymbol{b} = \operatorname{curl} \boldsymbol{A}, \tag{7.3}$$

$$\boldsymbol{E} = -(\partial \boldsymbol{A}/\partial t) - \operatorname{grad} \tilde{V}. \tag{7.4}$$

We have seen at various occasions that the Hamiltonian of a classical system is a good starting point for quantization. We therefore ask ourselves the question whether the equation of motion (7.1) can be derived from a Hamiltonian via Hamiltonian equations in much the same way as we have treated the motion of a harmonic oscillator in section 3.3. In the present case, finding a suitable Hamiltonian is not an entirely simple task. Therefore we write down the resulting Hamiltonian right away

$$H = \frac{1}{2m}(\boldsymbol{p} - e\boldsymbol{A})^2 + e\tilde{V}. \tag{7.5}$$

We want to convince ourselves that eq. (7.5) leads via the Hamiltonian equations to the equations of motion (7.1). The Hamiltonian equations read as usual (compare section 3.3)

$$\frac{\mathrm{d}}{\mathrm{d}t} x_j = \frac{\partial H}{\partial p_j}, \tag{7.6}$$

$$\frac{\mathrm{d}}{\mathrm{d}t} p_j = -\frac{\partial H}{\partial x_j}. \tag{7.7}$$

Inserting eq. (7.5) into eq. (7.6) and performing the differentiation of H with respect to p_j we obtain

$$\frac{\mathrm{d}}{\mathrm{d}t} x_j = \frac{1}{m}(p_j - eA_j). \tag{7.8}$$

Similarly eq. (7.7) yields

$$\frac{\mathrm{d}}{\mathrm{d}t} p_j = -\frac{1}{m}(\boldsymbol{p} - e\boldsymbol{A})\cdot\left(-e\frac{\partial \boldsymbol{A}}{\partial x_j}\right) - e\frac{\partial \tilde{V}}{\partial x_j}. \tag{7.9}$$

Differentiating eq. (7.8) with respect to time yields

$$m\frac{\mathrm{d}^2 x_j}{\mathrm{d}t^2} = \frac{\mathrm{d}}{\mathrm{d}t} p_j - e\frac{\mathrm{d}A_j}{\mathrm{d}t}. \tag{7.10}$$

In it, we express $\mathrm{d}p_j/\mathrm{d}t$ by eq. (7.9) and use again the relation (7.8). This yields

$$m\frac{\mathrm{d}^2 x_j}{\mathrm{d}t^2} = \frac{\mathrm{d}\boldsymbol{x}}{\mathrm{d}t} e\frac{\partial \boldsymbol{A}}{\partial x_j} - e\frac{\mathrm{d}A_j}{\mathrm{d}t} - e\frac{\partial \tilde{V}}{\partial x_j}. \tag{7.11}$$

Some care must be exercised when we wish to perform the differentiation of A_j with respect to time. The vector potential $\boldsymbol{A}$ depends on space point $\boldsymbol{x}$ and time t. It is understood that we have to take that coordinate $\boldsymbol{x}$ where the particle is present at time t, i.e. $\boldsymbol{A}$ must be considered as the function

$$\boldsymbol{A} = \boldsymbol{A}(\boldsymbol{x}(t), t). \tag{7.12}$$

However, since with varying time $\boldsymbol{x}(t)$ also varies, we have to take into account that variation when differentiating A_j with respect to time. This leads us to the relation

$$\frac{\mathrm{d}A_j}{\mathrm{d}t} = \sum_i \frac{\partial A_j}{\partial x_i}\frac{\mathrm{d}x_i}{\mathrm{d}t} + \frac{\partial A_j}{\partial t}. \tag{7.13}$$

Using eq. (7.13) we evaluate eq. (7.11) further. As an example we choose $j = 1$ and write $x_1 = x$, $x_2 = y$, and $x_3 = z$. We then obtain

$$m\frac{\mathrm{d}^2x}{\mathrm{d}t^2} = e\left[\frac{\partial A_x}{\partial x}\frac{\mathrm{d}x}{\mathrm{d}t} + \frac{\partial A_y}{\partial x}\frac{\mathrm{d}y}{\mathrm{d}t} + \frac{\partial A_z}{\partial x}\frac{\mathrm{d}z}{\mathrm{d}t} - \frac{\partial A_x}{\partial x}\frac{\mathrm{d}x}{\mathrm{d}t} - \frac{\partial A_x}{\partial y}\frac{\mathrm{d}y}{\mathrm{d}t} - \frac{\partial A_x}{\partial z}\frac{\mathrm{d}z}{\mathrm{d}t}\right] \underbrace{-e\frac{\partial A_x}{\partial t} - e\frac{\partial \tilde{V}}{\partial x}}_{eE_x.} \tag{7.14}$$

Using eq. (7.4) we can identify the last two terms in eq. (7.14) with eE_x. The terms in square brackets can be rearranged in the form

$$[\quad] = \frac{\mathrm{d}y}{\mathrm{d}t}\underbrace{\left(\frac{\partial A_y}{\partial x} - \frac{\partial A_x}{\partial y}\right)}_{B_z} - \frac{\mathrm{d}z}{\mathrm{d}t}\underbrace{\left(\frac{\partial A_x}{\partial z} - \frac{\partial A_z}{\partial x}\right)}_{B_y} \tag{7.15}$$

and using eq. (7.3) in the form

$$[\quad] = \frac{\mathrm{d}y}{\mathrm{d}t}B_z - \frac{\mathrm{d}z}{\mathrm{d}t}B_y. \tag{7.16}$$

Equation (7.16) is nothing but the x component of the vector product of $\boldsymbol{v} \times \boldsymbol{B}$. Thus eq. (7.14) reduces to

$$m\frac{\mathrm{d}^2x}{\mathrm{d}t^2} = eE_x + e(\boldsymbol{v} \times \boldsymbol{B})_x. \tag{7.17}$$

Making the same steps with the other components $j = 2, j = 3$ we readily verify eq. (7.1).

So far we have been dealing with a classical particle moving in classical fields, whose potentials are $\tilde{V}$, $\boldsymbol{A}$. When we wish to treat the electromagnetic field as not being given but as a variable with its own degrees of freedom we have to add to the Hamiltonian eq. (7.5) that of the free electromagnetic field. We are thus led to a total Hamiltonian

$$H_{\text{tot}} = H + H_{\text{field}}. \tag{7.18}$$

It is possible to derive from eq. (7.18) new Hamiltonian equations including those of the electromagnetic field. We will not go into the details of this somewhat lengthy calculation which eventually leads us back to Maxwell's equations including currents and charges. We would rather go on to show how to perform the quantization. When quantizing the motion of a particle in section 3.1 we have seen that we have simply to replace the momentum $\boldsymbol{p}$ by the operator $(\hbar/i)\nabla$. As we know from comparison between theory and experiment, this quantization procedure holds in the presence of a magnetic field also. This leads us to the Hamiltonian

$$H = \frac{1}{2m}\left(\frac{\hbar}{i}\nabla - e\boldsymbol{A}\right)^2 + V, \qquad V = e\tilde{V}. \tag{7.19}$$

To evaluate eq. (7.19) further we make use of the fact that there is still some arbitrariness in choosing $\boldsymbol{A}$. We adopt again the choice eq. (5.46) of section 5.1, namely

$$\nabla \boldsymbol{A} \equiv \operatorname{div} \boldsymbol{A} = 0 \tag{7.20}$$

which is called the Coulomb gauge. Since ∇ is an operator acting on the particle coordinate $\boldsymbol{x}$ on which $\boldsymbol{A}$ depends, we must carefully conserve the sequence of operators when evaluating the square of the bracket in eq. (7.19). We then obtain

$$H = -\frac{\hbar^2}{2m}\Delta - \frac{e\hbar}{2mi}\nabla \boldsymbol{A} - \frac{e\hbar}{2mi}\boldsymbol{A}\nabla + \frac{e^2}{2m}\boldsymbol{A}^2 + V, \tag{7.21}$$

which, due to eq. (7.20), can be brought into the form

$$H = -\frac{\hbar^2}{2m}\Delta - \frac{e\hbar}{mi}\boldsymbol{A}\nabla + \frac{e^2}{2m}\boldsymbol{A}^2 + V. \tag{7.22}$$

We represent the Hamiltonian of the field energy, H_{field}, in the form derived previously in section 5.8. We are now in a position to write eq. (7.18) in the form

$$H_{\text{tot}} = \underbrace{-\frac{\hbar^2}{2m}\Delta + V(x)}_{H_{\text{el}}} + \underbrace{\sum_\lambda \hbar\omega_\lambda b_\lambda^+ b_\lambda}_{H_{\text{field}}} + H_{I,1} + H_{I,2} \tag{7.23}$$

where we have used the abbreviations

$$H_{I,1} = -\frac{e\hbar}{mi}\boldsymbol{A}\nabla = -\frac{e}{m}\boldsymbol{A}\cdot\boldsymbol{p}, \qquad \boldsymbol{p} = \frac{\hbar}{i}\nabla \tag{7.24}$$

and

$$H_{I,2} = \frac{e^2}{2m}\boldsymbol{A}^2. \tag{7.25}$$

In it $\boldsymbol{a}$ is still to be replaced by [cf. eq. (5.163)]

$$\boldsymbol{A}(\boldsymbol{x},t) = \sum_\lambda (b_\lambda^+ + b_\lambda)\sqrt{\hbar/(2\omega_\lambda\varepsilon_0)}\,\boldsymbol{u}_\lambda(\boldsymbol{x}). \tag{7.26}$$

Therefore we find in particular

$$H_{I,1} = \sum_\lambda (b_\lambda^+ + b_\lambda)\left(-\frac{e\hbar}{mi}\right)\sqrt{\hbar/(2\omega_\lambda\varepsilon_0)}\,\boldsymbol{u}_\lambda(\boldsymbol{x})\nabla. \tag{7.27}$$

The Hamiltonian eq. (7.23) often serves as a starting point from which the interaction between a particle and the quantized electromagnetic field can be treated. Many of the underlying processes become still more transparent when we use the formalism of the electron wave field quantization which we developed in chapter 6. At a first glance this may appear a little bit more complicated but the reader will be fully rewarded for his efforts. We will see that this method allows us to describe the interaction processes between field and matter in a very transparent way.

7.3. Interaction light field–electron wave field

We proceed exactly as in section 6.2 starting from the expression (6.7) for the expectation value $\hat{H}$ of the Hamiltonian H. Remember that H in eq. (6.7) was the Hamiltonian of the "ordinary" Schrödinger equation we got to know in section 3.1, formula (3.18). Now we use the more general Hamiltonian eq. (7.22) because it describes the electron's motion under the impact of a general electromagnetic field. Inserting eq. (7.22) into (6.7), we readily obtain

$$\hat{H} \equiv \hat{H}_{\text{el}} + \hat{H}_I, \tag{7.28}$$

where

$$\hat{H}_{\text{el}} = \int \psi^+(\boldsymbol{x}) \left\{ -\frac{\hbar^2}{2m} \Delta + V(\boldsymbol{x}) \right\} \psi(\boldsymbol{x}) \mathrm{d}^3 x \tag{7.29}$$

refers to the "unperturbed" motion of the electron.

$$\hat{H}_I = \hat{H}_{I,1} + \hat{H}_{I,2} \tag{7.30}$$

describes the interaction of the electron with fields (in addition to V), i.e. in the present context with the light field. In correspondence with eqs. (7.24) and (7.25), $\hat{H}_{I,1}$ and $\hat{H}_{I,2}$ read explicitly

$$\hat{H}_{I,1} = \int \psi^+(\boldsymbol{x}) \left\{ -\frac{e}{m} \boldsymbol{A}\boldsymbol{p} \right\} \psi(\boldsymbol{x}) \mathrm{d}^3 x, \tag{7.31}$$

$$\hat{H}_{I,2} = \int \psi^+(\boldsymbol{x}) \left\{ \frac{e^2}{2m} \boldsymbol{A}^2 \right\} \psi(\boldsymbol{x}) \mathrm{d}^3 x. \tag{7.32}$$

In complete analogy to chapter 6, section 6.2 we express $\psi(\boldsymbol{x})$ and $\psi^+(\boldsymbol{x})$ as a superposition of unperturbed wave functions φ_j where we choose

$$H_{\text{el}} \varphi_j = W_j \varphi_j; \qquad H_{\text{el}} = -\frac{\hbar^2}{2m} \Delta + V(\boldsymbol{x}). \tag{7.33}$$

We thus have

$$\psi(\boldsymbol{x}) = \sum_j a_j \varphi_j(\boldsymbol{x}), \qquad \psi^+(\boldsymbol{x}) = \sum_j a_j^+ \varphi_j^*(\boldsymbol{x}). \tag{7.34, 35}$$

We now insert eqs. (7.34) and (7.35) jointly with eq. (7.26) into eq. (7.31). We have encountered such calculations several times, e.g. in section 6.2, so that we may readily write down the final result. In order not to overload our representation we represent only $\hat{H}_{I,1}$. In this new formalism the corresponding Hamiltonians thus read

$$\hat{H}_{\text{el}} = \sum_j W_j a_j^+ a_j, \tag{7.36}$$

$$\hat{H}_{I,1} = \hbar \sum_{j,k,\lambda} a_j^+ a_k g_{\lambda,jk}(b_\lambda + b_\lambda^+), \tag{7.37}$$

where the coefficients g are explicitly given by

$$g_{\lambda,jk} = -\frac{e}{m}\sqrt{\frac{1}{2\hbar\omega_\lambda\varepsilon_0}} \int \varphi_j^*(\boldsymbol{x})(\boldsymbol{u}_\lambda(\boldsymbol{x})\boldsymbol{p})\varphi_k(\boldsymbol{x})\mathrm{d}^3\boldsymbol{x}. \tag{7.38}$$

We leave it as an exercise to the reader to express eq. (7.32) in an analogous fashion by the operators a^+, a, b^+, b. In a number of practical applications the interaction term can still be simplified by means of the dipole approximation. We merely quote the result and will derive it later in section 7.5.

$$\hat{H}_{I,1} = \hbar \sum_{j,k,\lambda} a_j^+ a_k g'_{\lambda,jk}(b_\lambda - b_\lambda^+), \tag{7.39}$$

where

$$g'_{\lambda,jk} = -i\sqrt{\frac{1}{2\hbar\varepsilon_0\omega_\lambda}}\;\frac{W_j - W_k}{\hbar}\boldsymbol{u}_\lambda(\boldsymbol{x}_0)\boldsymbol{\vartheta}_{jk} \tag{7.39a}$$

and

$$\boldsymbol{\vartheta}_{jk} = \int \varphi_j^*(\boldsymbol{x})(e\boldsymbol{x})\varphi_k(\boldsymbol{x})\mathrm{d}^3\boldsymbol{x}. \tag{7.39b}$$

$\boldsymbol{x}_0$ is the coordinate of the center of the atom. $\boldsymbol{\vartheta}_{jk}$ is the dipole moment connected with the transition $k \to j$. Now we have to do the last step. As we have seen in section 5, the light field possesses its own degrees of freedom and its own field energy which, in quantum theory, gives rise to the (field-) Hamiltonian

$$\hat{H}_{\text{field}} = \sum_\lambda \hbar\omega_\lambda b_\lambda^+ b_\lambda \tag{7.40}$$

(where we have dropped the zero point energy). In analogy to classical theory, where the Hamiltonian is a sum of those of the electron, the field, and their mutual interactions, we have in quantum theory

$$\hat{H}_{\text{tot}} = \hat{H}_{\text{el}} + \hat{H}_I + \hat{H}_{\text{field}}, \tag{7.41}$$

where the suffix "tot" stands for "total system" (= electron wave field +

light field). In all practical calculations to follow we will use $\hat{H}_{I,1}$ instead of $\hat{H}_I$. $\hat{H}_{el}$, $\hat{H}_{I,1}$ and $\hat{H}_{field}$ are given by eqs. (7.36), (7.37), (7.40), respectively. This implies that we will neglect that term of the Hamiltonian which stems from $\boldsymbol{A}^2$. One may show that this term is small compared to the other terms of the Hamiltonian provided we deal with atoms (whose electron clouds are rather localized) and with not too high field strengths, i.e. light from conventional light sources. In nonlinear optics, where we often deal with high light intensities, the term $\boldsymbol{A}^2$ can become important.

When inspecting the different Hamiltonians we discover that they contain two types of operators, namely the Bose operators b_λ^+, b_λ and the "Fermi" operators a_j^+, a_j. Their commutation relations are given by eqs. (5.154)–(5.156) and (6.15), respectively. However, what about mutual commutation relations between as and bs? At least in a "gedanken experiment", we may decouple the electron wave field and the light field from each other by letting the electric charge go to zero. Since the two fields are then entirely independent of each other it is suggestive to require that the as and bs commute. This requirement is part of a theory which is verified up to a very high order (e.g. Lamb shift, see below) so that we adopt the relations

$$a_j^+ b_\lambda^+ - b_\lambda^+ a_j^+ = 0, \qquad a_j b_\lambda - b_\lambda a_j = 0,$$
$$a_j^+ b_\lambda - b_\lambda a_j^+ = 0, \qquad a_j b_\lambda^+ - b_\lambda^+ a_j = 0. \tag{7.42}$$

After all this, the reader may guess from his experience from previous sections what will be the end of this section: We write down the Schrödinger equation

$$i\hbar \frac{\mathrm{d}\Phi}{\mathrm{d}t} = \hat{H}_{tot}\Phi. \tag{7.43}$$

The reader will soon recognize that solving the Schrödinger equation, (7.43), is quite a different thing from writing it down. Indeed, a good deal of the rest of this book is devoted to extracting some of the physical content from eq. (7.43). Before doing so, we advise the reader to pay attention to the following exercises which will greatly help him to get used to the formalism. After having treated some exercises he will most probably enjoy working with operators. He may even experience a "God-like feeling", by being able to create and destroy photons and particles at will by merely applying the operators!

Exercises on section 7.3.

(1) Calculate $g_{\lambda,jk}$ eq. (7.39a) numerically for the two lowest states $j \equiv (n = 1, l = 0, m = 0)$ and $k \equiv (n = 2, l = 1, m = 0)$ of the hydrogen atom and $\boldsymbol{u}_\lambda(\boldsymbol{x}_0) = 1/\sqrt{V}(0, 0, 1)$, V: volume, $= 1\ m^3$.

Hint: Use energies, wave functions, and dipole moment as given in section 3.4 and its exercises, respectively. Put $\omega_\lambda = (W_j - W_k)/\hbar$.

$$\varepsilon_0 = 8.854 \cdot 10^{-12} \text{ farads/m},$$

$$e = 1.602 \cdot 10^{-19} \text{ Coulomb}.$$

Use MKSA units.

(2) Show that the solutions of

$$(h\omega b^+ b + W a^+ a)\Phi = W_{\text{tot}}\Phi$$

are given by

$$\Phi = \frac{1}{\sqrt{n!}}(b^+)^n \Phi_0,\ n = 0, 1, 2, \ldots \text{ with } W_{\text{tot}} = n\hbar\omega,$$

where $a\Phi_0 = b\Phi_0 = 0(*)$, i.e. $\Phi_0 =$ vacuum state, and

$$\Phi = \frac{1}{\sqrt{n!}}(b^+)^n a^+ \Phi_0,\ W_{\text{tot}} = W + n\hbar\omega.$$

Hint: Use the commutation relations for the as and bs and $(*)$.

(3) Show that the solutions of

$$\left(\sum_\lambda \hbar\omega_\lambda b_\lambda^+ b_\lambda + \sum_j W_j a_j^+ a_j\right)\Phi = W_{\text{tot}}\Phi$$

are given by

$$\Phi = \frac{1}{\sqrt{n_1! \cdots n_\lambda! \ldots}}(b_1^+)^{n_1} \ldots (b_\lambda^+)^{n_\lambda} \ldots (a_1^+)^{m_1}(a_2^+)^{m_2}\Phi_0,$$

with

$$W_{\text{tot}} = \sum_\lambda n_\lambda \hbar\omega_\lambda + \sum_j m_j W_j,$$

$$n_\lambda \geqslant 0, \quad \text{integer}; \quad m_j = 0 \text{ or } 1. \quad (a_j^+)^0 = 1, \quad (b_\lambda^+)^0 = 1.$$

Hint: Same as in exercise (2).

(4) Two-level atom interacting with a single mode:
Solve:

$$(\hbar\omega b^+ b + W_1 a_1^+ a_1 + W_2 a_2^+ a_2 + \hbar g b^+ a_1^+ a_2 + \hbar g b a_2^+ a_1)\Phi = W_{\text{tot}}\Phi, \quad (*)$$

g real.

Hint: Try $\Phi^{(0)} = a_1^+ \Phi_0$ and

$$\Phi^{(n+1)} = c_1 a_1^+ \frac{1}{\sqrt{(n+1)!}}(b^+)^{n+1}\Phi_0 + c_2 a_2^+ \frac{1}{\sqrt{n!}}(b^+)^n \Phi_0, \quad n \geqslant 0 \quad (**)$$

and determine the unknown coefficients by comparing the two sides of (*) when (* *) is inserted into (*). What are the energies W_{tot}? Note: for each $n \geqslant 0$ there are two pairs of solution for c_1, c_2. Plot the two branches of $W_{tot}^{(n)}$ as a function of ω for fixed $W_2 - W_1$ and fixed g. Discuss how $W_{tot}^{(n)}$ depends on $\hbar\omega$, W_2, W_1 and g.

(5) Solve the time-dependent Schrödinger equation belonging to (*) under the initial condition

$$\Phi(0) = a_2^+\Phi_0.$$

Hint: Try a wave packet of the pair of solutions (* *) for $n = 0$. Determine the free constants of the wave packet by the requirement $\Phi(0) = a_2^+\Phi_0$. Discuss the time-dependence of the resulting wave function. After which time is the initial state $a_2^+\Phi_0$ restored?

7.4. The interaction representation

In this section we get to know a procedure which is not restricted to the interaction between field and matter but which can be applied to many other cases, too. In it the Schrödinger equation reads

$$(H_0 + H_I)\Phi = i\hbar\frac{d\Phi}{dt}, \tag{7.44}$$

where we have evidently split the total Hamiltonian into two Hamiltonians H_0 and H_I. Of course, we may identify this decomposition with the one we encountered in chapter 7, section 7.3. Note that from now on we shall in general omit the "hat" ^ from the Hamiltonian operators to simplify the writing. There may be no misunderstanding because we will represent the Hamiltonians explicitly whenever it is necessary. Now let us forget for the moment being that $H_0 + H_I$ are operators. Then one may try to apply tricks well known from the theory of ordinary differential equations. So let us make the substitution

$$\Phi = U\tilde{\Phi}, \tag{7.45}$$

where U and $\tilde{\Phi}$ are in general both time dependent. Differentiating eq. (7.45) with respect to time yields

$$\frac{d\Phi}{dt} = \frac{dU}{dt}\tilde{\Phi} + U\frac{d\tilde{\Phi}}{dt}. \tag{7.46}$$

Since we have in mind to facilitate the solution of eq. (7.44) by means of eq. (7.45) we subject U to the equation

$$\frac{dU}{dt} = -\frac{i}{\hbar}H_0U. \tag{7.47}$$

We now remind the reader of a result we derived in section 5.6. There we saw that an equation of the shape (7.47) can be solved in a formal manner. In the present case one readily verifies that eq. (7.47) is solved by

$$U = \exp[-iH_0 t/\hbar]. \tag{7.48}$$

We now observe that all the steps we have done so far can be done not only for numbers but equally well for operators, provided we always keep the sequence of these operators. Inserting eqs. (7.46) and (7.47) into eq. (7.44), we immediately realize that H_0 drops out so that eq. (7.44) reduces to

$$H_I U\tilde{\Phi} = i\hbar U \frac{\mathrm{d}\tilde{\Phi}}{\mathrm{d}t}. \tag{7.49}$$

To obtain an equation for $\tilde{\Phi}$ which has the form of the usual Schrödinger equation we multiply both sides of eq. (7.49) by U^{-1}.

$$U^{-1}H_I U\tilde{\Phi} = i\hbar \frac{\mathrm{d}\tilde{\Phi}}{\mathrm{d}t}. \tag{7.50}$$

This equation can be given a particularly simple form when we introduce the abbreviation

$$U^{-1}H_I U = \tilde{H}_I(t). \tag{7.51}$$

We then find

$$\tilde{H}_I(t)\tilde{\Phi} = i\hbar \frac{\mathrm{d}\tilde{\Phi}}{\mathrm{d}t}. \tag{7.52}$$

From our procedure it follows that we have replaced the original equation (7.44) which contains both H_0 and H_I by an equation which is simpler because it contains only a single Hamiltonian, namely eq. (7.51). However, we must not forget that $\tilde{H}_I$ still contains H_0 through U. Thus it may seem that our problem is not really simplified. However, in many cases, such as those given in the following examples, $\tilde{H}_I$ can be explicitly calculated, and in these cases, a considerable simplification results.

Example 1 – (*displaced harmonic oscillator*). We choose b^+, b as the usual Bose operators and H_0, H_I in the form

$$H_0 = \hbar\omega b^+ b, \qquad H_I = \gamma^* b + \gamma b^+. \tag{7.53, 54}$$

U reads

$$U = \exp[-i\omega b^+ b t]. \tag{7.55}$$

To evaluate eq. (7.51) we must know the expressions

$$U^{-1}bU, \qquad U^{-1}b^+U. \tag{7.56, 57}$$

They are identical with the expressions we evaluated by formulas (5.74) and (5.75) of section 5.3, namely when we put $\alpha = \omega t$

$$U^{-1} b U = b \exp[-i\omega t], \tag{7.58}$$

$$U^{-1} b^+ U = b^+ \exp[i\omega t]. \tag{7.59}$$

Thus the Hamiltonian of the interaction representation can be explicitly evaluated and yields

$$\tilde{H}_I = \gamma^* b \exp[-i\omega t] + \gamma b^+ \exp[i\omega t], \tag{7.60}$$

and the Schrödinger equation in the interaction representation reads

$$i\hbar \frac{\mathrm{d}\tilde{\Phi}}{\mathrm{d}t} = \{\gamma^* b^+ \exp[i\omega t] + \gamma b \exp[-i\omega t]\}\tilde{\Phi}. \tag{7.61}$$

Example 2 – *Two-level atom interacting with a given classical electric field.* The operators are now Fermi–Dirac operators obeying the commutation relations of chapter 6, section 6.2. The unperturbed Hamiltonian H_0 reads

$$H_0 = W_1 a_1^+ a_1 + W_2 a_2^+ a_2. \tag{7.62}$$

The interaction with the external electric field is described by

$$H_I = \hbar(g a_2^+ a_1 + g^* a_1^+ a_2), \tag{7.63}$$

where we assume that g contains the external field. The transformation operator U reads

$$U = \exp[-i(W_1 a_1^+ a_1 + W_2 a_2^+ a_2)t/\hbar]. \tag{7.64}$$

We leave the evaluation of the interaction Hamiltonian $\tilde{H}_I$ as an exercise to the reader and give only the result

$$\tilde{H}_I = \hbar\{g a_2^+ a_1 \exp[i(W_2 - W_1)t/\hbar] + g^* a_1^+ a_2 \exp[-i(W_2 - W_1)t/\hbar]\}. \tag{7.65}$$

For hints see exercise at the end of this section.

Example 3 – *Interaction of a two-level atom with a single quantized field mode.* We specialize the Hamiltonian eq. (7.41) to the case where the electron can occupy only two levels, i.e. $j = 1, 2$ and where only a single field mode is present. Dropping all unnecessary indices in the Hamiltonian, the Schrödinger equation reads (see fig. 7.1).

$$i\hbar \frac{\mathrm{d}\Phi}{\mathrm{d}t} = \{W_1 a_1^+ a_1 + W_2 a_2^+ a_2 + \hbar\omega b^+ b\}\Phi + \hbar\{g a_1^+ a_2 (b + b^+) + g^* a_2^+ a_1 (b + b^+)\}\Phi. \tag{7.66}$$

We identify H_0 with the expression

$$H_0 = W_1 a_1^+ a_1 + W_2 a_2^+ a_2 + \hbar\omega b^+ b, \tag{7.67}$$

and the rest of the Hamiltonian with the interaction Hamiltonian. U now

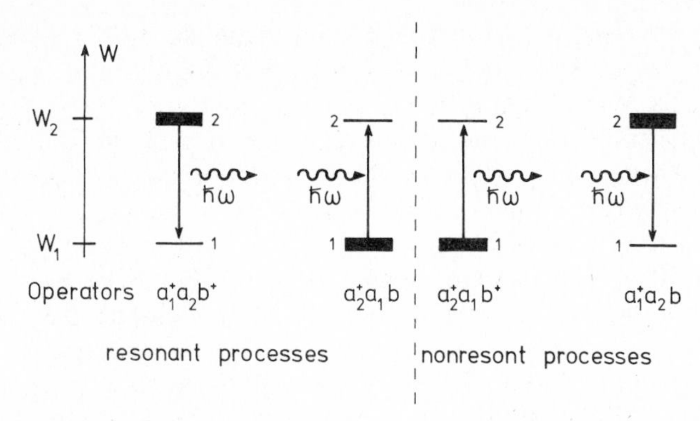

Fig. 7.1. This figure shows how we may visualize the individual terms occurring in the Hamiltonian eq. (7.66). The left-hand side represents resonant processes, namely the emission or the absorption of a photon. Here energy is exactly or nearly exactly conserved. The right-hand side shows nonresonant processes where the transition of an electron from its lower level to its upper level is accompanied by the creation of a photon, or where the transition of the electron from its upper level to its lower level is accompanied by the annihilation of the photon. These two processes clearly violate the energy conservation law. They are called virtual processes.

acquires the form

$$U = \exp\left[-i(W_1 a_1^+ a_1 + W_2 a_2^+ a_2 + \hbar\omega b^+ b)t/\hbar\right]. \tag{7.68}$$

Since as and bs commute with each other we may evaluate expressions of the form

$$U^{-1} a_1^+ a_2 b U \tag{7.69}$$

by the decomposition

$$U = U_{\text{atom}} U_{\text{field}}, \tag{7.70}$$

and thus

$$U^{-1} a_1^+ a_2 b U = U_{\text{atom}}^{-1} a_1^+ a_2 U_{\text{atom}} U_{\text{field}}^{-1} b U_{\text{field}}. \tag{7.71}$$

We can therefore immediately apply the results of the preceding two examples. Using the abbreviations

$$\bar{\omega} = \omega_{21} \equiv (W_2 - W_1)/\hbar, \tag{7.72}$$

we find as resulting Schrödinger equation in the interaction representation

$$\begin{aligned}
&\{g a_1^+ a_2 b^+ \exp[i(\omega - \bar{\omega})t] + g^* a_2^+ a_1 b \exp[-i(\omega - \bar{\omega})t] \\
&\quad + g a_1^+ a_2 b \exp[-i(\omega + \bar{\omega})t] + g^* a_2^+ a_1 b^+ \exp[i(\omega + \bar{\omega})t]\}\tilde{\Phi} \\
&= i\frac{\mathrm{d}}{\mathrm{d}t}\tilde{\Phi}.
\end{aligned} \tag{7.73}$$

By means of eq. (7.73), we can easily explain the rotating wave approximation which we came across on section 4.6, when dealing with classical fields. When looking at the left-hand side of (7.73) we find two types of exponential functions, namely those that depend on $(\bar{\omega} + \omega)t$ and $(\bar{\omega} - \omega)t$. When $\bar{\omega}$ is close to ω, the exponential functions containing $(\bar{\omega} + \omega)t$ oscillate much more rapidly than those containing $(\bar{\omega} - \omega)t$. As explained in section 4.6, we may then neglect the rapidly oscillating terms in a very good approximation compared to the slowly varying ones. In the frame of our present description we can easily describe the physical meaning of the corresponding processes. The term containing $a_2^+ a_1 b$ describes the annihilation of a photon where the atom goes from the state 1 to the state 2. This is an energy-conserving transition which is expressed by the fact $\Delta W = \hbar\omega$. The inverse process in which an electron goes from the upper state 2 to the lower state 1, creating a photon, is described by $a_1^+ a_2 b^+$. (We have assumed that the energies are such that $W_2 - W_1 > 0$.) It is now evident that the other terms $a_1^+ a_2 b$ and $a_2^+ a_1 b^+$ describe processes which are physically "unreasonable" because they won't conserve energy.

There are two problems within our present discussion. First of all, while the energy-conserving processes seem to be obvious we must derive the proper meaning of energy conservation in more detail which we will do in the subsequent sections. Furthermore the reader must be warned against believing that the non-energy conserving processes are unreasonable or non-existent. We will see, quite on the contrary, that they give rise to important effects such as the Lamb shift which we will discuss in section 7.8.

Let us briefly return to eq. (7.73) to which we now apply the rotating wave approximation. In the case of resonance, the Hamiltonian becomes time independent!

Example 4 – A (rather) realistic case: interaction of a two-level atom with many field modes. This example is quite similar to the preceding one. Our starting point is again the Hamiltonian eq. (7.41). We keep only the necessary indices $j = 1, 2$ for the atomic states and λ for the field modes. Furthermore, we know from example (3) that terms of the form $a_1^+ a_2 b$ and $a_2^+ a_1 b^+$ can be dropped within the rotating wave approximation. Adopting these simplifications, the Schrödinger equation reads

$$i\hbar \frac{\mathrm{d}\Phi}{\mathrm{d}t} = \{W_1 a_1^+ a_1 + W_2 a_2^+ a_2\}\Phi + \sum_\lambda \hbar\omega_\lambda b_\lambda^+ b_\lambda \Phi$$

$$+ \hbar \sum_\lambda (g_\lambda a_1^+ a_2 b_\lambda^+ + g_\lambda^* a_2^+ a_1 b_\lambda)\Phi, \tag{7.74}$$

where, according to eq. (7.38)

$$g_\lambda \equiv g_{\lambda,12} = -\frac{e}{m}(2\hbar\omega_\lambda\varepsilon_0)^{-1/2}\int \varphi_1^*(\boldsymbol{x})\boldsymbol{u}_\lambda(\boldsymbol{x})\,\boldsymbol{p}\varphi_2(\boldsymbol{x})\mathrm{d}^3\boldsymbol{x}. \tag{7.75}$$

We identify H_0 with the expression

$$H_0 = W_1 a_1^+ a_1 + W_2 a_2^+ a_2 + \sum_\lambda \hbar\omega_\lambda b_\lambda^+ b_\lambda \tag{7.76}$$

and the rest of the Hamiltonian of eq. (7.74) with the interaction Hamiltonian. U reads

$$U = \exp\left\{-i\left(W_1 a_1^+ a_1 + W_2 a_2^+ a_2 + \sum_\lambda \hbar\omega_\lambda b_\lambda^+ b_\lambda\right)t/\hbar\right\} \tag{7.77}$$

Owing to the fact that not only the a_js and a_j^+ s commute with the b_λs and b_λ^+ s but also the expression $b_\lambda^+ b_\lambda$ with all other $b_{\lambda'}^+$, $b_{\lambda'}$, $\lambda \neq \lambda'$, we may decompose U into the form of a product of commuting operators U_{atom} and U_λ

$$U = U_{\text{atom}} U_{\lambda_1} U_{\lambda_2} \dots U_{\lambda_N}. \tag{7.78}$$

From now on, everything goes through as in example (3) and we find

$$\tilde{H}_I = \hbar\sum_\lambda \{g_\lambda a_1^+ a_2 b_\lambda^+ \exp[i(\omega_\lambda - \bar{\omega})t]$$

$$+ g_\lambda^* a_2^+ a_1 b_\lambda \exp[-i(\omega_\lambda - \bar{\omega})t]\}, \tag{7.79}$$

where

$$\bar{\omega} = (W_2 - W_1)/\hbar. \tag{7.79a}$$

Exercises on section 7.4

These two exercises are for the more mathematically minded reader:

(1) Show:

$a_1^+ a_1$ and $a_2^+ a_2$	commute, i.e. $[a_1^+ a_1, a_2^+ a_2] = 0$	$(*)$
$a_j^+ a_j$ and a_k^+, $j \neq k$	commute,	
$a_j^+ a_j$ and a_k, $j \neq k$	commute.	

U_1 and U_2, where

$$U_j = \exp[-iW_j a_j^+ a_j t/\hbar] \quad \text{commute.} \tag{$**$}$$

$U = \exp[-i(W_1 a_1^+ a_1 + W_2 a_2^+ a_2)t/\hbar]$ can be written in the form

$$U = U_1 U_2 \tag{$***$}$$

and

$$U^{-1} = U_2^{-1} U_1^{-1}.$$

U_1 commutes with a_2^+ and a_2, U_2 commutes with a_1^+ and a_1.
Hints: Use the commutation relations (6.15) to prove ($*$). Expand ($* *$) into a power series of $(a_j^+ a_j)$, insert them into U_1, U_2 and use ($*$). To prove ($* * *$) expand both sides of ($* * *$) into a power series of $(a_j^+ a_j)$ and rearrange terms.

(2) Show that

$$U^{-1} a_2^+ a_1 U = a_2^+ a_1 \exp\left[i(W_2 - W_1)t/\hbar \right], \qquad (*)$$

where U is given in exercise (1) (or 7.64).
Hints: Use the results of exercise (1) and write the l.h.s. of (2) ($*$) as

$$U^{-1} a_2^+ U U^{-1} a_1 U$$

Write further for example

$$U^{-1} a_2^+ U = U_2^{-1} U_1^{-1} a_2^+ U_1 U_2 = U_2^{-1} a_2^+ U_2. \qquad (* *)$$

To determine ($* *$), proceed in analogy to section 5.3, equations (5.63) to (5.75).

7.5. The dipole approximation*

In this book, we have described the interaction between an electron and the light field in two different manners. In chapter 4 we described the field by its electric field strength and introduced the term

$$\boldsymbol{E}(-e\boldsymbol{x}) \qquad (7.80)$$

as interaction Hamiltonian. In the present section, on the other hand, we introduce the interaction between the light field and electron by means of the vector potential $\boldsymbol{a}$, which led to two terms in the classical Hamiltonian, namely

$$\frac{e}{m} \boldsymbol{A} \cdot \boldsymbol{p} \quad \text{and} \quad \frac{e^2}{2m} \boldsymbol{A}^2. \qquad (7.81, 82)$$

(note the change of sign of e, because we are now dealing with the electron with charge $-e$). This second formulation, which in classical physics describes the electromagnetic forces on a moving charge properly, forms the basis for the quantum theoretical formulation presented in this chapter. This leads us to the question, whether the hypothesis (7.80) was only some sort of a model, or if it can be justified properly.

We want to show that eq. (7.80) is a well-defined approximation. We do this in the frame of the quantum mechanical treatment. First we remark

*This section is rather technical and can be skipped.

that for fields produced by thermal sources the quadratic term, (7.82), which reappears in $\hat{H}_{I,2}$, eq. (7.32), can usually be neglected compared to the first term. We will show that the term $\hat{H}_{I,1}$, eqs. (7.31) or (7.39), can be transformed into an expression entirely corresponding to (7.80) in our present formulation.

We start from the Hamiltonian [cf. eq. (7.41)]

$$\hat{H}_{\text{tot}} = \hat{H}_{\text{el}} + \hat{H}_{\text{field}} + \hat{H}_{I,1}, \tag{7.83}$$

where we inspect $\hat{H}_{I,1}$, eq. (7.31), more closely. We assume that the vector potential varies in its space-coordinate more slowly than the electronic wave functions under consideration. This allows us to put $\boldsymbol{A}$ in front of the integral eq. (7.31), where we choose $\boldsymbol{A}$ at the atomic center, $\boldsymbol{x}_0$,

$$(7.31) = -\frac{e}{m}\boldsymbol{A}(\boldsymbol{x}_0)\int \psi^+(\boldsymbol{x})\boldsymbol{p}\psi(\boldsymbol{x})\mathrm{d}^3\boldsymbol{x}; \qquad \boldsymbol{p} = \frac{\hbar}{i}\nabla . \tag{7.84}$$

All essential points of our argument can be seen when we consider a two-level atom. According to section 6.2, we therefore use the decompositions

$$\psi(\boldsymbol{x}) = a_1\varphi_1(\boldsymbol{x}) + a_2\varphi_2(\boldsymbol{x}),$$

$$\psi^+(\boldsymbol{x}) = a_1^+\varphi_1^*(\boldsymbol{x}) + a_2^+\varphi_2^*(\boldsymbol{x}). \tag{7.85}$$

Inserting eq. (7.85) into (7.84) we find expressions of the form

$$\int \varphi_j^*(\boldsymbol{x})\boldsymbol{p}\varphi_k(\boldsymbol{x})\mathrm{d}^3\boldsymbol{x} = \boldsymbol{p}_{jk}. \tag{7.86}$$

We will assume as usual

$$\boldsymbol{p}_{jj} = 0 \tag{7.87}$$

(though our following approach also works if eq. (7.87) is not fulfilled). Under these assumptions we find

$$(7.31) = -\frac{e}{m}\{\boldsymbol{A}^{(+)}(\boldsymbol{x}_0)\boldsymbol{p}_{21}a_2^+a_1 + \boldsymbol{A}^{(-)}(\boldsymbol{x}_0)\boldsymbol{p}_{12}a_1^+a_2\}. \tag{7.88}$$

To proceed further we introduce the interaction representation where we identify

$$\hat{H}_{\text{el}} + \hat{H}_{\text{field}} \to H_0, \qquad \hat{H}_{I,1} \to H_I. \tag{7.89}$$

As we know from the detailed examples from the foregoing section,

$$U = \exp\{-iH_0t/\hbar\} \tag{7.90}$$

transforms $\boldsymbol{A}(\boldsymbol{x}_0)$ into

$$U^{-1}\boldsymbol{A}(\boldsymbol{x}_0)U = \boldsymbol{A}(\boldsymbol{x}_0, t). \tag{7.91}$$

We split $\boldsymbol{A}(\boldsymbol{x}, t)$ into its positive and negative frequency parts

$$\boldsymbol{A}(\boldsymbol{x}_0, t) = \boldsymbol{A}^{(+)} + \boldsymbol{A}^{(-)} \tag{7.92}$$

where $\boldsymbol{A}^{(+)}$ and $\boldsymbol{A}^{(-)}$ are superpositions of annihilation and creation operators

$$\begin{aligned} \boldsymbol{A}^{(+)}: &\quad b_\lambda \exp\{-i\omega_\lambda t\}, \\ \boldsymbol{A}^{(-)}: &\quad b_\lambda^+ \exp\{i\omega_\lambda t\}. \end{aligned} \tag{7.93}$$

Simultaneously, $a_2^+ a_1$ is transformed into

$$(a_2^+ a_1)_t = a_2^+ a_1 \exp\{i\bar{\omega}t\}, \tag{7.94}$$

where

$$\bar{\omega} \equiv \omega_{21} = (W_2 - W_1)/\hbar. \tag{7.95}$$

We will now show that the matrix elements $\boldsymbol{p}_{jk}$ of the momentum operator, $\boldsymbol{p}$, can be transformed into those of the dipole moment, $e\boldsymbol{x}$. To this end we start from the relation

$$[H_{\text{el}}, \boldsymbol{x}] = \frac{\hbar}{mi}\boldsymbol{p}, \tag{7.96}$$

where

$$H_{\text{el}} = -\frac{\hbar^2}{2m}\Delta + V(\boldsymbol{x}). \tag{7.97}$$

The relation (7.96) was derived in exercise 6 of section 3.2. Using eq. (7.96) we form

$$\int \varphi_2^* \boldsymbol{p} \varphi_1 \mathrm{d}^3x = \frac{mi}{\hbar} \int \varphi_2^* \{H_{\text{el}}\boldsymbol{x} - \boldsymbol{x}H_{\text{el}}\} \varphi_1 \mathrm{d}^3x. \tag{7.98}$$

The first term on the r.h.s. of this equations can be given another form because H_{el} is a hermitian operator

$$\int \varphi_2^* H_{\text{el}} \boldsymbol{x} \varphi_1 \mathrm{d}^3x = \int \boldsymbol{x} \varphi_1 (H_{\text{el}} \varphi_2^*) \mathrm{d}^3x. \tag{7.99}$$

Furthermore using

$$H_{\text{el}}\varphi_2^* = W_2\varphi_2^*, \qquad H_{\text{el}}\varphi_1 = W_1\varphi_1 \tag{7.100, 101}$$

we obtain the r.h.s. of eq. (7.98) in the form

$$\frac{mi}{\hbar}(W_2 - W_1) \int \varphi_2^* \boldsymbol{x} \varphi_1 \mathrm{d}^3x. \tag{7.102}$$

Introducing the abbreviation (7.95), we are eventually led to the important relation, announced above:

$$p_{21} = mi\bar{\omega}x_{21}. \tag{7.103}$$

We insert eqs. (7.103) and (7.92) into eq. (7.88) and apply the rotating wave approximation (cf. section 7.4, example 3) i.e. we keep only terms close to resonance and neglect antiresonant terms.

$$\begin{aligned}(7.31) &= (7.88)\\ &= -e\left(i\bar{\omega}A^{(+)}(x_0,t)x_{21}(a_2^+a_1)_t + \text{Herm. conj.}\right)\end{aligned} \tag{7.104}$$

Now we can do the final steps. Remember that in classical theory as well as in the interaction representation

$$E(x,t) = -\frac{\partial A(x,t)}{\partial t}, \tag{7.105}$$

where we use for A the decomposition eq. (7.92) with eq. (7.93). Since close to resonance, $\omega_\lambda = \bar{\omega}$, the relevant contributions to eq. (7.104) are those, for which

$$\frac{\partial A^{(\pm)}(x_0,t)}{\partial t} = \mp i\bar{\omega}A^{(\pm)}(x_0,t). \tag{7.106}$$

This latter assumption has a further implication which can best be judged upon when we interpret the operators b_λ, b_λ^+ as classical time-dependent amplitudes, $\beta_\lambda(t)$, $\beta_\lambda^*(t)$. Our assumption eq. (7.106) implies that $\beta_\lambda(t)$, $\beta_\lambda^*(t)$ change much more slowly in time than their exponential factors $\exp(-i\omega_\lambda t)$ and $\exp(i\omega_\lambda t)$, respectively. This conclusion can be carried over to the operators b_λ, b_λ^+ when we use the Heisenberg picture (cf. section 5.6). In general we may assume that this slowness condition is fulfilled, though, for instance in experiments dealing with very short light pulses it might become necessary to reconsider the approximations just made. A careful study of these assumptions is also necessary in the case of real or virtual multiphoton processes. By taking eqs. (7.106) and (7.105) into account, we can cast eq. (7.104) into the form

$$\tilde{H}_{I,1} = -E^{(+)}(x_0,t)\vartheta_{21}(a_2^+a_1)_t - E^{(-)}(x_0,t)\vartheta_{12}(a_1^+a_2)_t, \tag{7.107}$$

where we put, as usual

$$ex_{21} = \vartheta_{21}. \tag{7.108}$$

When we insert the decomposition

$$E^{(-)}(x,t) = -i\sum_\lambda b_\lambda^+\sqrt{\hbar\omega_\lambda/(2\varepsilon_0)}\,u_\lambda(x) \tag{7.109}$$

assuming level 2 higher energy

and a corresponding one for $\boldsymbol{E}^{(+)}$ [cf. eq. (5.149)] into eq. (7.107) and return to the Schrödinger picture, we are led to $\hat{H}_{I,1}$, as given in equation (7.39). This completes our derivation. Our derivation shows that the terms $\boldsymbol{Ap}$ and $\boldsymbol{xE}$ are approximately equivalent. In processes such as spontaneous and stimulated emission and absorption the assumptions we made above are well fulfilled. On the other hand when dealing with multiphoton processes the reader is well advised to check these assumptions—or still better, use right away the formulation by means of the vector potential $\boldsymbol{A}$. We shall base our following chapters on that latter formulation.

7.6. Spontaneous and stimulated emission and absorption

For the treatment of these problems it is convenient to start with the Schrödinger equation in the interaction representation (compare section 7.4). We consider a single atom with two levels, 1 and 2, whose upper level, 2, is occupied in the beginning ($t = 0$). Because spontaneous and stimulated emission as well as absorption are processes in which only one light quantum is created or annihilated, it suffices to take into account only that part of the Hamiltonian, which is linear in the light-field operators b_λ^+, b_λ.

$$i\hbar \frac{\mathrm{d}\tilde{\Phi}}{\mathrm{d}t} = \tilde{H}_I \tilde{\Phi}. \tag{7.110}$$

Keeping only resonant terms (cf. the discussion following, eq. (7.73)) we find eq. (7.79) i.e.

$$\begin{aligned}\tilde{H}_I = \hbar \sum_\lambda \{ & g_\lambda a_1^+ a_2 b_\lambda^+ \exp[i(\omega_\lambda - \bar{\omega})t] \\ & + g_\lambda^* a_2^+ a_1 b_\lambda \exp[-i(\omega_\lambda - \bar{\omega})t]\},\end{aligned} \tag{7.111}$$

where, according to eq. (7.75)

$$\begin{aligned} g_\lambda \equiv g_{\lambda,12} &= -\frac{e}{m}(2\hbar\omega_\lambda\varepsilon_0)^{-1/2} \int \varphi_1^*(\boldsymbol{x}) \boldsymbol{u}_\lambda(\boldsymbol{x}) \boldsymbol{p} \varphi_2(\boldsymbol{x}) \mathrm{d}V, \\ \boldsymbol{p} &= \frac{\hbar}{i}\nabla. \end{aligned} \tag{7.112}$$

To solve eq. (7.110), we use perturbation theory. In a first still exact step we integrate both sides over time τ from $\tau = 0$ till $\tau = t$:

$$\tilde{\Phi}(t) = \tilde{\Phi}(0) + \left(\frac{-i}{\hbar}\right) \int_0^t \tilde{H}_I(\tau) \tilde{\Phi}(\tau) \mathrm{d}\tau. \tag{7.113}$$

$\tilde{\Phi}(0)$ is the quantum state of the system at the initial time. We shall specify

$\tilde{\Phi}(0)$ below. In the spirit of perturbation theory, we approximate $\tilde{\Phi}(\tau)$ under the integral on the r.h.s. of eq. (7.113) by $\tilde{\Phi}(0)$. We then obtain an improved function $\tilde{\Phi}^{(1)}(t)$ by the formula

$$\tilde{\Phi}^{(1)}(t) = \tilde{\Phi}(0) + \left(\frac{-i}{\hbar}\right)\int_0^t \tilde{H}_I(\tau)\mathrm{d}\tau\tilde{\Phi}(0). \tag{7.114}$$

This approach is called time-dependent perturbation theory in first order. In this chapter we shall base our analysis on eq. (7.114).

To treat (a) *spontaneous emission* we assume an initial state for the total system: light field + one electron, in which there is no photon present, and the electron is in its excited state:

$$\tilde{\Phi}(0) = a_2^+\Phi_0, \tag{7.115}$$

Φ_0: vacuum state.

Inserting eqs. (7.115) and (7.111) into (7.114) we are led to evaluate expressions of the form

$$a_2^+ a_1 b_\lambda a_2^+\Phi_0 \tag{7.116}$$

and

$$a_1^+ a_2 b_\lambda^+ a_2^+\Phi_0. \tag{7.117}$$

The first expression vanishes, because b_λ commutes with a_2^+ and

$$b_\lambda\Phi_0 = 0.$$

Using the commutation relation (6.15)

$$a_2 a_2^+ + a_2^+ a_2 = 1, \tag{7.118}$$

and the fact that b's and a's commute, we reduce eq. (7.117) to

$$a_1^+ b_\lambda^+\Phi_0, \tag{7.119}$$

where use was made of $a_2\Phi_0 = 0$.

After inserting these results into eq. (7.114), we may perform the integration, and obtain

$$\tilde{\Phi}(t) = a_2^+\Phi_0 + \sum_\lambda g_\lambda \frac{1}{\Delta\omega_\lambda}(1 - \exp[i\Delta\omega_\lambda t])a_1^+ b_\lambda^+\Phi_0, \tag{7.120}$$

where

$$\Delta\omega_\lambda = \omega_\lambda - \bar{\omega}. \tag{7.121}$$

From eq. (7.120) we see that the original wave function $a_2^+\Phi_0$ is replaced by a sum over terms of the form eq. (7.119). Thus our formalism describes a process in which a state

electron in excited state + no photon

is replaced by a state

electron in ground state + one photon.

The absolute square of the coefficient connected with the normalized wave function

$$a_1^+ b_\lambda^+ \Phi_0,$$

i.e.

$$\left|\frac{g_\lambda}{\Delta\omega_\lambda}\right|^2 |1 - \exp[i\Delta\omega_\lambda t]|^2,$$

gives the probability of finding a photon of the mode λ and the electron in the ground-state. In general, it makes sense to single out a specific mode connected with spontaneous emission only if the dimensions of the cavity (with closed walls) are of the same order as the wavelength of the emitted electromagnetic wave. On the other hand, in infinite space or in a cavity with open sides, there is a continuum of modes, so that we have to sum up (integrate) over a total range of final states:

$$\sum_\lambda \left|\frac{g_\lambda}{\Delta\omega_\lambda}\right|^2 |1 - \exp[i\Delta\omega_\lambda t]|^2. \tag{7.122}$$

By differentiation of this expression with respect to the time we obtain for the transition probability per second

$$P = \sum_\lambda |g_\lambda|^2 2 \frac{\sin \Delta\omega_\lambda t}{\Delta\omega_\lambda}. \tag{7.123}$$

We encountered an entirely analogous expression in section 4.2, formula (4.33). Having in our mind to replace the sum by an integral, we use the relation (cf. also the mathematical appendix)

$$\lim_{t\to\infty} \frac{\sin \omega t}{\omega} = \pi\delta(\omega). \tag{7.124}$$

The meaning of eq. (7.124) is explained in fig. 7.2. $\delta(\omega)$ is Dirac's δ-function which is defined by the properties

$$\delta(\omega) = 0 \quad \text{for} \quad \omega \neq 0 \tag{7.125}$$

and

$$\int_{-\varepsilon}^{+\varepsilon} \delta(\omega)\, d\omega = 1, \quad \varepsilon > 0, \quad \text{arbitrary}. \tag{7.126}$$

Using eq. (7.124) in eq. (7.123) we obtain

$$P = 2\pi \sum_\lambda |g_\lambda|^2 \delta(\omega_\lambda - \bar{\omega}). \tag{7.127}$$

We calculate first the sum over λ considering only those photons, which are emitted into a certain group of modes. In order to be specific, we treat

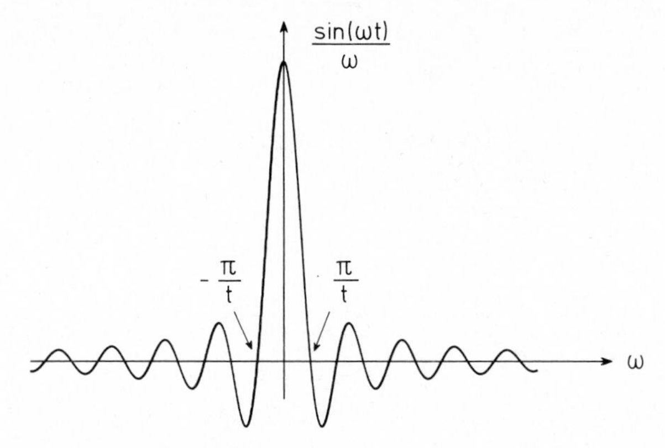

Fig. 7.2. This figure shows us how we may understand the properties of the Dirac δ-function which we obtain of $\sin \omega t/\omega$ in the limit $t \to \infty$. $\sin \omega t/\omega$ is plotted as a function of the frequency. The maximum of this function tends to infinity when t goes to infinity. On the other hand the first zero approaches the ordinate more and more when t goes to infinity. The area below the curve, however, is independent of t.

plane waves, so that

$$\boldsymbol{u}_\lambda(\boldsymbol{x}) = \frac{1}{\sqrt{V}} \boldsymbol{e} \exp[\,i\boldsymbol{k}_\lambda \boldsymbol{x}\,]. \tag{7.128}$$

$\boldsymbol{e}$: vector of polarization, V = volume of the box, with $V \to \infty$. We consider photons emitted into a space angle $\mathrm{d}\Omega$. To transform the sum Σ_λ into an integral we have to count the number of modes $\mathrm{d}N$ in the volume V and in the wave number (or frequency) interval $\mathrm{d}k$ ($\mathrm{d}\nu$). We have seen in chapter 2, section 2.3, formulas (2.54)–(2.56), how this can be done. Applying similar arguments to plane waves eq. (7.128), whose wave vectors lie within the space angle $\mathrm{d}\Omega$, we obtain

$$\mathrm{d}N = \frac{V}{(2\pi)^3} k^2 \,\mathrm{d}k \,\mathrm{d}\Omega \tag{7.129}$$

and therefore

$$\sum_\lambda \cdots = \frac{V}{(2\pi)^3} \int \cdots k^2 \,\mathrm{d}k \,\mathrm{d}\Omega. \tag{7.130}$$

Inserting eq. (7.130) into eq. (7.127) we find

$$P(\mathrm{d}\Omega) \equiv a^1_{2,\boldsymbol{e}} \,\mathrm{d}\Omega$$

$$= \frac{e^2 \overline{\omega}}{8\pi^2 \hbar \varepsilon_0 m^2 c^3} \left| \int \varphi_1^*(\boldsymbol{x}) \exp[\,i\boldsymbol{k}_{\lambda_0} \boldsymbol{x}\,] \boldsymbol{e}\boldsymbol{p} \varphi_2(\boldsymbol{x}) \,\mathrm{d}V \right|^2 \mathrm{d}\Omega \tag{7.131}$$

where

$$ck_{\lambda_0} = \bar{\omega}. \tag{7.132}$$

$a^1_{2,e}$ in eq. (7.131) is the Einstein coefficient for spontaneous emission of a light quantum of polarization $\boldsymbol{e}$ into $d\Omega$.

In the dipole approximation [compare eq. (7.39) or section 7.5], eq. (7.131) simplifies to

$$P(d\Omega) = \frac{\bar{\omega}^3}{8\pi^2\hbar\varepsilon_0 c^3}|\boldsymbol{e}\boldsymbol{\vartheta}_{21}|^2\, d\Omega, \tag{7.133}$$

where

$$\boldsymbol{\vartheta}_{21} = e\int \varphi_2^*(\boldsymbol{x})\boldsymbol{x}\varphi_1(x)\, dV \tag{7.134}$$

and we have used

$$\boldsymbol{p}_{21} = im\bar{\omega}\boldsymbol{x}_{21}. \tag{7.135}$$

In order to get the total transition probability for the emission of any photon, we integrate eq. (7.133) over $d\Omega$ and sum up over the two directions of polarization:

$$P = \frac{\bar{\omega}^3}{3\pi\hbar\varepsilon_0 c^3}|\boldsymbol{e}\cdot\boldsymbol{\vartheta}_{21}|^2. \tag{7.136}$$

As it will transpire below, $P = 1/\tau$, where τ is the lifetime of the upper state.

(b) *Stimulated emission.* We assume first that there is a definite number of light quanta in a definite mode λ_0 initially present. The normalized initial state is then

$$\tilde{\Phi}(0) = a_2^+ \frac{1}{\sqrt{n!}}(b_{\lambda_0}^+)^n\Phi_0. \tag{7.137}$$

We insert again $\Phi(0)$ into the right-hand side of eq. (7.114) and find immediately that there are two kinds of final states, depending on whether the index λ in Σ_λ equals λ_0 or not:

(α)

$$\lambda \neq \lambda_0: \qquad a_1^+\frac{1}{\sqrt{n!}}(b_{\lambda_0}^+)^n b_\lambda^+\Phi_0, \tag{7.138}$$

(β)

$$\lambda = \lambda_0: \qquad a_1^+\frac{1}{\sqrt{n!}}(b_{\lambda_0}^+)^{n+1}\Phi_0. \tag{7.139}$$

In case (α), a quantum of another mode λ has been emitted spontaneously. In case (β), a light quantum has been added to the mode under consideration. Under the action of the light field the atom is forced to emit a quantum by "stimulated emission". In order to obtain the transition probability, we first naively repeat the above steps, which yields:

$$P = 2\pi |g_{\lambda_0}|^2 \delta(\bar{\omega} - \omega_{\lambda_0})(n+1) + 2\pi \sum_{\lambda \neq \lambda_0} |g_\lambda|^2 \delta(\bar{\omega} - \omega_\lambda). \quad (7.140)$$

The first term stems from (β), while the second one comes from (α). We split the first term $(n+1)$ into n and 1 and combine the expressions connected with "1" with the second expression in eq. (7.140), so that we obtain with this combination just the spontaneous emission into all modes. The remaining expression

$$P_{st} = 2\pi n |g_{\lambda_0}|^2 \delta(\bar{\omega} - \omega_{\lambda_0}) \quad (7.141)$$

is then the stimulated emission rate. In it no summation over λ appears. On the other hand, it is necessary to integrate over a continuum in order that the δ-function can be evaluated. Thus the formalism forces us to start with a more realistic initial state, which is formed as a wave packet. We assume, that it is built up out of plane waves within a region Δk_x, Δk_y, Δk_z with a corresponding frequency spread $\Delta\omega = c\Delta k$. If there are M modes the normalized wave function reads

$$\tilde{\Phi}(0) = \frac{1}{\sqrt{M}} \frac{1}{\sqrt{n!}} \sum_{\Delta k} a_2^+ (b_k^+)^n \Phi_0 \quad (7.142)$$

(in order to be quite clear we use $\boldsymbol{k}$ instead of λ). With this initial state we readily obtain

$$P_{st} = 2\pi \frac{n}{M} \sum_{\Delta k} |g_k|^2 \delta(\bar{\omega} - \omega_k). \quad (7.143)$$

With

$$M = m_x m_y m_z, \qquad \Delta k_i = \frac{2\pi m_i}{L} \quad (7.144)$$

(where L: length of normalization box) we find

$$M = \frac{L^3}{(2\pi)^3} \Delta k_x \Delta k_y \Delta k_z = \frac{V}{(2\pi)^3} k^2 \Delta k \, \mathrm{d}\Omega. \quad (7.145)$$

Using further eqs. (7.130) and (7.141) we obtain for the transition probability for stimulated emission of photons into a space angle $\mathrm{d}\Omega$:

$$P_{st}(\mathrm{d}\Omega) = \frac{\pi e^2 n}{m^2 \hbar \varepsilon_0 \bar{\omega} \Delta\omega V \mathrm{d}\Omega} \left| \int \varphi_1^*(\boldsymbol{x}) \exp[i\boldsymbol{k}_{\lambda_0} \boldsymbol{x}] \boldsymbol{e}\boldsymbol{p} \varphi_2(\boldsymbol{x}) \, \mathrm{d}V \right|^2 \mathrm{d}\Omega. \quad (7.146)$$

We write $P_{st}(\mathrm{d}\Omega)$ in the form

$$P_{st}(\mathrm{d}\Omega) = \rho_e(\bar{\omega}, \mathrm{d}\Omega) b^1_{2,e}\, \mathrm{d}\Omega, \tag{7.147}$$

where

$$b^1_{2,e} = \frac{\pi e^2}{\hbar^2 \varepsilon_0 m^2 \bar{\omega}^2} \left| \int \varphi_1^*(\boldsymbol{x}) \exp\left[i\boldsymbol{k}_{\lambda_0}\boldsymbol{x} \right] \boldsymbol{e}\boldsymbol{p}\varphi_1(\boldsymbol{x})\, \mathrm{d}V \right|^2 \tag{7.148}$$

is the Einstein coefficient for stimulated emission of photons with polarization $\boldsymbol{e}$ into $\mathrm{d}\Omega$.

$$\rho_e(\bar{\omega}, \mathrm{d}\Omega) = \frac{n\hbar\omega}{\Delta\omega\, \mathrm{d}\Omega V} \tag{7.149}$$

is the total energy of the n photons, divided by the frequency spread, the space angle, and the volume, or in other words: ρ is the energy density per unit frequency interval, unit space angle, and unit volume. A comparison between eqs. (7.131) and (7.148) yields one of Einstein's relations:

$$\frac{a^1_{2,e}}{b^1_{2,e}} = \frac{\hbar\bar{\omega}^3}{(2\pi)^3 c^3}. \tag{7.150}$$

In the dipole approximation the transition probability P_{st} becomes:

$$P_{st}(\mathrm{d}\Omega) = \rho_e(\bar{\omega}, \mathrm{d}\Omega) b^1_{2,e}\, \mathrm{d}\Omega, \tag{7.151}$$

where ρ is given by eq. (4.38) and

$$b^1_{2,e} = \frac{\pi}{\hbar^2 \varepsilon_0} |\boldsymbol{e}\boldsymbol{\vartheta}_{21}|^2. \tag{7.152}$$

(c) *Comparison between spontaneous and stimulated emission rates.* We already know of one connection, namely the one between Einstein's coefficients eq. (7.150). This connection can be given another appearance. We determine the spontaneous emission rate, $\tilde{P}$, per number of modes (not photons!) in the volume V, the angle $\mathrm{d}\Omega$ and the frequency range $\Delta\omega$ $(\bar{\omega}, \bar{\omega} + \Delta\omega)$ which we had considered just now. By dividing eq. (7.131) by this number

$$N_m = \frac{k^2 \Delta k V\, \mathrm{d}\Omega}{8\pi^3} = \frac{\bar{\omega}^2 \Delta\omega V\, \mathrm{d}\Omega}{8\pi^3 c^3}, \tag{7.153}$$

we find

$$\tilde{P} = \frac{P}{N_m} = \frac{\pi e^2}{\hbar \varepsilon_0 m^2 \bar{\omega} \Delta\omega V} \left| \int \varphi_1^*(\boldsymbol{x}) \exp\left[i\boldsymbol{k}_{\lambda_0}\boldsymbol{x} \right] \boldsymbol{e}\boldsymbol{p}\varphi_2(\boldsymbol{x})\, \mathrm{d}V \right|^2 = \frac{1}{n} P_{st}, \tag{7.154}$$

so that the ratio of stimulated emission rate to the spontaneous emission rate is $= n$ (= total number of photons in this range).

(d) *Absorption*. The calculation of the transition probability P_{abs} for the atom going from its ground state to the excited state by absorption of a quantum out of a wave-packet which propagates within a space angle $d\Omega$ can be done in complete analogy to the stimulated emission. The only difference is that we start now with an electron in its ground state instead of one in its excited state:

$$P_{abs} = \rho_e(\bar{\omega}, d\Omega) b_{1,e}^2 \, d\Omega, \tag{7.155}*$$

where $b_{1,e}^2 = b_{2,e}^1$ so that the Einstein coefficient for absorption is equal to that for stimulated emission. The absorption rate is proportional to the energy density, ρ, of the incident light. In order to derive the absorption coefficient α, we introduce the energy flux density $I(\omega) = \rho_e(\bar{\omega}, d\Omega) c \, d\Omega$ (energy flux per s, per unit area) into eq. (7.155) so that

$$P_{abs} = I(\omega) \frac{b_{1,e}^2}{c}. \tag{7.156}$$

The decrease of the photon number, n, per s is equal P_{abs} for a single atom. If there are N atoms, we find

$$\frac{dn}{dt} = -P_{abs} N. \tag{7.157}$$

Introducing

$$I(\omega) = \frac{\bar{n} \hbar \omega c}{\Delta\omega} \tag{7.158}$$

($\bar{n} = n/V$: photon density) into eq. (7.157) yields

$$\frac{dI(\omega)}{dt} = -I(\omega) \frac{b_{1,e}^2}{c} \frac{N \hbar \omega}{V \Delta\omega}. \tag{7.159}$$

In writing $dx = c\,dt$ we obtain the spatial absorption equation

$$\frac{dI(\omega)}{dx} = -I(\omega)\alpha, \tag{7.160}$$

where the absorption coefficient

$$\alpha = \frac{b_{1,e}^2 N \hbar \omega}{c V \Delta\omega}. \tag{7.161}$$

Since P_{abs} and P_{st} play a completely symmetrical role, we find quite generally for a system of noninteracting, partially inverted atoms:

$$\frac{dI}{dt} = I \frac{N_2 - N_1}{N} \alpha. \tag{7.162}$$

*Do not read b squared, but b one–two, because "2" is an index as is "1".

We conclude with an important remark about coherence properties of light created by spontaneous or stimulated emission. The above transition rates were calculated for an experimental setup, in which the photon number is measured. Such a measurement, however, destroys any phase information (compare section 5.9). An appropriate treatment which retains phase information will be given in section 8.4.

7.7. Perturbation theory and Feynman graphs

In the foregoing section we have discussed processes in which the absorption or emission of a single photon was involved. Both, in the fundamental theory of radiation processes as well as in many practical applications of nonlinear optics, it will be necessary to study effects in which several or even many photons play a role. For a qualitative and, in many cases, even a quantitative discussion higher-order perturbation theory is useful and necessary. To get an overview of the possible processes, the Feynman technique of graphs is particularly useful. Therefore we present here some of its basic ideas. Let us start with the spontaneous emission of a photon. As shown at the beginning of section 7.6 this process is described by the wave function eq. (7.120). This result can be interpreted as follows. The initial state in which an electron in the upper state 2 and no photon are present is transformed under the influence of the electron field interaction into a new state. In it the electron is now in state 1 while a photon with wave vector $\boldsymbol{k}(\leftrightarrow\lambda)$ has been created. This process can be interpreted by means of a graph. Here and in the following we shall read such graphs from right to left. This may be somewhat inconvenient for the beginner but it is most useful for later applications of the rules which we will develop now. We represent the incoming electron by a solid line with an arrow pointing to the left. The interaction of the electron with the photon field is indicated by a vertex. The outgoing electron after the interaction is represented again by a solid line with an arrow. Furthermore the outgoing photon is indicated by a wavy line with an arrow (compare fig. 7.3).

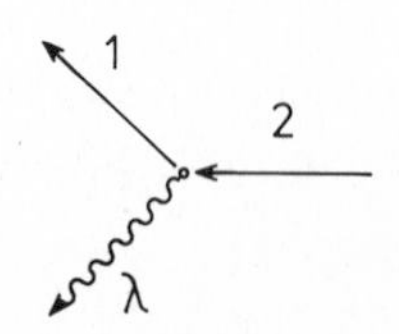

Fig. 7.3. Simple Feynman diagram in which an incoming electron emits a photon of wavelength λ and is hereby scattered into a new state 1.

These and similar graphs enable us to describe all processes, and furthermore, to calculate the function for the final state by means of precise rules. To this end we write the function for the final state in the form

$$\tilde{\Phi}(t) = \sum_{\lambda} c_{\lambda}(t) a_1^+ b_{\lambda}^+ \Phi_0 + a_2^+ \Phi_0. \tag{7.163}$$

It is then the task of the theory to calculate the coefficients $c_{\lambda}(t)$. The prescription for this calculation now reads as follows:

Table 1

Incoming electron wave	$\exp(-i\varepsilon_2 t)$	$\overset{2}{\rightarrow}\circ$
Outgoing electron wave	$\exp(i\varepsilon_1 t)$	$\circ\rightarrow$
Outgoing photon	$\exp(i\omega_{\lambda} t)$	$\circ\leftsquigarrow$ λ
Vertex	$-ig_{\lambda}$	$\circ$

For sake of convenience (to get rid of too many $\hbar$s!) we have used the abbreviation

$$\varepsilon_j = W_j/\hbar, \qquad j = 1, 2, \ldots \tag{7.164}$$

so that ε_j has the dimension of a frequency. The functions in the middle of our scheme must be multiplied with each other and finally we have to integrate the product from an initial time $t_0 = 0$ until the final time t. This prescription yields the coefficient

$$c_{\lambda} = -g_{\lambda} \frac{(\exp[i(\varepsilon_1 - \varepsilon_2 + \omega_{\lambda})t] - 1)}{\varepsilon_1 - \varepsilon_2 + \omega_{\lambda}}. \tag{7.165}$$

The further discussion can now proceed again as in section 7.6 before and we do not wish to repeat that here. Instead we now reinterpret the stimulated emission of a photon. It can be represented by a graph of a structure given in fig. 7.4. The table which must now be applied reads

Table 2

Incoming electron wave	$\exp(-i\varepsilon_2 t)$	$\overset{2}{\rightarrow}\circ$
n incoming photons	$\exp(-in\omega_{\lambda_0} t)$	$\lambda_0(n)$ $\circ\leftsquigarrow$
Outgoing electron wave	$\exp(i\varepsilon_1 t)$	$\overset{1}{\circ\rightarrow}$
$(n+1)$ outgoing photons	$\exp(i(n+1)\omega_{\lambda_0} t)$	$\lambda_0(n+1)$ $\rightsquigarrow$
Vertex	$-ig_{\lambda_0}\sqrt{n+1}$	$\circ$

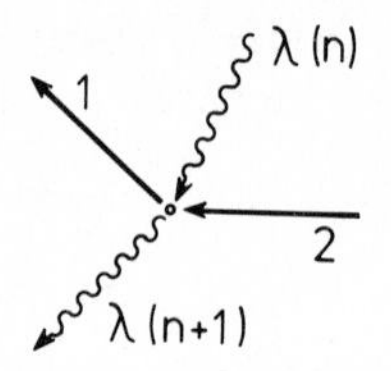

Fig. 7.4. A diagram showing a process in which an electron and n photons of wavelength λ interact whereby an additional photon of the same wavelength is emitted and the electron is scattered into the state 1.

the coefficients of the corresponding expansion of the wave functions in terms of photons and electron states can now be achieved by multiplying the functions on the right-hand side of table 2 with each other and eventually integrating the product over time. We leave it as an exercise to the reader to establish a corresponding diagram and rules for the absorption of a photon.

In our above treatment we have established rules for calculating the time-dependent coefficients. In this treatment, time is distinguished with respect to spatial coordinates. We mention that it is also possible to derive rules for Feynman graphs treating space and time in a symmetric fashion as it is required by the theory of relativity. Indeed, Feynman's original graphs were relativistically symmetric. However, since the most important results in our context can be visualized by means of time dependent Feynman graphs we will stick to our simpler description.

So far we have been treating processes which occur in first order perturbation theory. We now discuss some results of second order perturbation theory and will conclude this chapter by discussing a general case. To start with, let us reconsider formula (7.113) of section 7.6. There we realized that we could integrate Schrödinger's equation in a formal way which led us to the result

$$\tilde{\Phi}(t) = \tilde{\Phi}(0) + \frac{1}{i\hbar}\int_0^t \tilde{H}_I(\tau)\tilde{\Phi}(\tau)\,\mathrm{d}\tau. \tag{7.166}$$

The state function $\tilde{\Phi}$ and the interaction Hamiltonian are taken in the interaction picture. Since the solution of the integral equation (7.166) is as difficult as that of the Schrödinger equation we have tried, in section 7.6, to solve it by an approximation. To this end we had replaced $\tilde{\Phi}(\tau)$ by the initial state $\tilde{\Phi}(0)$. This resulted in an improved state function given by

$$\tilde{\Phi}^{(1)}(t) = \tilde{\Phi}(0) + \frac{1}{i\hbar}\int_0^t \tilde{H}_I(\tau)\tilde{\Phi}(0)\,\mathrm{d}\tau. \tag{7.167}$$

Now we may repeat this procedure replacing $\tilde{\Phi}(\tau)$ in eq. (7.166) by the improved state function $\tilde{\Phi}^{(1)}$. Thus we obtain the following more improved state function $\tilde{\Phi}^{(2)}$

$$\tilde{\Phi}^{(2)}(t) = \tilde{\Phi}(0) + \frac{1}{i\hbar}\int_0^t \tilde{H}_I(\tau)\tilde{\Phi}^{(1)}(\tau)\,\mathrm{d}\tau. \tag{7.168}$$

By introducing the explicit representation of $\tilde{\Phi}^{(1)}$ as given by eq. (7.167) into eq. (7.168) we obtain an explicit representation of $\tilde{\Phi}^{(2)}$

$$\tilde{\Phi}^{(2)}(t) = \tilde{\Phi}(0) + \frac{1}{i\hbar}\int_0^t \mathrm{d}\tau_1 \tilde{H}_I(\tau_1)\tilde{\Phi}(0)$$

$$+ \left(\frac{1}{i\hbar}\right)^2 \int_0^t \mathrm{d}\tau_2 \int_0^{\tau_2} \mathrm{d}\tau_1 \tilde{H}_I(\tau_2)\tilde{H}_I(\tau_1)\tilde{\Phi}(0). \tag{7.169}$$

This formulation means that on the right-hand side we have to use the initial wave function and to apply certain operations to it. Finally, we have to evaluate the integrals.

We now show, by means of an example, how to evaluate the right-hand side. Again we use as explicit representation for $\tilde{H}_I$ the interaction Hamiltonian atom field. We have encountered the first two terms of equation (7.169) before so that we focus our attention on the last term. In it $\tilde{H}_I$ must be applied twice in succession to the initial state. We choose as initial state and treat only virtual processes where $\varepsilon_2 - \varepsilon_1 \neq \omega_\lambda$.

$$a_2^+\Phi_0. \tag{7.170}$$

When we apply $\tilde{H}_I$ first on (7.170), we obtain terms of the form

$$b_\lambda^+ a_1^+ a_2 \mid a_2^+\Phi_0 = b_\lambda^+ a_1^+\Phi_0. \tag{7.171}$$

Thus we are dealing with the process depicted in fig. 7.3. When we now apply $\tilde{H}_I$ at time τ_2 for a second time we obtain two processes. One process describes the emission of a second photon, the other one the absorption of the photon already present. We treat the reabsorption of a photon. Taking the result (7.171) into account we must now calculate

$$b_\lambda a_2^+ a_1 \mid b_\lambda^+ a_1^+\Phi_0. \tag{7.172}$$

By using our well-known rules for field and atomic operators we immediately obtain the final state $a_2^+\Phi_0$, i.e. again exactly the initial state. The subsequent action of the two interaction Hamiltonians at times τ_1 and τ_2 can be represented by the diagram of fig. 7.5. We now consider more closely that part of eq. (7.169) which yields the original state on account of the emission and absorption processes just discussed

$$c_2(t)a_2^+\Phi_0 + a_2^+\Phi_0. \tag{7.173}$$

It will be our task to calculate the coefficient $c_2(t)$ explicitly. To this end

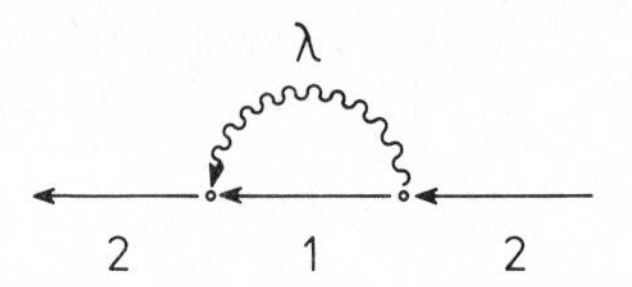

Fig. 7.5. Feynman diagram giving a contribution to the self-energy of the electron. An incoming electron in state 2 emits a photon which is eventually reabsorbed restoring the initial state of the electron.

we have only to remember that the vertices stem each time from an application of the interaction operator $\tilde{H}_I$. The factor in front of the operator of formula (7.171) reads

$$-ig_\lambda \int_0^{\tau_2} \exp\{i(\varepsilon_1 + \omega_\lambda - \varepsilon_2)\tau_1\}\, \mathrm{d}\tau_1. \tag{7.174}$$

In it we have made use of the fact that we have to integrate over τ_1 from 0 till τ_2. Furthermore we will have to remember that we still have to perform the sum over λ. At vertex 2 that factor comes into play which stands in the interaction Hamiltonian describing the absorption process. By using eq. (7.172) we obtain from the second vertex the expression

$$-ig_\lambda^* \int_0^t \mathrm{d}\tau_2 \exp\{i(\varepsilon_2 - \omega_\lambda - \varepsilon_1)\tau_2\}. \tag{7.175}$$

In it we have to use the same photon wave vector $\boldsymbol{k}$ ($\hat{=}$ index λ) as in formula (7.174). In putting eqs. (7.174) and (7.175) together, the following explicit representation of the coefficient $c_2(t)$ of formula (7.173) results

$$c_2(t) = \int_0^t \mathrm{d}\tau_2 \exp[i\varepsilon_2\tau_2] \sum_\lambda (-ig_\lambda^*) \int_0^{\tau_2} \exp[-i(\varepsilon_1 + \omega_\lambda)(\tau_2 - \tau_1)]$$
$$\times (-ig_\lambda)\exp[-i\varepsilon_2\tau_1]\, \mathrm{d}\tau_1. \tag{7.176}$$

The result of our little calculation can again be described by rules which are connected with the graph of fig. 7.5. Eventually we have to integrate over the times τ_1, τ_2, where $0 \leqslant \tau_1 \leqslant \tau_2 \leqslant t$, and to sum over λ. The reader will recognize that the rules described by table 3 are a systematic continuation of the rules given above in connection with perturbation theory of first order.

Table 3

Vertex 1	
Incoming free electron	$\exp[-i\varepsilon_2\tau_1]$
Emission of a photon	$-ig_\lambda$
Propagation of electron	
and photon from τ_1 to τ_2	$\exp[-i(\omega_\lambda + \varepsilon_1)(\tau_2 - \tau_1)]$
Vertex 2	
Absorption of a photon	$-ig_\lambda^*$
Outgoing free electron	$\exp[i\varepsilon_2\tau_2]$

Now it is not very difficult to imagine what such rules look like when we deal with still more complicated processes. We will get to know a further example in the next paragraph.

After the integration over τ_1 we obtain

$$c_2(t) = \sum_\lambda |g_\lambda|^2 \int_0^t \frac{1 - \exp\{i(\varepsilon_2 - \varepsilon_1 - \omega_\lambda)\tau_2\}}{i(\varepsilon_2 - \varepsilon_1 - \omega_\lambda)} \, \mathrm{d}\tau_2. \tag{7.177}$$

The number 1 under the integral gives rise to a factor t. The exponential function itself leads to a term oscillatory in time. For sufficiently large times only the term which increases proportional to t is important so that we may neglect the oscillatory term. Thus we obtain an expression of the form

$$c_2(t) = -it\Delta\varepsilon, \tag{7.178}$$

where we have used the abbreviation

$$\Delta\varepsilon = \sum_\lambda \frac{|g_\lambda|^2}{\varepsilon_2 - \varepsilon_1 - \omega_\lambda}. \tag{7.179}$$

Using this result we find the wanted state function (7.173) in the form

$$\tilde{\Phi}(t) = (1 - i\Delta\varepsilon t) a_2^+ \Phi_0. \tag{7.180}$$

This result looks rather strange. It tells us that the process under study, namely emission and reabsorption of a photon, does not lead to any change of the original state but that its coefficient seems to grow with time more and more.

A result making sense can be obtained only when we consider also the higher-order terms of perturbation theory. We will do this in the next section and anticipate here its result. According to that analysis it turns out that $1 - i\Delta\varepsilon t$ are just the first two terms of an expansion of the exponential function. Therefore we are allowed to write the state function (7.180) in the form

$$\tilde{\Phi}(t) = \exp\{-i\Delta\varepsilon t\} a_2^+ \Phi_0. \tag{7.181}$$

We will now assume that the form (7.181) is correct and we want to study what the time-dependent factor in eq. (7.181) means. To this end we go back from the interaction picture [state function $\tilde{\Phi}(t)$] which we have been using to the Schrödinger picture [state function $\Phi(t)$]. According to section 7.4, $\Phi(t)$ and $\tilde{\Phi}(t)$ are connected by means of the unperturbed Hamiltonian H_0 [cf. eqs. (7.45) and (7.48)]. Using the relations

$$\Phi(t) = U\tilde{\Phi}(t)$$

$$U = \exp\{-iH_0 t/\hbar\} \tag{7.182}$$

and (7.181), and making a small rearrangement of terms we obtain

$$\Phi(t) = \exp\{-i\Delta\varepsilon t\} \underbrace{Ua_2^+ U^{-1}}_{I} \underbrace{U\Phi_0}_{II}. \tag{7.182a}$$

However, we know expressions I and II from our earlier chapters, namely

$$I = a_2^+ \exp\{-i\varepsilon_2 t\}, \qquad II = \Phi_0. \tag{7.183}$$

We thus obtain as a final result

$$\Phi(t) = \exp\{-i(\varepsilon_2 + \Delta\varepsilon)t\}\Phi_0. \tag{7.184}$$

Comparing this result with the stationary solution of the Schrödinger equation containing the total Hamiltonian $H = H_0 + H_I$

$$\Phi(t) = \exp\{-iWt/\hbar\}\Phi(0) \tag{7.185}$$

we readily obtain

$$W = \hbar\varepsilon_2 + \hbar\Delta\varepsilon. \tag{7.186}$$

This relation yields the wanted interpretation of $\Delta\varepsilon$. $\hbar\Delta\varepsilon$ is an energy shift of the electron in state 2. In the frame of second order perturbation theory this shift is explicitly given by eq. (7.179). This energy shift stems from the emission and subsequent absorption of a photon. Since energy is not conserved during the intermediate state in which a photon is present the corresponding processes are called virtual emission of a photon. We are thus led to a result which is also of fundamental importance in the theory of elementary particles. It tells us that the energy of an electron is shifted by the virtual emission and reabsorption of quanta, in our case of photons.

So far we have described these processes by dealing explicitly with the bound electron of an atom which can occupy the states 1 or 2. Exactly the same discussion can be performed when we deal with free electrons with the momentum $\boldsymbol{p}$. The only generalization of that case as compared to the case here consists of the rule that in the intermediate state we have also to sum up over all $\boldsymbol{p}$ vectors of the electron which obey, according to momentum conservation, the rule

$$\boldsymbol{p} = \boldsymbol{p}_{\text{initial}} + \hbar\boldsymbol{k}, \tag{7.187}$$

where $\hbar\boldsymbol{k}$ now denotes the photon momentum. In this more general case, which we will come to in section 7.8, the energy shift depends on $\boldsymbol{p}^2$. This is usually called self-energy. Now let us briefly determine what the $\boldsymbol{p}$-dependence of $\Delta\varepsilon$ means. For a free electron its energy depends on $\boldsymbol{p}$ in the form

$$W_k^0 = W_0^0 + \frac{\boldsymbol{p}^2}{2m}, \tag{7.188}$$

where $\boldsymbol{p} = \boldsymbol{p}_{\text{initial}}$ and m is the electron mass. It can be shown further (compare also section 7.8) that the energy shift can be expressed in the form

$$\Delta W = \hbar \Delta \varepsilon = \Delta W_0 - Cp^2. \tag{7.189}$$

To reveal the significance of C we write it in the form

$$C = \frac{\delta}{2m} \tag{7.190}$$

where δ is assumed to be a small quantity. By now adding eq. (7.189) to eq. (7.188) we are led to formula

$$W = W_0 + \frac{\boldsymbol{p}^2}{2m^*}, \tag{7.191}$$

where, by comparison, we find

$$\frac{1}{m^*} = \frac{1}{m} - \frac{\delta}{m}. \tag{7.192}$$

Since δ was assumed small, eq. (7.192) is equivalent to

$$m^* = m(1 + \delta). \tag{7.193}$$

This result puts the significance of the $\boldsymbol{p}$-dependent energy shift in evidence. We realize that by virtual emission processes the mass of the electron is changed.

That change of mass due to virtual processes has played an important role in the modern development of quantum electrodynamics. Since the change of self-energy and mass give rise to observable effects, the corresponding measurements have been a test of quantum electrodynamics. We will discuss these points in the following section 7.8.

By means of Feynman graphs and the rules connected with them we can study further processes which lead to a genuine change of the initial state in second-order perturbation theory. Such processes involve two photons. Figure 7.6 represents the subsequent spontaneous emission of two photons. Similarly all higher-order processes can be described by Feynman graphs.

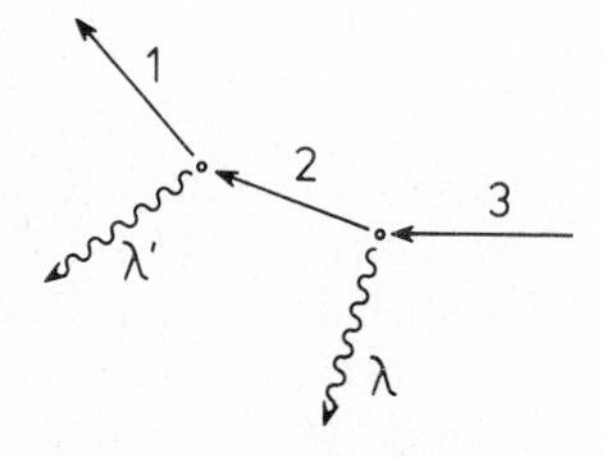

Fig. 7.6. An example in which an electron consecutively emits two photons with wavelengths λ and λ′, respectively.

In concluding this chapter we now deal with perturbation theory in arbitrary order. Again we start from the exact expression (7.166) which we presented at the beginning of this chapter. As has been seen above we may try to solve this equation by an iteration procedure. In such a procedure we determine the nth approximation on the left-hand side of eq. (7.166) by inserting the $(n-1)$th approximation on the right-hand side of eq. (7.166)

$$\tilde{\Phi}^{(n)}(t) = \tilde{\Phi}(0) + \frac{1}{i\hbar}\int_0^t \tilde{H}_I(\tau_n)\,\mathrm{d}\tau_n \tilde{\Phi}^{(n-1)}(\tau_n). \tag{7.194}$$

To have a concise formula we put in the 0th approximation

$$\tilde{\Phi}^{(0)} = \tilde{\Phi}(0) = \Phi(0). \tag{7.195}$$

It will be our goal to express the nth approximation directly by $\Phi(0)$. To this end we express consecutively $\Phi^{(n)}$ by $\Phi^{(n-1)}$, then $\Phi^{(n-1)}$ by $\Phi^{(n-2)}$, etc. To recognize the general structure of the resulting expression first look at formula (7.194) for $n = 1$ and for $n = 2$. To obtain $\tilde{\Phi}^{(3)}$ let us insert $\hat{\Phi}^{(2)}$ into eq. (7.194). One easily convinces oneself that the resulting state function reads

$$\tilde{\Phi}^{(3)}(t) = \tilde{\Phi}^{(2)} + \left(\frac{1}{i\hbar}\right)^3 \int_0^t \tilde{H}_I(\tau_3)\,\mathrm{d}\tau_3 \int_0^{\tau_3} \tilde{H}_I(\tau_2)\,\mathrm{d}\tau_2 \int_0^{\tau_2} \tilde{H}_I(\tau_1)\,\mathrm{d}\tau_1 \Phi(0). \tag{7.196}$$

As can be seen from our examples, each step further yields an additional term. Such an additional term is an n-fold integral over the product of interaction Hamiltonians taken at subsequent times. Thus we are directly led to the following explicit expression for the state function in the nth approximation

$$\begin{aligned}\tilde{\Phi}^{(n)}(t) &= \Phi(0)\\ &\quad + \sum_{\nu=1}^{n} (i\hbar)^{-\nu} \int_0^t \tilde{H}_I(\tau_\nu)\,\mathrm{d}\tau_\nu \int_0^{\tau_\nu} \tilde{H}_I(\tau_{\nu-1})\,\mathrm{d}\tau_{\nu-1} \cdots\\ &\quad \times \int_0^{\tau_2} \tilde{H}_I(\tau_1)\,\mathrm{d}\tau_1 \Phi(0).\end{aligned} \tag{7.197}$$

Thus the problem of finding $\tilde{\Phi}^{(n)}$ is reduced to an evaluation of eq. (7.197). One should not forget, however, that eq. (7.197) contains an enormous number of different processes involving the absorption and emission of photons. In the present context we confine ourselves to the following process. We consider a sequence of virtual emission and absorption processes as depicted in fig. 7.7. Exactly speaking the diagram of fig. 7.7 corresponds to a term with $\nu = 8$ in eq. (7.197). Our former results allow us

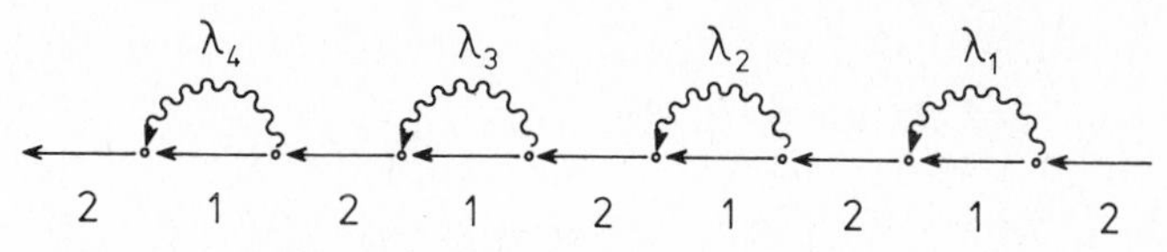

Fig. 7.7. A Feynman diagram of a sequence of virtual photon emission and reabsorption processes (compare text).

to determine the corresponding coefficients in a simple way. Each emission and subsequent absorption process introduces a factor $(-ig_\lambda^*)(-ig_\lambda)$. At each time we have to sum up over the wave vectors of the virtual photons in the intermediate states. Finally we have to perform the successive integrations. Thus we have to evaluate the following expression

$$\sum_{\lambda_1 \ldots \lambda_4} (-1)^4 |g_{\lambda_4}|^2 \ldots |g_{\lambda_1}|^2 \int_0^t \mathrm{d}\tau 8 \int_0^{\tau 8} \mathrm{d}\tau_7 \ldots \int_0^{\tau_2} \mathrm{d}\tau_1 \exp\{\ldots\}, \tag{7.198}$$

where we have indicated the time-dependent factors by dots. We still have to determine the time-dependent functions under the integrals. These functions may be obtained by means of the rules:

Incoming electron	$\exp[-i\varepsilon_2\tau]$
Outgoing electron	$\exp[i\varepsilon_1\tau]$
Outgoing photon	$\exp[i\omega_\lambda\tau]$.

By multiplying these functions with each other, we obtain the factor which occurs at vertex 1

$$\exp\{i\Delta_1 t\}, \tag{7.199}$$

where we have used the abbreviation

$$\Delta_1 = \varepsilon_1 + \omega_\lambda - \varepsilon_2. \tag{7.200}$$

At vertex 1 we identify τ with τ_1 and integrate from 0 till τ_2

$$\int_0^{\tau_2} \exp\{i\Delta_1\tau_1\}\, \mathrm{d}\tau_1 = (i\Delta_1)^{-1}(\exp\{i\Delta_1\tau_2\} - 1). \tag{7.201}$$

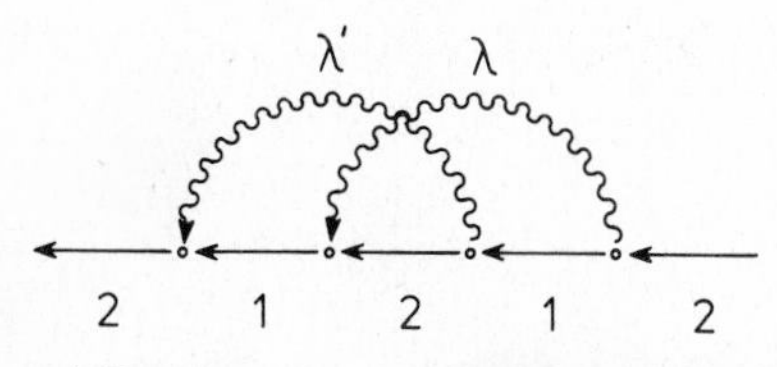

Fig. 7.8. A more complicated Feynman diagram giving a contribution to the self-energy of the electron. Shown is the virtual emission and reabsorption of two photons.

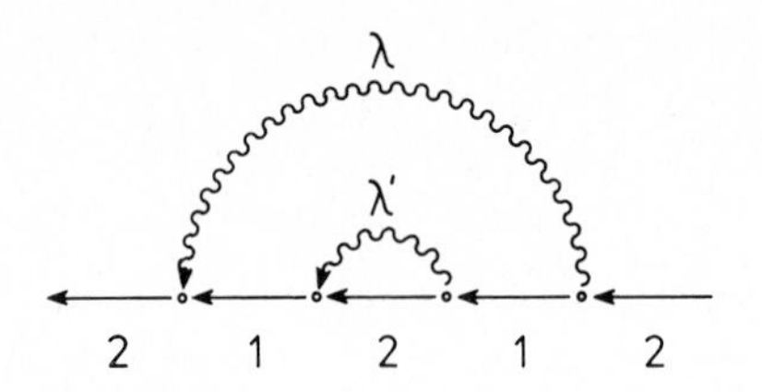

Fig. 7.9. Similar Feynman diagram as in fig. 7.8, but the sequence of reabsorption processes of the two photons is exchanged.

At vertex point 2 the opposite process occurs introducing a factor

$$\exp\{-i\Delta_2\tau_2\}. \tag{7.202}$$

We now have to multiply eq. (7.201) by (7.202) and to integrate over τ_2. This yields

$$(i\Delta_1)^{-1}\int_0^{\tau_3}\exp\{-i\Delta_2\tau_2\}(\exp\{i\Delta_1\tau_2\}-1)\,\mathrm{d}\tau_2=(i\Delta_1)^{-1}\tau_3. \tag{7.203}$$

As usual we neglect oscillatory terms here and in the following because they are inessential for large times. At the next vertex we multiply the thus obtained results by a factor of the type (7.199) with $\tau=\tau_3$ and integrate

$$(i\Delta_1)^{-1}\int_0^{\tau_4}\exp\{i\Delta_2\tau_3\}\tau_3\,\mathrm{d}\tau_3=\frac{\tau_4(\exp\{i\Delta_2\tau_4\}-1)}{i\Delta_1 i\Delta_2}. \tag{7.204}$$

By multiplying the result of eq. (7.204) by the corresponding factor of vertex 4 and again integrating we obtain

$$\int_0^{\tau_5}\exp\{-i\Delta_2\tau_4\}\,\mathrm{d}\tau_4\frac{\tau_4(\exp\{i\Delta_2\tau_4\}-1)}{i\Delta_1 i\Delta_2}=\frac{\tau_5^2}{2}\cdot\frac{1}{i\Delta_1}\cdot\frac{1}{i\Delta_2}. \tag{7.205}$$

It is now obvious how to continue this procedure. After the last integration we obtain

$$\int_0^{t}\exp\{-i\Delta_4\tau_8\}\tau_8^3\,\mathrm{d}\tau_8\frac{1}{3!i\Delta_1 i\Delta_2 i\Delta_3}=\frac{t^4}{4!i\Delta_1 i\Delta_2 i\Delta_3 i\Delta_4}. \tag{7.206}$$

The coefficient of the wave function corresponding to diagram fig. 7.7 with four virtual emission processes can now be formulated by use of eqs. (7.198) and (7.206)

$$\frac{(-it)^4}{4!}\left(\sum_\lambda\frac{|g_\lambda|^2}{\varepsilon_2-\varepsilon_1-\omega_\lambda}\right)^4=\frac{1}{4!}(-i\Delta\varepsilon t)^4. \tag{7.207}$$

The whole wave function $\tilde{\Phi}(t)$ is composed of contributions of different order. These different contributions can be represented by a sum of diagrams of the type of fig. 7.7. We thus have a one-to-one correspondence between diagrams and wave functions:

$$a_2^+\Phi_0\left[1+(-i\Delta\varepsilon t)+\frac{1}{2!}(-i\Delta\varepsilon t)^2+\cdots+\frac{1}{n!}(-i\Delta\varepsilon t)^n+\cdots\right]. \tag{7.208}$$

In each of these diagrams the total wave function remains unaltered after the total process except for the time-dependent factor given in eq. (7.208). In the limit $n\to\infty$ the summation gives us

$$\tilde{\Phi}_2(t)=a_2^+\Phi_0\exp\{-it\Delta\varepsilon\}. \tag{7.209}$$

Thus we have proven that the processes considered give rise to an energy shift as discussed above.

Let us conclude this chapter with an important remark. Besides the diagrams (graphs) considered here quite different kinds of diagrams may also occur. To this end we need only to consider processes described by the succession of interaction operators in eq. (7.197). This discussion may best be performed again by a graphical representation. For instance diagrams of figs. 7.8 and 7.9 can also occur. As can be shown, such diagrams give rise to coefficients proportional to time t so that these processes lead again to an energy shift. The total energy shift can be obtained by summing up all such contributions which are called connected graphs.

Exercises on section 7.7

(1) Consider a three-level atom interacting with the light field (decomposed into modes λ). Let its interaction Hamiltonian be given by

$$g_{12}a_1^+\,a_2b_{\lambda_1}\exp\left[i(\varepsilon_1-\varepsilon_2-\omega_{\lambda_1})t\right]$$

$$+\,g_{23}a_2^+\,a_3b_{\lambda_2}\exp\left[i(\varepsilon_2-\varepsilon_3-\omega_{\lambda_2})t\right]+\text{h.c.}$$

Draw all Feynman graphs of second order taking into account the following initial states

$$\Phi=a_j^+\,\Phi_0,\qquad j=1,2,3;\qquad \Phi=a_j^+\,b_{\lambda_1}^+b_{\lambda_2}^+\Phi_0,\qquad j=1,2,3.$$

(2) Determine the coefficients of

$$a_1^+\,\Phi_0$$

in second order perturbation theory belonging to exercise (1) and to the initial state $a_3^+ b_{\lambda_1}^+ b_{\lambda_2}^+ \Phi_0$. Discuss these coefficients for the case

$$\varepsilon_1 - \varepsilon_2 - \omega_{\lambda_1} = 0, \qquad \varepsilon_2 - \varepsilon_3 - \omega_{\lambda_2} = 0.$$

(3) Is the interaction Hamiltonian eq. (7.111) in the rotating wave approximation sufficient to give rise to the diagrams of figs. 7.8 and 7.9 or must antiresonant terms be included, i.e. the full Hamiltonian $H_{I,1}$ eq. (7.31) (or its equivalent in the interaction representation) be used?

7.8. Lamb shift

In the preceding section we came across the term self-energy. We now turn to the question of the explicit computation of these self-energies. To come closer to reality in this section we abandon the model of a two-level atom and take into account all levels of an atom. In particular, we will be interested in the hydrogen atom. We distinguish the wave functions of the different levels by an index n. Note that in the case of the hydrogen atom n denotes not only the principle quantum number but the total set of all three quantum numbers of the electron. We denote the initial state, which was formerly denoted by 2, by n, and the intermediate states, formerly denoted by 1, by n'. The interaction Hamiltonian in the Schrödinger picture is now taken in the form

$$H_{I,1} = \hbar \sum_{\lambda,n,n'} \{ g_{\lambda,n,n'} b_\lambda^+ a_n^+ a_{n'} + g_{\lambda,n,n'}^* b_\lambda a_{n'}^+ a_n \}, \tag{7.210}$$

which generalizes the two-level Hamiltonian of sections 7.6 and 7.7. Taking into account that we have to sum up over the intermediate states of both the virtually emitted photon and the electron, the energy shift is given by

$$\Delta W_n/\hbar = \sum_{n',\lambda} \frac{|g_{\lambda,n,n'}|^2}{\varepsilon_n - \varepsilon_{n'} - \omega_\lambda}; \tag{7.211}$$

where, as before

$$\hbar\varepsilon_n = W_n.$$

In the following we need the explicit form of the optical matrix elements g which we therefore write down

$$g_{\lambda,n,n'} = -\frac{e}{m}(2\hbar\omega_\lambda\varepsilon_0 V)^{-1/2} \int \varphi_n^*(\boldsymbol{x})(\boldsymbol{e}_\lambda \hat{\boldsymbol{p}})\varphi_{n'}(\boldsymbol{x})\, \mathrm{d}^3\boldsymbol{x}. \qquad (7.212)^*$$

We have used the symbol $\hat{\boldsymbol{p}} = (\hbar/i)\nabla$ for the momentum operator instead

*Note that ε_0 eq. (2.212) is the dielectric constant and is not to be confused with ε_n in eq. (7.211).

of the $\boldsymbol{p}$ used formerly to avoid confusion with the classical momentum $\boldsymbol{p}$. In this we have assumed that the field is expanded into plane waves and that the atom is situated at the origin of the coordinate system. Furthermore the dipole approximation has been made in which we assume that the field amplitude is constant over the extension of the electronic wave functions. To save space we will use Dirac's notation which means that we denote the integral occurring in eq. (7.212) by

$$\int \ldots = \langle n | \boldsymbol{e}_\lambda \hat{\boldsymbol{p}} | n' \rangle . \tag{7.213}$$

Furthermore we have to distinguish the different field modes more carefully. To this end we replace the general mode index λ by the wave vector $\boldsymbol{k}$ and an index j indicating one of the two directions of polarization. Furthermore we use the relation between frequency and wave number

$$\omega_\lambda \equiv \omega_k = ck . \tag{7.214}$$

While we initially start with waves normalized in a volume V we will eventually go over to an integration which is done by the well-known rule

$$\frac{1}{V} \sum_\lambda \to \sum_{j=1,2} \int \frac{\mathrm{d}^3 \boldsymbol{k}}{(2\pi)^3} . \tag{7.215}$$

[Compare eq. (7.130)]. Using eqs. (7.212), (7.213), (7.214), and (7.215) we may cast (7.211) into the form

$$\Delta W_n / \hbar = C \int \mathrm{d}^3 \boldsymbol{k} \frac{1}{\omega_k} \sum_{n',j} \frac{|\langle n | \boldsymbol{e}_j \hat{\boldsymbol{p}} | n' \rangle|^2}{\varepsilon_n - \varepsilon_{n'} - \omega_k} . \tag{7.216}$$

In it we use the abbreviation

$$C = \frac{1}{(2\pi)^3} \frac{e^2}{2m^2 \hbar \varepsilon_0} . \tag{7.217}$$

For further evaluation we split the integral over $\boldsymbol{k}$-space into one over the space angle Ω and one over the magnitude of $\boldsymbol{k}$, in k,

$$\int \mathrm{d}^3 \boldsymbol{k} = \int k^2 \mathrm{d}k \int \mathrm{d}\Omega . \tag{7.218}$$

We then first perform the integration over the space angle and sum up over the two directions of polarizations. Since the evaluation is purely formal we immediately write down the result

$$\int \mathrm{d}\Omega \sum_j |\langle n | \boldsymbol{e}_j \hat{\boldsymbol{p}} | n' \rangle|^2 = 4\pi \tfrac{2}{3} |\langle n | \hat{\boldsymbol{p}} | n' \rangle|^2 . \tag{7.219}$$

This leads us to the following result for the self-energy

$$\Delta W_n/\hbar = \frac{1}{(2\pi)^2}\frac{2e^2}{3m^2\hbar\varepsilon_0 c^3}\int_0^\infty \omega\,\mathrm{d}\omega \sum_{n'} \frac{|\langle n|\,\hat{\boldsymbol{p}}|n'\rangle|^2}{\varepsilon_n - \varepsilon_{n'} - \omega}. \tag{7.220}$$

A detailed discussion of the sum over n' reveals that this sum certainly does not vanish more strongly than $1/\omega$. We thus immediately recognize that the integral over ω in eq. (7.220) diverges which means that the energy shift is infinitely great. This seemingly absurd result presented a great difficulty to theoretical physics. It was overcome by ideas of Bethe, Schwinger and Weisskopf which we will now explain.

When we do similar calculations for free electrons we again find an infinite result, which can be seen as follows. We repeat the whole calculation above but instead of eigenfunctions φ_n of the hydrogen atom we use the wave functions of free electrons

$$\varphi_n(\boldsymbol{x}) \to \varphi_{\boldsymbol{p}} = \mathcal{N}\exp[\,i\boldsymbol{p}\boldsymbol{x}/\hbar\,]. \tag{7.221}$$

Note that in this formula $\boldsymbol{p}$ is a usual vector whereas $\hat{\boldsymbol{p}}$ occurring for instance in eqs. (7.212) and (7.220) is the momentum operator $\hbar/i\,\nabla$. Instead of matrix elements, which were between the eigenstates of the hydrogen atom, we now have to evaluate matrix elements between plane waves. We immediately obtain

$$\langle\, \boldsymbol{p}'|\,\hat{\boldsymbol{p}}|\,\boldsymbol{p}\rangle = \mathcal{N}^2\int \exp[\,-i\boldsymbol{p}'\boldsymbol{x}/\hbar\,]\,\hat{\boldsymbol{p}}\exp[\,i\boldsymbol{p}\boldsymbol{x}/\hbar\,]\,\mathrm{d}^3\boldsymbol{x}, \tag{7.222}$$

and

$$\langle\, \boldsymbol{p}'|\,\hat{\boldsymbol{p}}|\,\boldsymbol{p}\rangle = (\hbar/i)\,\boldsymbol{p}\delta_{\boldsymbol{p},\boldsymbol{p}'}. \tag{7.223}$$

Furthermore, we have to make the substitution

$$\varepsilon_n - \varepsilon_{n'} \to \varepsilon_{\boldsymbol{p}} - \varepsilon_{\boldsymbol{p}'}, \tag{7.224}$$

but we immediately find

$$\varepsilon_{\boldsymbol{p}} - \varepsilon_{\boldsymbol{p}'} = 0 \tag{7.225}$$

on account of eq. (7.223). By putting all the results together we obtain the self-energy of a free electron in the form

$$\Delta W_{\boldsymbol{p}}/\hbar = -\frac{1}{(2\pi)^2}\frac{2e^2}{3m^2\hbar\varepsilon_0 c^3}\,\boldsymbol{p}^2\int_0^\infty \mathrm{d}\omega. \tag{7.226}$$

We notice that the self-energy of a free electron of momentum $\boldsymbol{p}$ is proportional to $\boldsymbol{p}^2$. In adopting our results of section 7.7 we may immediately state that eq. (7.226) can be interpreted as giving rise to a shift in the mass of the electron.

For the sake of clarity, we repeat our former arguments. The energy of the free electron without interaction with electromagnetic field "bare electron" reads

$$W_p = p^2/(2m_0). \tag{7.227}$$

In it, m_0 is the "bare" mass. The energy shift just calculated is

$$\Delta W_p = -\frac{1}{(2\pi)^2}\frac{2e^2}{3m^2\hbar\varepsilon_0 c^3}p^2\int_0^\infty d\omega. \tag{7.228}$$

Thus the total energy reads

$$W_p + \Delta W_p = \frac{p^2}{2m}. \tag{7.229}$$

While the mass of the "bare" electron neglecting electromagnetic interaction is m_0, taking this interaction into account the electron mass is m. Note that in this type of considerations one uses m and not m_0 in eq. (7.228). This follows from the "renormalization" procedure we will describe now.

Since we always make observations on free electrons with the electromagnetic interaction present, eq. (7.229) must be just the expression we normally write down for the energy of a free electron where m is the observed mass. Thus we can make the identification

$$\frac{1}{m} = \frac{1}{m_0} - \frac{1}{(2\pi)^2}\frac{4e^2}{3m^2\hbar\varepsilon_0 c^3}\int_0^\infty d\omega \equiv \frac{1}{m_0} - 2\tilde{a}, \tag{7.230}$$

where $2\tilde{a}$ is merely an abbreviation of the last term of the middle part of equation (7.230). The electromagnetic self-energy can be interpreted as a shift of the mass of an electron from its "bare" value to its observed value m. This shift is called renormalization of the mass.

The argument used in renormalization theory is now as follows. The reason that the result eq. (7.220) is infinite lies in the fact that it includes an infinite energy change that is already counted when we use the observed mass in the Hamiltonian rather than the bare mass. In other words, in fact, we should start with the Hamiltonian for the hydrogen atom in the presence of the radiation field given by

$$H = \frac{p^2}{2m_0} - \frac{e^2}{4\pi\varepsilon_0 r} + H_{\text{int}}. \tag{7.231}$$

Then using eq. (7.230) we can rewrite H as

$$H = \frac{p^2}{2m} - \frac{e^2}{4\pi\varepsilon_0 r} + \{H_{\text{int}} + \tilde{a}p^2\}. \tag{7.232}$$

Thus if we use the observed free particle mass in the expression for the kinetic energy (which we always do) we should not count that part of H_{int} that produces the mass shift, i.e. we should regard

$$H_{\text{int}} + \tilde{a}\boldsymbol{p}^2 \tag{7.233}$$

as the effective interaction of an electron of a renormalized mass m with the radiation field.

Returning then to the calculation of the Lamb shift we see that to first order in $e^2/\hbar c$ we must add the expectation value of the second term in (7.233) to (7.220) in order to avoid counting the electromagnetic interaction twice, once in m and once in H_{int}. Thus more correctly the shift of the level n is given by

$$\Delta W_n'/\hbar = a\int_0^\infty \omega\,\mathrm{d}\omega \left[\sum_{n'} \frac{|\langle n'|\,\hat{\boldsymbol{p}}\,|n\rangle|^2}{\varepsilon_n - \varepsilon_{n'} - \omega} + \frac{\langle n|\,\hat{\boldsymbol{p}}^2\,|n\rangle}{\omega}\right], \tag{7.234}$$

where we used the abbreviation

$$a\int_0^\infty \mathrm{d}\omega = \tilde{a} \equiv \frac{1}{(2\pi)^2}\,\frac{2e^2}{3m^2\hbar\varepsilon_0 c^3}\int_0^\infty \mathrm{d}\omega. \tag{7.235}$$

The second term under the integral in eq. (7.234) can be brought into a form similar to the first term under the integral by means of the relation

$$\langle n|\,\hat{\boldsymbol{p}}^2\,|n\rangle = \sum_{n'} \langle n|\,\hat{\boldsymbol{p}}\,|n'\rangle\langle n'|\,\hat{\boldsymbol{p}}\,|n\rangle. \tag{7.236}$$

In order not to interrupt the main discussion we will postpone the proof of this relation to the end of this section. Using eqs. (7.236) and (7.234) we find after a slight rearrangement of terms

$$\Delta W_n'/\hbar = a\sum_{n'} |\langle n'|\,\hat{\boldsymbol{p}}\,|n\rangle|^2 \int_0^\infty \mathrm{d}\omega \frac{\varepsilon_n - \varepsilon_{n'}}{\varepsilon_n - \varepsilon_{n'} - \omega}. \tag{7.237}$$

We note that the integral over ω is still divergent, however, only logarithmically. This divergence is not present in a more sophisticated relativistic calculation. Such a calculation yields a result quite similar to eq. (7.237), but with an integrand falling off more rapidly at high frequencies $\hbar\omega \gg mc^2$. We can mimic the result of such a calculation by cutting off the integral at $\omega = mc^2/\hbar$. The integral can be immediately performed and yields

$$\Delta W_n'/\hbar = a\sum_{n'} |\langle n'|\,\hat{\boldsymbol{p}}\,|n\rangle|^2 (\varepsilon_{n'} - \varepsilon_n) \ln\left|\frac{mc^2}{W_{n'} - W_n}\right|, \tag{7.238}$$

where we have neglected $|W_n - W_{n'}|$ compared to mc^2. The further evaluation of eq. (7.238) must be done numerically. We first cast eq. (7.238) into a more transparent form by interpreting the sum over n' as an averaging procedure over energy differences. More precisely, we introduce the expression

$$A = \sum_{n'} |\langle n'|\,\hat{\boldsymbol{p}}\,|n\rangle|^2(\varepsilon_{n'} - \varepsilon_n)\ln\left|\frac{W_{n'} - W_n}{mc^2}\right|, \tag{7.239}$$

which we interprete as an average of $\ln|(W_{n'} - W_n)/(mc^2)|$ by means of the weight function

$$|\langle n'|\,\hat{\boldsymbol{p}}\,|n\rangle|^2(\varepsilon_{n'} - \varepsilon_n). \tag{7.240}$$

Note that in general this "weight" may be also negative. With this idea in mind, we may formulate the following average

$$\ln\left|\frac{W_{n'} - W_n}{mc^2}\right|_{\text{av}} = \frac{A}{\sum_{n'} |\langle n'|\,\hat{\boldsymbol{p}}\,|n\rangle|^2(\varepsilon_{n'} - \varepsilon_n)}. \tag{7.241}$$

Clearly, we may now rewrite equation (7.238) in the form

$$\Delta W_n'/\hbar = a\ln\left|\frac{mc^2}{W_{n'} - W_n}\right|_{\text{av}} \sum_{n'} |\langle n'|\,\hat{\boldsymbol{p}}\,|n\rangle|^2(\varepsilon_{n'} - \varepsilon_n). \tag{7.242}$$

To simplify eq. (7.242) further we use the relation

$$\sum_{n'} |\langle n'|\,\hat{\boldsymbol{p}}\,|n\rangle|^2(W_{n'} - W_n) = -\tfrac{1}{2}\langle n|[\,\hat{\boldsymbol{p}},[\,\hat{\boldsymbol{p}},H_0]]|n\rangle, \tag{7.243}$$

which we will prove at the end of this section. The double commutator on the right-hand side of eq. (7.243) is most useful. Such a commutator can be easily evaluated. We assume H_0 in the form

$$H_0 = -\frac{\hbar^2}{2m}\Delta + V(\boldsymbol{x}). \tag{7.244}$$

We readily obtain (cf. exercise 6 on section 3.2)

$$[\,\hat{p}_x, H_0] = \frac{\hbar}{i}\frac{\partial V(\boldsymbol{x})}{\partial x} \tag{7.245}$$

and in a similar fashion

$$[\,\hat{\boldsymbol{p}},[\,\hat{\boldsymbol{p}},H_0]] = -\hbar^2\Delta V(\boldsymbol{x}). \tag{7.246}$$

Using for V the Coulomb potential of the electron in the hydrogen atom $V = -[e^2/(4\pi\varepsilon_0|\boldsymbol{x}|)]$ we can readily evaluate the right-hand side of eq. (7.246). Using a formula well known from electrostatics (potential of a point charge!) we find

$$\Delta\frac{1}{|\boldsymbol{x}|} = -4\pi\delta(\boldsymbol{x}), \tag{7.247}$$

where $\delta(\boldsymbol{x})$ is Dirac's function in three dimensions. Using this result and the definition of bra and kets (compare section 3.2), we readily obtain

$$(7.243) = \frac{\hbar^2}{2}\left\langle n\left|\Delta\frac{-e^2}{4\pi\varepsilon_0|x|}\right|n\right\rangle = \frac{e^2\hbar^2}{2\varepsilon_0}\int|\varphi_n(\boldsymbol{x})|^2\delta(\boldsymbol{x})\,\mathrm{d}^3\boldsymbol{x} \tag{7.248}$$

and making use of the properties of the δ-function

$$(7.248) = \frac{e^2\hbar^2}{2\varepsilon_0}|\varphi_n(0)|^2. \tag{7.249}$$

We are now in a position to write down the final formula for the renormalized self-energy shift by inserting the result (7.246) with (7.243) and (7.249) into (7.242). We then obtain

$$\Delta W_n'/\hbar = \frac{1}{(2\pi)^2}\frac{2e^2}{3m^2\hbar\varepsilon_0 c^3}\ln\left|\frac{mc^2}{W_{n'}-W_n}\right|_{\text{av}}\frac{e^2\hbar^2}{2\varepsilon_0}|\varphi_n(0)|^2. \tag{7.250}$$

To obtain final numerical results we have to calculate numerically the average eq. (7.241) as well as $|\varphi_n(0)|^2$. For the hydrogen atom $|\varphi_n(0)|^2$ is well known and is nonzero only for s-states. The average value was calculated by Bethe for the $2S$ level. Inserting all the numerical values we eventually find $\Delta W_n'/\hbar = 1040$ megacycles. According to these considerations a shift between an S and a P level must be expected. Such a shift was first discovered between the $2S_{1/2}$ and $2P_{1/2}$ level of hydrogen by Lamb and Retherford. A fully quantum-dynamic calculation gives excellent agreement with the experiments. While Lamb and Retherford used microwaves to measure the splitting, more recently Hänsch was able to measure the splitting by optical high resolution spectroscopy.

At first sight, it may seem strange that it is possible to obtain reasonable results by a subtraction procedure in which two infinitely large quantities are involved. However, it has turned out that such a subtraction procedure can be formulated in the frame of a beautiful theory, called renormalization, and such procedures are now a legitimate part of theoretical physics giving excellent agreement between theory and experiment. Unfortunately it is beyond the scope of our introductory book to cover the details of these renormalization techniques.

In conclusion we briefly present the proof of two auxiliary formulas which we have used above. To this end we write the bracket symbols again as integrals and start with the right-hand side of equation (7.236).

$$\sum_{n'}\int\varphi_n^*(\boldsymbol{x})\frac{\hbar}{i}\nabla\varphi_{n'}(\boldsymbol{x})\,\mathrm{d}^3\boldsymbol{x}\int\varphi_{n'}^*(\boldsymbol{x}')\frac{\hbar}{i}\nabla'\varphi_n(\boldsymbol{x}')\,\mathrm{d}^3\boldsymbol{x}'. \tag{7.251}$$

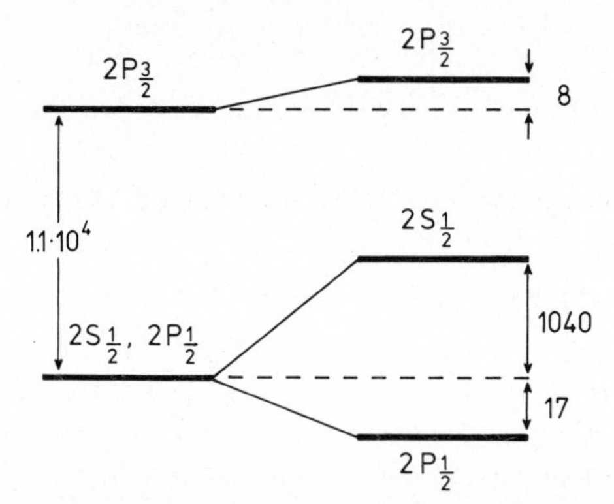

Fig. 7.10. Lamb-shift for the electron states with $n = 2$ in hydrogen. The left part shows the splitting due to spin-orbit coupling. According to Dirac's equation, the energy levels belonging to the $2S_{1/2}$ and $2P_{1/2}$ states should coincide. In reality, they are split (Lamb-shift). The r.h.s. of the figure shows the splitting based on calculations which take into account the mass renormalization and some further, though minor effects treated by quantum electrodynamics. The numbers indicate the corresponding frequencies in MHz.

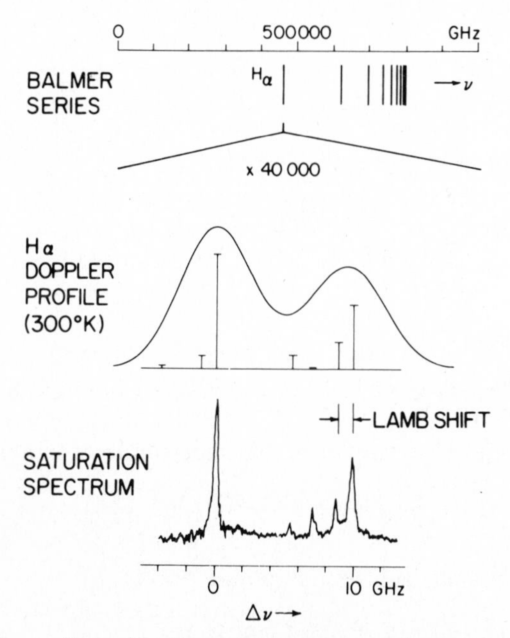

Fig. 7.11. By modern methods of nonlinear spectroscopy, namely saturation spectroscopy (cf. the 3rd volume), the Lamb-shift can be measured optically. This figure shows the saturation spectrum of the red Balmer line in atomic hydrogen. The splitting of the levels is indicated in the spectrum. [After T. W. Hänsch, I. S. Shahin,and A. L. Schawlow, Nature 235 (1972) 63; T. W. Hänsch, A. L. Schawlow and P. Toschek, IEEE J. Quant. Electr. QE-8 (1972) 802.]

Rearranging summation and integration we obtain

$$\int\int \mathrm{d}^3x\,\mathrm{d}^3x'\varphi_n^*(\boldsymbol{x})\frac{\hbar}{i}\nabla\left\{\sum_{n'}\varphi_{n'}(\boldsymbol{x})\varphi_{n'}^*(\boldsymbol{x}')\right\}\frac{\hbar}{i}\nabla'\varphi_n(\boldsymbol{x}'). \tag{7.252}$$

Since the functions φ_n form, in the mathematical sense, a complete set, the following relation is proved in mathematics

$$\sum_{n'}\varphi_{n'}(\boldsymbol{x})\varphi_{n'}^*(\boldsymbol{x}') = \delta(\boldsymbol{x}-\boldsymbol{x}'), \tag{7.253}$$

where δ is Dirac's function. This relation is often called "completeness" relation. Inserting eq. (7.253) into (7.252) allows us to perform the integration over $\boldsymbol{x}'$ immediately using the δ function so that (7.252) reduces to

$$\int \mathrm{d}^3x\,\varphi_n^*(\boldsymbol{x})\left(\frac{\hbar}{i}\nabla\right)^2\varphi_n(\boldsymbol{x})\,\mathrm{d}V = \langle n|\,\hat{\boldsymbol{p}}^2|n\rangle. \tag{7.254}$$

But this is evidently the left-hand side of equation (7.236).

We now prove eq. (7.243) and study the expression

$$\sum_{n'}\langle n|\,\hat{\boldsymbol{p}}|n'\rangle\langle n'|\,\hat{\boldsymbol{p}}|n\rangle W_{n'}. \tag{7.255}$$

We use the fact that $\varphi_{n'}$ is eigenfunction to the Hamiltonian H_0

$$H_0\varphi_{n'} = W_{n'}\varphi_{n'}. \tag{7.256}$$

With its aid we can immediately evaluate

$$\begin{aligned}\langle n|\,\hat{\boldsymbol{p}}H_0|n'\rangle &= \int\varphi_n^*(\boldsymbol{x})\,\hat{\boldsymbol{p}}H_0\varphi_{n'}(\boldsymbol{x})\,\mathrm{d}^3x\\ &= W_{n'}\langle n|\,\hat{\boldsymbol{p}}|n'\rangle,\end{aligned} \tag{7.257}$$

which allows us to write (7.255) in the form

$$\sum_{n'}\langle n|\,\hat{\boldsymbol{p}}|n'\rangle\langle n'|\,\hat{\boldsymbol{p}}|n\rangle W_{n'} = \sum_{n'}\langle n|\,\hat{\boldsymbol{p}}H_0|n'\rangle\langle n'|\,\hat{\boldsymbol{p}}|n\rangle. \tag{7.258}$$

Using the completeness relation again quite similarly as above we obtain

$$\sum_{n'}\langle n|\,\hat{\boldsymbol{p}}H_0|n'\rangle\langle n'|\,\hat{\boldsymbol{p}}|n\rangle = \langle n|\,\hat{\boldsymbol{p}}H_0\,\hat{\boldsymbol{p}}|n\rangle. \tag{7.259}$$

In the same way the completeness relation yields

$$\sum_{n'}\langle n|\,\hat{\boldsymbol{p}}|n'\rangle\langle n'|\,\hat{\boldsymbol{p}}|n\rangle W_n = W_n\langle n|\,\hat{\boldsymbol{p}}^2|n\rangle. \tag{7.260}$$

Thus we can rewrite the left-hand side of eq. (7.243) in the form

$$\langle n|\,\hat{\boldsymbol{p}}(H_0 - W_n)\,\hat{\boldsymbol{p}}|n\rangle. \tag{7.261}$$

Furthermore, by using the definition of brackets and using partial integrations we can readily establish the following relations

$$\left.\begin{aligned}\langle n|\,\hat{p}W_n\,\hat{p}|n\rangle &= \int \varphi_n^*(x)\frac{\hbar}{i}\nabla\, W_n\frac{\hbar}{i}\nabla\,\varphi_n(x)\,\mathrm{d}^3x \\ &= \int\left[\left(-\frac{\hbar}{i}\nabla\right)^2\varphi_n^*(x)\right]W_n\varphi_n(x)\,\mathrm{d}^3x \\ &= \int\left[\left(-\frac{\hbar}{i}\nabla\right)^2\varphi_n^*(x)\right]H_0\varphi_n(x)\,\mathrm{d}^3x \\ &= \int \varphi_n^*(x)\,\hat{p}^2H_0\varphi_n(x)\,\mathrm{d}^3x \\ &= \langle n|\,\hat{p}^2H_0|n\rangle .\end{aligned}\right\} \tag{7.262}$$

Similarly, we obtain

$$\begin{aligned}\int (H_0\varphi_n^*(x))\,\hat{p}^2\varphi_n(x)\,\mathrm{d}^3x &= \int \varphi_n^*(x)H_0\,\hat{p}^2\varphi_n(x)\,\mathrm{d}^3x \\ &= \langle n|H_0\,\hat{p}^2|n\rangle .\end{aligned} \tag{7.263}$$

Writing out the double commutators on the right-hand side of eq. (7.243) we obtain

$$-\tfrac{1}{2}\langle n|\,\hat{p}^2H_0 + H_0\,\hat{p}^2|n\rangle + \langle n|\,\hat{p}H_0\,\hat{p}|n\rangle . \tag{7.264}$$

Using the results in eqs. (7.259), (7.260), (7.262), and (7.263) we can easily verify the relation (7.243).

7.9. Once again spontaneous emission: Damping and line–width

In section 7.6 we solved the Schrödinger equation of a single atom interacting with the radiation field by first-order perturbation theory. Let us again inspect our former solution (7.120), which has the form

$$\tilde{\Phi}(t) = a_2^+\Phi_0 + \sum_\lambda c_\lambda(t)a_1^+b_\lambda^+\Phi_0 . \tag{7.265}$$

In the foregoing sections we focussed our attention on the sum over λ. Let us now consider the first term, $a_2^+\,\Phi_0$. According to it, the electron ought to remain all the time in the upper level, 2. Obviously, this result cannot be correct! We know experimentally that the number of electrons of an ensemble of atoms in the upper state, 2, (in quantum mechanics we always deal with ensembles) decays exponentially with time. This makes it necessary to seek a more exact solution of the Schrödinger equation. We choose, as in section 7.6, a two-level atom, but work in the Schrödinger representation. The corresponding Schrödinger equation which we have already met in section 7.5, eq. (7.74), reads

$$i\hbar \frac{\mathrm{d}\Phi}{\mathrm{d}t} = \left\{ W_1 a_1^+ a_1 + W_2 a_2^+ a_2 + \sum_\lambda \hbar\omega_\lambda b_\lambda^+ b_\lambda \right.$$
$$\left. + \hbar \sum_\lambda \left(g_\lambda a_1^+ a_2 b_\lambda^+ + g_\lambda^* a_2^+ a_1 b_\lambda \right) \right\} \Phi. \tag{7.266}$$

We now try the ansatz

$$\Phi(t) = A(t) a_2^+ \Phi_0 \exp[-i\varepsilon_2 t]$$
$$+ \sum_\lambda c_\lambda(t) b_\lambda^+ a_1^+ \Phi_0 \exp[-i(\omega_\lambda + \varepsilon_1)t], \quad \varepsilon_j = W_j/\hbar. \; A(t) \tag{7.267}$$

and $c_\lambda(t)$ are considered as unknown coefficients still to be determined. By inserting it into eq. (7.266) and comparing the coefficients of the linearly independent components $a_2^+\Phi_0$ and $b^+a_1^+\Phi_0$ on both sides yields

$$\frac{\mathrm{d}A}{\mathrm{d}t} = -i \sum_\lambda c_\lambda g_\lambda^* \exp[i(\bar{\omega} - \omega_\lambda)t], \tag{7.268}$$

$$\frac{\mathrm{d}c_\lambda}{\mathrm{d}t} = -i g_\lambda A \exp[-i(\bar{\omega} - \omega_\lambda)t], \tag{7.269}$$

where

$$\bar{\omega} \equiv \omega_{21} = \varepsilon_2 - \varepsilon_1. \tag{7.269a}$$

Since we expect that the occupation number of the upper level decays exponentially, we put $A = \exp[-\gamma t]$. By inserting this into eq. (7.269), we can perform the integration and find

$$c_\lambda = -i g_\lambda \frac{1 - \exp[i(\omega_\lambda - \bar{\omega})t - \gamma t]}{\gamma - i(\omega_\lambda - \bar{\omega})}. \tag{7.270}$$

By inserting this result and $A = \exp[-\gamma t]$ into eq. (7.268) we obtain

$$\gamma = \sum_\lambda |g_\lambda|^2 \frac{\exp[\gamma t - i(\omega_\lambda - \bar{\omega})t] - 1}{\gamma - i(\omega_\lambda - \bar{\omega})}. \tag{7.271}$$

This result seems to imply a contradiction, because γ on the l.h.s. is a constant, whereas the r.h.s. still depends on time. This puzzle can be solved as follows. Since we are only interested in time intervals $t \gg 1/\bar{\omega}$ and since it will turn out afterwards that $\gamma \ll \bar{\omega}$, we may use on the r.h.s. of eq.

(7.271) the mathematical relation (cf. also the mathematical appendix)

$$\lim_{t\to\infty} \frac{1-\exp[i(\bar{\omega}-\omega)t]}{\bar{\omega}-\omega} = P\frac{1}{\bar{\omega}-\omega} - i\pi\delta(\bar{\omega}-\omega). \tag{7.272}$$

This is a formal relation making sense only under an integral. δ is again Dirac's function, defined in section 7.6. P means principal value. It is defined with respect to an integral as follows

$$P\int_{-\infty}^{\infty} \frac{1}{\bar{\omega}-\omega} f(\omega)\,\mathrm{d}\omega = \lim_{\varepsilon\to 0}\left\{\int_{-\infty}^{\bar{\omega}-\varepsilon} \frac{f(\omega)\,\mathrm{d}\omega}{\bar{\omega}-\omega} + \int_{\bar{\omega}+\varepsilon}^{\infty} \frac{f(\omega)\,\mathrm{d}\omega}{\bar{\omega}-\omega}\right\}. \tag{7.273}$$

It thus cuts the "point" $\omega = \bar{\omega}$ out of the integral. Inserting the δ-function part into eq. (7.271) yields an expression identical with that for the total transition probability P, eq. (7.127), besides a factor of 2, so that

$$2\gamma = P. \tag{7.274}$$

We thus find that the inverse lifetime τ of the upper level is connected with the transition probability P by

$$\frac{1}{\tau} = P \tag{7.275}$$

We shall see later, in section 8.3, that the exponential decay of the upper level causes a finite width of the spontaneous emission line.

The principal part of eq. (7.272) causes an imaginary contribution to γ, i.e. a level shift of the upper atomic level, 2. This shift is proportional to

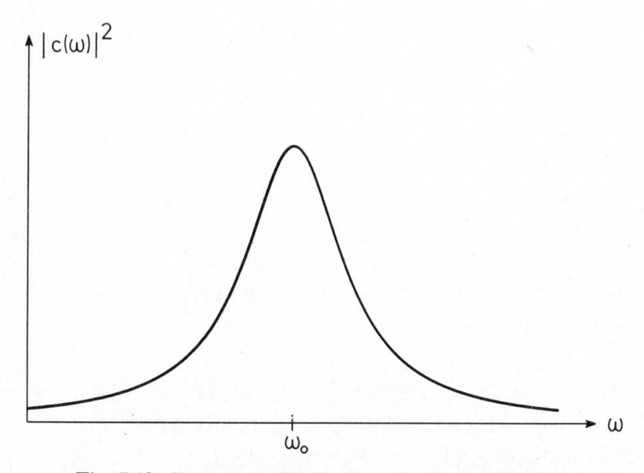

Fig. 7.12. Frequency distribution of a Lorentzian line.

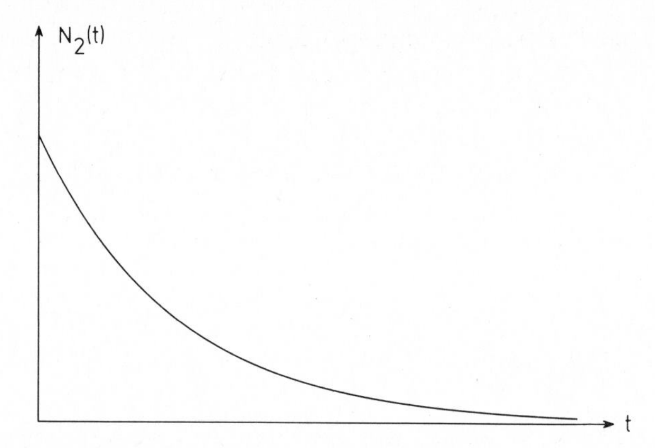

Fig. 7.13. Experimental decay of the excited electronic level as a function of time. The curves of fig. 7.12 and 7.13 are Fourier transforms of each other.

$|g_\lambda|^2$ and thus stems from the emission and reabsorption of a photon (compare sections 7.7 and 7.8). This shift is a contribution to the Lamb shift. It is only a contribution, because in reality transitions to other atomic levels are also involved.

Exercises on section 7.9

(1) Verify the statements made in section 1.10, ch. 1 on the line width and photon energy measurements by means of the results of the present section.
Hint: Use eq. (7.270), and the probability interpretation of quantum theory. Let $t \to \infty$.

(2) Verify the statements made in section 1.13 on the decay of the mean occupation number $N_2 = \langle \Phi | a_2^+ a_2 | \Phi \rangle$.

(3) Discuss section 1.12 using the results of the present section.

7.10. How to return to the semiclassical approach. Example: A single mode, absorption and emission

In this book we have encountered two ways of dealing with optical transitions in atoms. In chapter 4 we treated the atom quantum mechanically and the field classically. In that treatment the field was an externally prescribed quantity. On the other hand, in this chapter, 7, we treated both the atoms and the field according to quantum theory. We expect that in the limit of large photon numbers the results of both treatments must coincide. As we have seen at several occasions earlier a quantum mechanical coherent field has many properties in common with a classical field. Therefore we expect that a formulation using coherent fields will enable us to establish this connection we have in mind. Quite generally, in the interaction representation we have

$$\frac{\mathrm{d}}{\mathrm{d}t}\tilde{\Phi} = \frac{1}{i\hbar}\tilde{H}_I\tilde{\Phi}. \tag{7.276}$$

As an example we use the interaction Hamiltonian for a single light mode and a 2-level atom

$$\tilde{H}_I = a_2^+ a_1 bg \exp[i(\overline{\omega} - \omega)t] + a_1^+ a_2 b^+ g^* \exp[-i(\overline{\omega} - \omega)t],$$
$$\overline{\omega} \equiv \omega_{21}. \tag{7.277}$$

We now make the substitution for Φ

$$\tilde{\Phi} = U\hat{\Phi}, \tag{7.278}$$

where U is given by

$$U = \exp[\beta b^+ - \beta^* b]. \tag{7.279}$$

When U acts on the vacuum state this new state describes the coherent field. By inserting eq. (7.278) into eq. (7.276) we obtain

$$U\frac{\mathrm{d}}{\mathrm{d}t}\hat{\Phi} = \frac{1}{i\hbar}\tilde{H}_I U\hat{\Phi}, \tag{7.280}$$

or after multiplication of (7.280) on both sides by U^{-1}

$$\frac{\mathrm{d}}{\mathrm{d}t}\hat{\Phi} = \frac{1}{i\hbar}\hat{H}_I\hat{\Phi}. \tag{7.281}$$

In it we have used the abbreviation

$$\hat{H}_I = U^{-1}\tilde{H}_I U. \tag{7.282}$$

In section 5.3 we derived the relations

$$\exp[-\beta b^+ + \beta^* b]\begin{pmatrix} b \\ b^+ \end{pmatrix}\exp[\beta b^+ - \beta^* b] = \begin{pmatrix} b + \beta \\ b^+ + \beta^* \end{pmatrix}. \tag{7.283}$$

Making use of them in eq. (7.282) we can already evaluate $\hat{H}_I$ and find

$$\hat{H}_I = a_2^+ a_1 \beta g \exp[i(\overline{\omega} - \omega)t]$$
$$+ a_1^+ a_2 \beta^* g^* \exp[-i(\overline{\omega} - \omega)t] + \hat{H}_q. \tag{7.784}$$

Here $\hat{H}_q$ is identical with the original quantum mechanical interaction Hamiltonian $\hat{H}_q \equiv \tilde{H}_I$. The transformation eq. (7.279) thus transforms the original interaction Hamiltonian into two parts. Now one contains a classical field amplitude β, β^* and the other is still quantum mechanical. When we consider the definition of g etc. one can be convinced after some calculations that the following relations hold

$$\beta g \exp[i\bar{\omega}t] = \vartheta_{21}\boldsymbol{E}^{(+)}(t)$$

$$\beta^* g^* \exp[-i\bar{\omega}t] = \vartheta_{21}^* \boldsymbol{E}^{(-)}(t), \tag{7.285}$$

where ϑ_{21} is the optical dipole matrix element and $\boldsymbol{E}^{(+)}, \boldsymbol{E}^{(-)}$ are the positive and negative frequency parts, respectively, of the classical electric field strength. Thus we may rewrite the first part of eq. (7.284) in the form

$$\hat{H}_{\mathrm{el}} = a_2^+ a_1 \vartheta_{21} \boldsymbol{E}^{(+)}(t) + a_1^+ a_2 \vartheta_{21}^* \boldsymbol{E}^{(-)}(t). \tag{7.286}$$

We now see that we can do a perturbation theory quite similarly as in chapter 4 where the electrons are treated quantum mechanically but the field occurs as classical quantity. In this approach it is assumed that the field amplitude is so high that we can neglect the quantum fluctuations described by the quantum mechanical part of eq. (7.284), i.e. $\hat{H}_q$. A complete discussion of the impact of transformations such as eq. (7.278) on the physical content must include not only the Schrödinger equation but also expectation values and matrix elements. Because the field operators b^+, b commute with the electron operators a^+, a, one readily establishes that all quantities involving only electronic matrix elements are unchanged under the transformation (7.278). We leave it to the reader as an exercise to extend these considerations to many field modes.

7.11. The dynamic Stark effect*

The purpose of this chapter is as follows. A two-level atom is exposed to a coherent resonant driving field. What happens with the spectrum of its spontaneous emission? We shall see that under high enough driving fields the atomic line will split into three parts (cf. fig. 7.14). The splitting of spectral lines under the influence of static fields is well known. For instance a high enough static electric field causes the Stark effect. In the present case, the line splitting is caused by an oscillating electromagnetic field, so that the term "dynamic Stark" was coined to describe this

*The topic treated in this section is still under intense experimental and theoretical study. While some authors (see references at the end of this book) use methods similar to those presented in this section, others use the density matrix approach (for more details on density matrices see section 9.5). While these approaches give the same results concerning the size of the line-splitting, the results differ what the ratio of the heights of the intensity peaks is concerned. Possibly, the assumption made below, after eq. (7.298), might be still too strong.

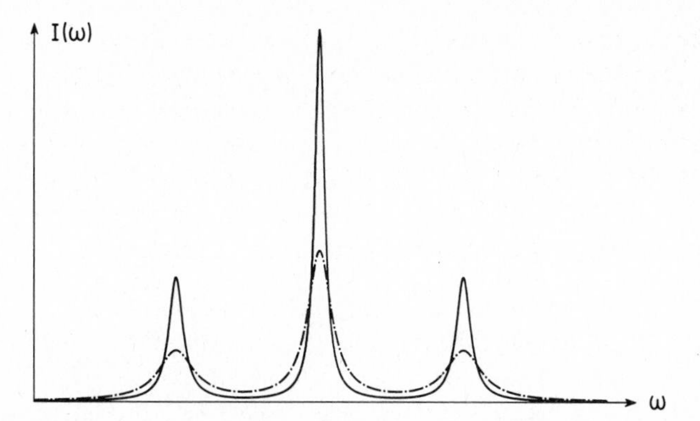

Fig. 7.14a. Typical spectra of the emitted radiation for the dynamic Stark-effect. The solid and dot-dashed lines correspond to smaller and larger atomic line widths, respectively. (Theoretical results by Gardiner, private communication.)

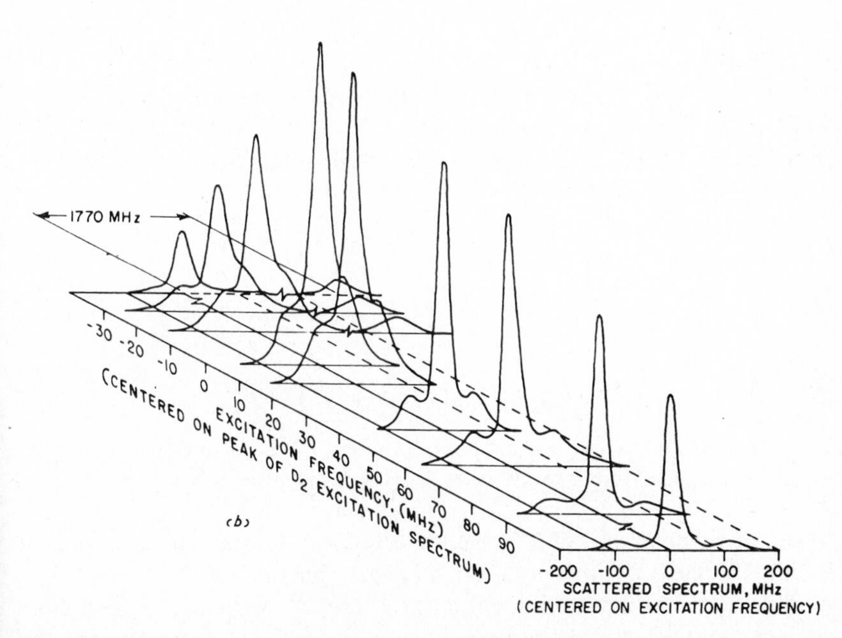

Fig. 7.14b. This figure shows the intensity (ordinate) of the emitted light versus frequency (abscissa to the right) for various detunings of the exciting laser light. The experimental results were obtained for the hyperfine transition $F = 2 \rightarrow 3$ of the sodium D_2 line by F. Schuda, C. R. Stroud, Jr. and M. Hercher, J. Phys. B7 (1974) L 198.

phenomenon. This terminology is sometimes considered not as adequate because the mechanisms of the two effects are rather different.

We start with the Schrödinger equation in the interaction representation in agreement with section 7.4. Since the mode of the driving field will be highly excited we write the corresponding Schrödinger equation a little bit differently by treating the creation and annihilation operator of the driving field mode, i.e. b_0^+, b_0, separately from the other modes

$$\frac{\mathrm{d}}{\mathrm{d}t}\tilde{\Phi} = -ig\left\{a_1^+ a_2\left(\sum_\lambda{}' b_\lambda^+ \exp[i\Delta_\lambda t] + b_0^+\right) + a_2^+ a_1\left(\sum_\lambda{}' b_\lambda \exp[-i\Delta_\lambda t] + b_0\right)\right\}\tilde{\Phi}, \tag{7.287}$$

where

$$\Delta_\lambda = \omega_\lambda - \bar{\omega}. \tag{7.288}$$

For simplicity, but without losing any essentials, we choose g independent of λ and real. The prime at the sums indicates that the sums are taken without $\lambda = 0$. As in the foregoing chapter we put

$$\tilde{\Phi} = U\hat{\Phi}, \tag{7.289}$$

with

$$U = \exp[b_0^+ \beta - \beta^* b_0]. \tag{7.290}$$

Making use of the relations (cf. section 5.3 including exercise 3)

$$U^+ b_0 U = b_0 + \beta \tag{7.291}$$

and

$$U^+ b_0^+ U = b_0^+ + \beta^* \tag{7.292}$$

we immediately find

$$\frac{\mathrm{d}}{\mathrm{d}t}\hat{\Phi} = -ig\left\{a_1^+ a_2\left(\sum_\lambda b_\lambda^+ \exp[i\Delta_\lambda t] + \beta^*\right) + a_2^+ a_1\left(\sum_\lambda b_\lambda \exp[-i\Delta_\lambda t] + \beta\right)\right\}\hat{\Phi}. \tag{7.293}$$

When we compare eq. (7.293) with the original Schrödinger equation of section 7.6 where we treated the spontaneous emission of a two-level atom, we recognize that new additional terms β and β^* occur which stem from the external field. It is tempting to make a hypothesis for $\hat{\Phi}$ similar to the Weisskopf-Wigner theory which we treated in section 7.9. However, some care must be exercised here. To see this, let us start with an initial wave function with no photon present and the electron in the excited state

$$a_2^+ \Phi_0. \tag{7.294}$$

When we insert (7.294) for $\hat{\Phi}$ in the sense of perturbation theory and let the operators in front of $\hat{\Phi}$ act on it, we immediately realize, that (7.294) is transformed into linear combinations of the function

$$a_1^+ \Phi_0 \tag{7.295}$$

and of functions

$$a_1^+ b_\lambda^+ \exp[i\Delta_\lambda t] \Phi_0. \tag{7.296}$$

In the next step we insert (7.295) and (7.296) into the right-hand side of (7.293). We then see that (7.296) is transformed into

$$a_2^+ b_\lambda^+ \exp[i\Delta_\lambda t] \Phi_0, \tag{7.297}$$

which means that we now also find states in which the electron has returned to its upper state but a photon is present in addition. In the same way the function (7.297) gives rise to new terms of the form

$$a_1^+ b_\lambda^+ b_{\lambda'}^+ \exp[i\Delta_\lambda t + i\Delta_{\lambda'} t] \Phi_0. \tag{7.298}$$

Here we have to deal with two additional photons. When continuing this procedure we will obtain wave functions with an increasing number of photons. Thus the system of equations is not closed. To turn the problem into one which we can solve explicitly we will neglect the terms of the form of eq. (7.298) and those containing still more photons. This procedure can be justified when we assume that the production rate of photons is not too big and if the photons can escape quickly enough out of the system under consideration. As we have seen, an initial function (7.294) generates additional functions (7.295), (7.296), and (7.297). This leads us to make the following hypothesis for the solution of eq. (7.293):

$$\begin{aligned} \hat{\Phi} = \Big\{ & D_1 a_1^+ + \sum_\lambda D_{1,\lambda} \exp[i\Delta_\lambda t] b_\lambda^+ a_1^+ \\ & + D_2 a_2^+ + \sum_\lambda D_{2,\lambda} \exp[i\Delta_\lambda t] b_\lambda^+ a_2^+ \Big\} \Phi_0. \end{aligned} \tag{7.299}$$

In it, the coefficients D are still unknown functions of time. Our further procedure is now quite analogous to that of section 7.9. We insert eq. (7.299) into (7.293) and let the operators act on $\hat{\Phi}$. Then we collect terms describing the same electron-photon state. Since the corresponding states are orthogonal to each other we can fulfill the corresponding equations only by putting the coefficients of each function equal to zero. This yields the equations

$$\frac{\mathrm{d}}{\mathrm{d}t} D_1 = -igD_2\beta^*, \tag{7.300}$$

$$\frac{\mathrm{d}}{\mathrm{d}t} D_2 = -igD_1\beta - ig\sum_{\lambda} D_{1,\lambda}, \tag{7.301}$$

$$\frac{\mathrm{d}}{\mathrm{d}t} D_{1,\lambda} = -i\Delta_\lambda D_{1,\lambda} - igD_2 - igD_{2,\lambda}\beta^*, \tag{7.302}$$

$$\frac{\mathrm{d}}{\mathrm{d}t} D_{2,\lambda} = -i\Delta_\lambda D_{2,\lambda} - igD_{1,\lambda}\beta. \tag{7.303}$$

Since we have an infinite number of modes λ, eqs. (7.300)–(7.303) represent a set of ordinary linear differential equations with constant coefficients. The reader, who is not so much interested in the details of the further calculations, may skip the following and continue after eq. (7.328).

Here we want to show how the equations (7.300)–(7.303) can be solved in an elegant fashion. To this end we make use of the Laplace transformation. Since not all readers might be familiar with this method, we briefly explain it. Let us consider a time-dependent function $x(t)$. We then define the Laplace transform $\mathfrak{L}_s$ by the relation

$$\mathfrak{L}_s[x] = \int_0^\infty \exp[-st]x(t)\,\mathrm{d}t. \tag{7.304}$$

It must be assumed, of course, that $x(t)$ guarantees the existence of the integral on the right-hand side. When we choose $x = 1$ we immediately find its Laplace transform to be

$$\mathfrak{L}_s[1] = \frac{1}{s}. \tag{7.305}$$

Furthermore we use the identity

$$x(t) = x(0) + \int_0^t \mathrm{d}t' \frac{\mathrm{d}x(t')}{\mathrm{d}t'} \tag{7.306}$$

which can be immediately verified by evaluating the integral. Taking the Laplace transform in eq. (7.306) on both sides we find

$$\mathfrak{L}_s[x] = \frac{x(0)}{s} + \int_0^\infty \mathrm{d}t \exp[-st] \int_0^t \mathrm{d}t' \frac{\mathrm{d}x(t')}{\mathrm{d}t'} \tag{7.307}$$

or, after a slight rearrangement

$$s\mathfrak{L}_s[x] = x(0) + \mathfrak{L}_s\left[\frac{\mathrm{d}x}{\mathrm{d}t}\right]. \tag{7.308}$$

This formula allows us immediately to determine the Laplace transform of the derivative of x by the Laplace transform of x.

To apply this formalism we have to identify x with any of the Ds occurring in eqs. (7.300)–(7.303). For abbreviation we further put

$$\mathfrak{L}_s[D_j] = d_{j,s}. \tag{7.309}$$

Taking then the Laplace transform of the equations (7.300)–(7.303) we readily obtain

$$sd_{1,s} = D_1(0) - ig\beta^* d_{2,s}, \tag{7.310}$$

$$sd_{2,s} = D_2(0) - ig\beta d_{1,s} - ig\sum_\lambda d_{1,\lambda,s}, \tag{7.311}$$

$$sd_{1,\lambda,s} = D_{1,\lambda}(0) - i\Delta_\lambda d_{1,\lambda,s} - igd_{2,s} - ig\beta^* d_{2,\lambda,s}, \tag{7.312}$$

$$sd_{2,\lambda,s} = D_{2,\lambda}(0) - i\Delta_\lambda d_{2,\lambda,s} - ig\beta d_{1,\lambda,s}. \tag{7.313}$$

As an initial condition, we choose that at time $t = 0$, the atom is in an excited state and no photon is present. This implies

$$D_2(0) = 1;\ D_1(0) = D_{1,\lambda}(0) = D_{2,\lambda}(0) = 0. \tag{7.314}$$

Equations (7.312) and (7.313) allow us to express $d_{1,\lambda,s}$ and $d_{2,\lambda,s}$ by $d_{2,s}$

$$d_{1,\lambda,s} = \frac{-ig(s + i\Delta_\lambda)}{(s + i\Delta_\lambda)^2 + g^2|\beta|^2} d_{2,s}, \tag{7.315}$$

$$d_{2,\lambda,s} = \frac{-ig(-ig\beta)}{(s + i\Delta_\lambda)^2 + g^2|\beta|^2} d_{2,s}. \tag{7.316}$$

By inserting eq. (7.315) into eq. (7.311) and then solving the eqs. (7.310) and (7.311) yields

$$d_{2,s} = \frac{sD_2(0)}{s(s + \Gamma(s)) + g^2|\beta|^2}. \tag{7.317}$$

Now (7.315) and (7.316) are determined explicitly. $\Gamma(s)$ is the following abbreviation

$$\Gamma(s) = \sum_\lambda \frac{g^2(s + i\Delta_\lambda)}{(s + i\Delta_\lambda)^2 + g^2|\beta|^2}. \tag{7.318}$$

$\Gamma(s)$ can be decomposed into

$$\Gamma(s) = \frac{g^2}{2} \sum_\lambda \left\{ \frac{1}{s + i\Delta_\lambda + ig|\beta|} + \frac{1}{s + i\Delta_\lambda - ig|\beta|} \right\}. \tag{7.319}$$

When we use the relation (A.19) of the mathematical appendix we see that the line width and frequency shift resulting from $\Gamma(s)$ are the same as those of eq. (7.271), where (7.272) is used, provided $s \approx 0$ and $g(\beta)$ is absorbed in Δ_λ by a change of the summation variable. As will turn out below in a self-consistent fashion, the important contributions to the final result, are those in which s is small compared to ω. Therefore in the following we will use

$$\Gamma(s) = \gamma + i\Omega, \tag{7.320}$$

where γ and Ω are the linewidth and frequency shift without the external field.

Thus we have determined the Laplace transform of the wanted functions $D(t)$ explicitly. We now indicate how we can return from the Laplace transform to the original time dependent functions. To this end let us write instead of s the complex quantity $z = a + ib$; a, b real. Let us assume for example that $\mathfrak{L}_s(f)$ has the form $1/(s - z_0)$ so that

$$\mathfrak{L}_s[f] = \frac{1}{s - z_0}. \tag{7.321}$$

One is readily convinced that

$$f(t) = \exp[z_0 t]; \qquad \mathrm{Re}(z_0) < 0 \tag{7.322}$$

when inserted into eq. (7.321) just yields the right-hand side. One may show that this solution is unique and in this way may establish a whole list of rules for evaluating Laplace transforms. For instance one may show that a factor s corresponds to the differentiation

$$s \to -\frac{\mathrm{d}}{\mathrm{d}t} \tag{7.323}$$

or a power of s to n-fold differentiation

$$s^n \to \left(-\frac{\mathrm{d}}{\mathrm{d}t}\right)^n. \tag{7.324}$$

Furthermore when a Laplace transform

$$\frac{1}{G(s)} \equiv \frac{1}{a(s - z_1)(s - z_2)\cdots(s - z_n)} \tag{7.325}$$

is given, one may decompose this fraction into partial fractions

$$\frac{1}{G(s)} = \frac{a_1}{s - z_1} + \frac{a_2}{s - z_2} + \cdots + \frac{a_n}{s - z_n}. \tag{7.326}$$

For each of the expressions of the right-hand side we know the original function of the Laplace transform.

It is now a straightforward task to decompose eqs. (7.315) and (7.316) with (7.317) into such partial fractions and to evaluate the final result.

For the following analysis we need the roots s_j of the polynomials in s occurring in eqs. (7.315), (7.316) with (7.317). These roots are the following quantities

$$s_{1,2} = -i\Delta_\lambda \pm i|g\beta|, \tag{7.327}$$

and

$$s_{3,4} = -\frac{\Gamma}{2} \pm \sqrt{\frac{\Gamma^2}{4} - |g\beta|^2}\,. \tag{7.328}$$

In order to make contact with experiments we wish to determine the probability to find a photon of mode λ at time t. This probability can be easily deduced from the hypothesis (7.299) and is given by

$$p_\lambda(t) = |D_{1,\lambda}(t)|^2 + |D_{2,\lambda}(t)|^2. \tag{7.329}$$

Since the corresponding expressions become rather lengthy we quote only a few important final results. For $t \to \infty$ one obtains

$$p_\lambda = \frac{g^2}{2}\frac{1}{|s_3 - s_4|^4}\left\{\begin{aligned} &\left|\frac{s_3}{\Delta_\lambda - |g\beta| - is_3} - \frac{s_4}{\Delta_\lambda - |g\beta| - is_4}\right|^2 \\ &+\left|\frac{s_3}{\Delta_\lambda + |g\beta| - is_3} - \frac{s_4}{\Delta_\lambda + |g\beta| - is_4}\right|^2\end{aligned}\right\}. \tag{7.330}$$

In the limit in which $|g\beta|$ is much bigger than the line width γ, i.e. for strong enough driving fields, p_λ acquires the form

$$\begin{aligned} p_\lambda = \frac{g^2}{2}\Bigg\{&\left[\frac{\gamma^2}{4} + \left(\bar{\omega} - \omega_\lambda + 2|g\beta| + \frac{\Omega}{2}\right)^2\right]^{-1} \\ &+\left[\frac{\gamma^2}{4} + \left(\bar{\omega} - \omega_\lambda - 2|g\beta| + \frac{\Omega}{2}\right)^2\right]^{-1} \\ &+2\left[\frac{\gamma^2}{4} + \left(\bar{\omega} - \omega_\lambda + \frac{\Omega}{2}\right)^2\right]^{-1}\Bigg\}. \end{aligned} \tag{7.331}$$

p_λ appears here as a function of the mode frequency ω_λ which belongs to a continuous spectrum. We thus can plot p as a function of the continuous frequency ω. Clearly the first term in the curly bracket represents a Lorentzian curve with the width $\gamma/2$ centered around $\omega = \bar{\omega} + 2|g\beta| + \Omega/2$. The second term describes the same line but now centered around $\omega = \bar{\omega} - 2|g\beta| + \Omega/2$. The last term finally represents a Lorentzian curve centered around $\omega = \bar{\omega} + \Omega/2$ and has double the intensity than the two other curves. According to this formula the original emission line of an unperturbed atom splits into three lines under the influence of an external electric field (dynamic Stark effect). The theoretical and experimental results are shown in figs. 7.14a and b, respectively. The dynamic Stark effect can also be observed by absorption of light which hits atoms subjected to a strong coherent resonant driving field.

8. Quantum theory of coherence

8.1. Quantum mechanical coherence functions

In section 2.2 we got acquainted with classical coherence functions. In particular we dealt with the following coherence function

$$\langle E_i^{(+)}(\boldsymbol{x}',t')E_j^{(-)}(\boldsymbol{x},t)\rangle \tag{8.1}$$

In it $\boldsymbol{x}$ and t were the space point and time, respectively, at which the field strength $\boldsymbol{E}$ is measured. The indices i and j in eq. (8.1) refer to the components of the field vector, whereas plus and minus indicate the positive and negative frequency part, respectively. As the reader may recall the brackets denote the time average over the measuring time which was assumed long compared to the inverse of the light frequency. Several times in this book we have had occasion to learn how to translate classical quantities into the corresponding quantum mechanical ones. As we have seen, the classical observable "field strength" becomes an operator. The measurable quantities are then derived from the operators by certain expectation values. When we try to apply this scheme to the translation of (8.1) into a quantum theoretical expression we encounter a difficulty, namely $\boldsymbol{E}^{(-)}$ and $\boldsymbol{E}^{(+)}$ don't commute with each other (cf. exercise below). That means we now have to take care of the correct sequence of these operators. It turns out that their sequence depends on the experiment by which we measure the coherence function. When we use light detectors which absorb light then the positive frequency part must stand on the right and the negative frequency part on the left. On the other hand, when light is detected by processes involving downward transitions, the opposite sequence holds. Since the upward transitions are the usual ones we will confine our analysis to this case.

We now wish to derive the coherence function just mentioned. To this end we consider a realistic system consisting of a light source, the light field and a detector. Each of these individual systems is described fully

Table 1

Source	H_s		
Interaction source-field:	H_{s-f}	$H_s + H_{s-f} + H_f$	
Field:	H_f		$H_s + H_{s-f} + H_f$ $+ H_{f-d} + H_d$
Interaction field-detector:	H_{f-d}	$H_f + H_{f-d} + H_d$	
Detector:	H_d		

quantum mechanically. The corresponding individual Hamiltonians are listed in table 1. Thus in principle we are confronted with the solution of the Schrödinger equation of the total system composed of source, field and detector.

Its Hamiltonian reads

$$H = \underbrace{H_s + H_{s-f} + H_f}_{H_F} + H_{f-d} + H_d. \tag{8.2}$$

Of course, in general we will not be able to solve a Schrödinger equation with such a complicated Hamiltonian. However, we want to show how we can reduce the problem in a way which allows for direct comparison with experiments. It is most convenient to transform eq. (8.2) into the interaction representation which we have explained in section 7.4. To this end we make the identifications

$$H_F + H_d \rightarrow H_0 \tag{8.3}$$

and

$$H_{F-d} \rightarrow H_{int} \tag{8.4}$$

so that the total Schrödinger equation to be solved reads

$$(H_0 + H_{\text{int}})\Phi = i\hbar \frac{\mathrm{d}\Phi}{\mathrm{d}t}. \tag{8.5}$$

Making the hypothesis

$$\Phi = \exp[-iH_0 t/\hbar]\tilde{\Phi} \tag{8.6}$$

we go over to the interaction representation which now reads

$$i\hbar \frac{\mathrm{d}\tilde{\Phi}}{\mathrm{d}t} = \tilde{H}_{\text{int}}\tilde{\Phi} \tag{8.7}$$

with

$$\tilde{H}_{\text{int}} = \exp[iH_0 t/\hbar] H_{\text{int}} \exp[-iH_0 t/\hbar] \tag{8.8}$$

as we know from section 7.4. Our first task will be to determine (8.8) more explicitly. To this end we formulate in a first step the Hamiltonian of the detector and the Hamiltonian of the interaction between field and detector explicitly. To avoid too many complications which would obscure the procedure we assume that the detector consists of a single, two-level atom. Its Hamiltonian reads

$$H_d = \hbar \bar{\omega} a_2^+ a_2. \tag{8.9}$$

$\bar{\omega} = \omega_{21}$ is the transition frequency of the atom, a_j, a_j^+ are, as usual, the annihilation and creation operator of the electron in level j. The wave functions of the corresponding two states $j = 1, 2$ are, as usual, denoted by

$$\varphi_j(\boldsymbol{x}). \tag{8.10}$$

Adopting the general result of section 7.2 we may write the Hamiltonian for the interaction field-detector as

$$H_{f-d} = a_2^+ a_1 \left(-\frac{e}{m}\right) \int \varphi_2^*(\boldsymbol{x}) \boldsymbol{A}^{(+)}(\boldsymbol{x}) \frac{\hbar}{i} \nabla \varphi_1(\boldsymbol{x}) d^3 x$$
$$+ \text{Hermitian conjugate.} \tag{8.11}$$

Our next task will be to evaluate (8.8) using (8.9)–(8.11). We note that we made the rotating wave approximation in (8.11), i.e. we are interested only in real transitions. To verify our statement we recall that the operator $a_2^+ a_1$ describes an upward transition of the detector atom, while $\boldsymbol{A}^{(+)}$ is a linear combination of photon annihilation operators.

Since the detector Hamiltonian and the Hamiltonian H_F describe different systems the corresponding Hamiltonian operators commute

$$[H_d, H_F] = 0. \tag{8.12}$$

This allows us to split the exponential functions containing H_0 in eq. (8.8) into two factors. Treating the factor containing H_d alone we readily obtain

$$\exp[iH_d t/\hbar] a_2^+ a_1 \exp[-iH_d t/\hbar] = a_2^+ a_1 \exp[i\bar{\omega} t], \tag{8.13}$$

where we use results obtained in section 7.4.

At a first sight it appears rather hopeless to perform a similar procedure with respect to that part of (8.8) which contains H_F. The reason for that is, of course, that we don't know yet anything about the source. Indeed by measuring light absorption by the detector we want to get information about the nature of the source. However, a formal trick helps us quite a lot. Namely, having in our mind again the decomposition of H_0 (8.3), we are led to consider an expression of the form

$$\exp[iH_F t/\hbar] \boldsymbol{A}^{(+)}(\boldsymbol{x}) \exp[-iH_F t/\hbar] \tag{8.14}$$

which occurs in eq. (8.8). But now we recall that we encountered expressions of such a type earlier, namely when we considered the change of operators in the Heisenberg picture (cf. section 5.6). In our case we may think that the operator of the vector potential $\boldsymbol{A}$ (positive frequency part) changes under the action of the full Hamiltonian H_F describing the dynamics of the systems field and source. We will see in a minute that we need not evaluate (8.14) explicitly to obtain our final result. We would rather make use of the fact that (8.14) just defines the operator $\boldsymbol{A}^{(+)}$ in the corresponding Heisenberg picture which allows us to put

$$(8.14) = \boldsymbol{A}^{(+)}(\boldsymbol{x}, t). \tag{8.15}$$

According to the positive and negative frequency parts of $\boldsymbol{A}$ in (8.11) we split the interaction Hamiltonian into the corresponding parts

$$\tilde{H}_{\text{int}} = \tilde{H}_{\text{int}}^{(+)} + \tilde{H}_{\text{int}}^{(-)}. \tag{8.16}$$

Our considerations just performed allow us to write now explicitly

$$\tilde{H}_{\text{int}}^{(+)} = a_2^+ a_1\left(-\frac{e}{m}\right)\exp[i\bar{\omega}t]\int \varphi_2^*(\boldsymbol{x})\boldsymbol{A}^{(+)}(\boldsymbol{x},t)\frac{\hbar}{i}\nabla\varphi_1(\boldsymbol{x})\mathrm{d}^3\boldsymbol{x}. \tag{8.17}$$

We now wish to deal with the interaction between the field or, more precisely speaking, between field and source with the detector. To this end we have to solve the Schrödinger equation (8.7). We treat a specific experimental situation. In it, at the initial time $t = 0$, the coupling between field and detector atom is switched on. At this time we still may decompose the initial wave function into a product of the wave functions of the detector atom and the field + source system

$$\tilde{\Phi}(0) = \Phi(0)_{\text{detector}}\underbrace{\Phi_F(0)}_{\text{field + source}}\quad . \tag{8.18}$$

To be more specific we assume that the detector atom is at that time in its groundstate

$$\Phi_{\text{detector}}(0) = a_1^+\Phi_{0,\text{atom}}. \tag{8.19}$$

The field acting on the detector atom may now cause transitions and then we may be able to verify by other means that the detector atom has made the transition into its upper state. So our next task will be to calculate the probability of finding after time t a detector atom in its upper state

$$P_2(t) = \langle\tilde{\Phi}(t)|a_2^+ a_2|\tilde{\Phi}(t)\rangle. \tag{8.20}$$

To evaluate eq. (8.20) we need $\tilde{\Phi}(t)$. We determine that function by means of first-order perturbation theory in the well known fashion (cf. section 7.7)

$$\tilde{\Phi}(t) = \tilde{\Phi}(0) + \left(-\frac{i}{\hbar}\right)\int_0^t \tilde{H}_{\text{int}}(\tau)\,\mathrm{d}\tau\,\tilde{\Phi}(0). \tag{8.21}$$

When we insert eq. (8.21) into (8.20) we are left with four terms. However, looking at the action of the atomic operators a_j, a_j^+ and taking into account the form of the initial state eq. (8.19) we may readily convince ourselves that eq. (8.20) reduces to

$$P_2(t) = \frac{1}{\hbar^2}\langle\tilde{\Phi}(0)|\int_0^t \tilde{H}_{\text{int}}^{(-)}(\tau)\,\mathrm{d}\tau \int_0^t \tilde{H}_{\text{int}}^{(+)}(\tau')\,\mathrm{d}\tau'|\tilde{\Phi}(0)\rangle. \tag{8.22}$$

We now use the explicit expression (8.17) for $\tilde{H}_{\text{int}}$. When we insert this expression into eq. (8.22) we obtain a rather lengthy formula which should not shock the reader too much. He will notice that the resulting formula contains two major parts. One part refers to the atom containing the electron wave function and the momentum operators, whereas another part contains the vector potential in the Heisenberg picture. This formula reads

$$P_2(t) = \frac{e^2}{m^2\hbar^2}\sum_{j,k=x,y,z}\int_0^t \mathrm{d}\tau \int_0^t \mathrm{d}\tau' \exp[i\bar{\omega}(\tau'-\tau)]$$
$$\times \int \mathrm{d}^3x \int \mathrm{d}^3x' \varphi_2^*(x)\left(\frac{\hbar}{i}\nabla_j\right)\varphi_1(x)\varphi_2^*(x')\left(\frac{\hbar}{i}\nabla_k'\right)\varphi_1(x')\langle\ldots\rangle. \tag{8.23}$$

In it the bracket $\langle\ldots\rangle$ is the abbreviation for

$$\langle\ldots\rangle = \langle\Phi_F(0)|A_j^{(-)}(x,\tau)A_k^{(+)}(x',\tau')|\Phi_F(0)\rangle. \tag{8.24}$$

This latter bracket is most important for our purpose. It shows us that the probability of finding the detector atom in its upper state is quite essentially determined by the correlation function (8.24) between positive and negative frequency parts of the vector potential. The connection with the coherence function quoted at the beginning of this section can be made apparent immediately. We could have performed the whole procedure in the dipole approximation in which case the electric field strength would have appeared instead of the vector potential. Repeating all the above steps our final result then would depend on $\boldsymbol{E}$, replacing $\boldsymbol{A}$ in eq. (8.24) everywhere by the corresponding $\boldsymbol{E}$.

$$\langle\Phi_F(0)|E_j^{(-)}(x,\tau)E_k^{(+)}(x',\tau')|\Phi_F(0)\rangle. \tag{8.25}$$

[Another way of going from eq. (8.24) to (8.25) can be achieved by taking the time derivatives of the $\boldsymbol{A}$s, compare sections 5.1 and 5.8.]

To be quite clear let us repeat some essential points. The $\boldsymbol{A}$s represent free fields in the interaction representation. Free refers to the interaction with the detector. On the other hand, when there is an interaction of $\boldsymbol{A}$ with the source, the $\boldsymbol{A}$s are the full Heisenberg operators with respect to the source. We will present later on a number of explicit examples how the $\boldsymbol{A}$s and the correlation function (8.24) can be explicitly determined. The quantum mechanical averages in eqs. (8.24) and (8.25) refer to the field and source variables but no longer to the detector. In our present model which represents the detector by a single atom, $\boldsymbol{x}$ and $\boldsymbol{x}'$ are very close together within the atomic dimensions. When we use a system of atoms with macroscopic dimensions, for instance crystals, one may also encounter macroscopic differences of $\boldsymbol{x}$ and $\boldsymbol{x}'$ but we shall not consider this case any further here. In the dipole approximation we can replace the coordinates $\boldsymbol{x}$ and $\boldsymbol{x}'$ in the $\boldsymbol{A}$'s by $\boldsymbol{x}_0$. To evaluate P_2 for practical cases one has to take into account an ensemble of atoms and to sum up over the individual contributions. Furthermore each of these atoms may have more than 2 states. One has then as usual to average over the initial states and to sum up over the final states of all these atoms. Whether the double time integration τ, τ' can be replaced by a single one depends on the spread of atomic energies (band width $\Delta\omega$ of the counter involved in the counting and detecting process). If $t > 1/\Delta\omega$ this can be achieved. These averaging processes are somewhat involved but do not give us a new insight at this moment. Therefore, we just state that we find that the detecting rate is directly given by functions of the form of eq. (8.24) or (8.25) where the other factors depend on the detector. As it transpires from our discussion just performed we may put $\boldsymbol{x} = \boldsymbol{x}'$, $t = t'$. Thus eqs. (8.24) and (8.25) are the quantum mechanical analogues to eqs. (2.24) with (2.25). In order to obtain correlation functions containing different times and space points we may proceed in complete analogy to sections 2.2 and 1.11.

We conclude this section by two comments: (1) it is simple to extend our procedure to processes in which a successive annihilation of photons occurs. In this case we have to apply $\tilde{H}_{\text{int}}^{(+)}$ several times on $\Phi(0)$, together with the corresponding sequence of time integrations. The probability that n photons are absorbed is then determined by correlation functions of the form

$$\langle\Phi_F(0)|A_{i_1}^{(-)}(\boldsymbol{x}_1,t_1)\cdots A_{i_n}^{(-)}(\boldsymbol{x}_n,t_n)$$

$$\times A_{j_n}^{(+)}(\boldsymbol{x}_n',t_n')\cdots A_{j_1}^{(+)}(\boldsymbol{x}_1',t_1')|\Phi_F(0)\rangle. \tag{8.26}$$

(2) One might also imagine processes by which light is detected not by upward transitions but by downward transitions. In our specific case in such an experiment the detector atom is initially in the upper state 2 and is then caused to make a transition towards the lower state 1. Repeating all our above steps one readily verifies that this has the effect of exchanging the role of the positive and negative frequency parts of the vector potential $\boldsymbol{A}$.

Exercise on section 8.1

(1) Show that $\boldsymbol{E}^{(-)}(\boldsymbol{x},t)$ and $\boldsymbol{E}^{(+)}(\boldsymbol{x}',t')$ do not commute.
Hint: Use eq. (5.149). b_λ, b_λ^+ are taken in the Heisenberg picture. Use further eq. (5.154) (in the Heisenberg picture).

(2) Discuss under which conditions (8.26) can be replaced by

$$\langle \Phi_F(0)|E_{i_1}^{(-)}(\boldsymbol{x}_1,t_1)\cdots E_{i_n}^{(-)}(\boldsymbol{x}_n,t_n)E_{j_n}^{(+)}(\boldsymbol{x}_n,t_n)\cdots E_{j_1}^{(+)}(\boldsymbol{x}_1,t_1)|\Phi_F(0)\rangle .$$

8.2. Examples of the evaluation of quantum mechanical coherence functions

In the following we will show how we can evaluate coherence functions. We shall do this for the following coherence functions

$$\langle \Phi|E_j^{(\pm)}(\boldsymbol{x},t)|\Phi\rangle, \tag{8.27}$$

$$\langle \Phi|E_i^{(-)}(\boldsymbol{x}',t')E_j^{(+)}(\boldsymbol{x},t)|\Phi\rangle, \tag{8.28}$$

$$\langle \Phi|E_i^{(-)}(\boldsymbol{x}',t')E_j^{(-)}(\boldsymbol{x}',t')E_j^{(+)}(\boldsymbol{x},t)E_i^{(+)}(\boldsymbol{x},t)|\Phi\rangle. \tag{8.29}$$

We decompose the positive and negative frequency parts of the electric field strength in the usual way (compare section 5.8)

$$E_i^{(+)}(\boldsymbol{x},t) = \sum_\lambda c_\lambda u_{\lambda,i}(\boldsymbol{x})b_\lambda, \tag{8.30}$$

$$E_i^{(-)}(\boldsymbol{x},t) = \sum_\lambda c_\lambda^* u_{\lambda,i}^*(\boldsymbol{x})b_\lambda^+ . \tag{8.31}$$

In it λ distinguishes the different light modes which may belong to a cavity or which might be modes of free space. b_λ^+, b_λ are as usual the creation and annihilation operators. $u_{\lambda,i}(\boldsymbol{x})$ are the classical field modes, in most cases of practical applications running waves. c_λ^*, c_λ are certain normalization factors (cf. 5.149). In the Schrödinger picture the Φs are time dependent and b_λ, b_λ^+ time independent. This picture is applicable only to (8.27), but not to (8.28) or (8.29), however. The reason for this lies in the fact that in (8.28) or (8.29) operators $\boldsymbol{E}$ occur which contain different times. Throughout this chapter we will therefore use the Heisenberg picture in which the whole time dependence is inherent in b_λ^+, b_λ. The brackets in (8.27), (8.28), and (8.29) again denote the quantum mechanical average which affects b_λ^+, b_λ, but leaves all other quantities in eqs. (8.30) and (8.31) unaffected. When we insert eqs. (8.30) and (8.31) into one of the coherence functions (8.27)–(8.29), we can take out of these coherence functions all parts of eqs. (8.30) and (8.31) except for b_λ^+, b_λ. Thus the evaluation of eqs.

(8.27)–(8.29) is reduced to the evaluation of expressions of the form

$$\langle \Phi | b_\lambda(t) | \Phi \rangle, \tag{8.32}$$

$$\langle \Phi | b^+_{\lambda'}(t') b_\lambda(t) | \Phi \rangle, \tag{8.33}$$

$$\langle \Phi | b^+_{\lambda_4}(t_4) b^+_{\lambda_3}(t_3) b_{\lambda_2}(t_2) b_{\lambda_1}(t_1) | \Phi \rangle. \tag{8.34}$$

We now show, by examples, how we can evaluate such expressions. We first start with *free fields* in which case the field operators obey the equations

$$\frac{\mathrm{d}b_\lambda}{\mathrm{d}t} = -i\omega_\lambda b_\lambda, \qquad \frac{\mathrm{d}b^+_\lambda}{\mathrm{d}t} = i\omega_\lambda b^+_\lambda, \tag{8.35, 36}$$

with the solutions

$$b_\lambda(t) = b_\lambda(0)\exp[-i\omega_\lambda t], \qquad b^+_\lambda(t) = b^+_\lambda(0)\exp[i\omega_\lambda t]. \tag{8.37, 38}$$

$b^+_\lambda(0)$ and $b_\lambda(0)$ are the operators taken at initial time $t = 0$. We assume that at that time the Heisenberg picture and the Schrödinger picture coincide. By inserting eqs. (8.37) and (8.38) into (8.32)–(8.34) we readily obtain

$$\langle \Phi | b_\lambda(t) | \Phi \rangle = \exp[-i\omega_\lambda t]\langle \Phi | b_\lambda(0) | \Phi \rangle, \tag{8.39}$$

$$\langle \Phi | b^+_{\lambda'}(t') b_\lambda(t) | \Phi \rangle = \exp[i\omega_{\lambda'} t' - i\omega_\lambda t]\langle \Phi | b^+_{\lambda'}(0) b_\lambda(0) | \Phi \rangle, \tag{8.40}$$

$$\begin{aligned}(8.34) &= \exp\left[i\omega_{\lambda_4} t_4 + i\omega_{\lambda_3} t_3 - i\omega_{\lambda_2} t_2 - i\omega_{\lambda_1} t_1\right] \\ &\quad \times \langle \Phi | b^+_{\lambda_4}(0) b^+_{\lambda_3}(0) b_{\lambda_2}(0) b_{\lambda_1}(0) | \Phi \rangle.\end{aligned} \tag{8.41}$$

Thus we have only to evaluate expectation values at the initial time which we can easily do in the Schrödinger picture.

We first choose for Φ eigenfunctions of the Schrödinger equation, i.e. eigenfunctions in the energy or photon number representation

$$\Phi_{\lambda_1\lambda_2\cdots\lambda_n} = \frac{1}{\sqrt{n_{\lambda_1}! n_{\lambda_2}! \cdots n_{\lambda_n}!}} (b^+_{\lambda_1})^{n_{\lambda_1}} (b^+_{\lambda_2})^{n_{\lambda_2}} \cdots \Phi_0. \tag{8.42}$$

Using methods developed in sections 3.3 and 5.3 we may immediately write down the results. We obtain

$$\langle \Phi | b_\lambda(0) | \Phi \rangle = 0, \tag{8.43}$$

$$\langle \Phi | b^+_{\lambda'}(0) b_\lambda(0) | \Phi \rangle = \delta_{\lambda\lambda'} n_\lambda, \tag{8.44}$$

$$\begin{aligned}&\langle \Phi | b^+_{\lambda_4}(0) b^+_{\lambda_3}(0) b_{\lambda_2}(0) b_{\lambda_1}(0) | \Phi \rangle \\ &= n_{\lambda_1} n_{\lambda_2}(\delta_{\lambda_3\lambda_2}\delta_{\lambda_4\lambda_1} + \delta_{\lambda_3\lambda_1}\delta_{\lambda_4\lambda_2}) - (n_{\lambda_1} + 1) n_{\lambda_1}\delta_{\lambda_1\lambda_2}\delta_{\lambda_3\lambda_2}\delta_{\lambda_4\lambda_1}.\end{aligned} \tag{8.45}$$

We obtained the result (8.43) earlier, namely (8.27) vanishes when we are dealing with free fields and the field being in an eigenstate of photon numbers. These results put us in a position to write down the explicit expressions for the coherence functions we wanted to calculate. As we just saw we obtain

$$\left.\begin{aligned}&\langle\Phi|E_j^{(-)}(\boldsymbol{x},t)|\Phi\rangle = 0,\\&\langle\Phi|E_j^{(-)}(\boldsymbol{x}',t')E_k^{(+)}(\boldsymbol{x},t)|\Phi\rangle\\&\quad=\sum_\lambda|c_\lambda|^2 n_\lambda u_{\lambda,j}^*(\boldsymbol{x}')u_{\lambda,k}(\boldsymbol{x})\exp[i\omega_\lambda(t'-t)]\end{aligned}\right\}. \tag{8.46}$$

Coherent initial fields. We now do the same steps for free fields but choose as initial field (which then, of course, persists for all times) a coherent field

$$\Phi = \mathfrak{N}\exp\left[\sum_\lambda \beta_\lambda b_\lambda^+\right]\Phi_0. \tag{8.47}$$

The normalization factor $\mathfrak{N}$ is given by

$$\mathfrak{N} = \exp\left[-\tfrac{1}{2}\sum_\lambda|\beta_\lambda|^2\right]. \tag{8.48}$$

Using the results of section 5.5 we readily obtain

$$\langle\Phi|b_\lambda^+|\Phi\rangle = \beta_\lambda^*, \tag{8.49}$$

$$\langle\Phi|b_{\lambda_2}^+ b_{\lambda_1}|\Phi\rangle = \beta_{\lambda_2}^*\beta_{\lambda_1}, \tag{8.50}$$

$$\langle\Phi|b_{\lambda_4}^+ b_{\lambda_3}^+ b_{\lambda_2} b_{\lambda_1}|\Phi\rangle = \beta_{\lambda_4}^*\beta_{\lambda_3}^*\beta_{\lambda_2}\beta_{\lambda_1}. \tag{8.51}$$

Again we may determine the coherence functions (8.27), (8.28), and (8.29) by using these intermediate results. Thus we obtain

$$\langle\Phi|E_j^{(-)}(\boldsymbol{x},t)|\Phi\rangle = \sum_\lambda c_\lambda^* u_{\lambda,j}^*(\boldsymbol{x})\exp[i\omega_\lambda t]\beta_\lambda^*. \tag{8.52}$$

Similarly we obtain

$$\begin{aligned}&\langle\Phi|E_j^{(-)}(\boldsymbol{x}',t')E_k^{(+)}(\boldsymbol{x},t)|\Phi\rangle\\&\quad=\sum_{\lambda\lambda'} c_{\lambda'}^* c_\lambda u_{\lambda',j}^*(\boldsymbol{x}')u_{\lambda,k}(\boldsymbol{x})\exp[i\omega_{\lambda'}t'-i\omega_\lambda t]\beta_{\lambda'}^*\beta_\lambda,\end{aligned} \tag{8.53}$$

which we can also cast into the form

$$(8.53) = \langle\Phi|E_j^{(-)}(\boldsymbol{x}',t')|\Phi\rangle\langle\Phi|E_k^{(+)}(\boldsymbol{x},t)|\Phi\rangle. \tag{8.54}$$

Thus, (8.53) factorizes into coherence functions of lower order.

So far we have evaluated coherence functions for free fields. With respect to our next chapters it is most illuminating to calculate coherence functions of fields driven by prescribed classical sources. For simplicity we consider a single field mode. As we saw in section 5.7, the operators b and b^+ of the field mode obey equations of the form

$$\frac{\mathrm{d}b}{\mathrm{d}t} = -i\omega b + f(t) \tag{8.55}$$

and

$$\frac{\mathrm{d}b^+}{\mathrm{d}t} = i\omega b^+ + f^*(t). \tag{8.56}$$

In it $f(t)$ describes the effect of a classical source on the field mode. $f(t)$ can be an arbitrary classical time dependent function, for instance it can describe the oscillations or currents of classical charges. The solution of eq. (8.55) reads

$$b(t) = \underbrace{\int_0^t \exp[-i\omega(t-\tau)] f(\tau)\,\mathrm{d}\tau}_{B(t)} + b(0)\exp[-i\omega t], \tag{8.57}$$

where we have chosen as a specific initial time $t = 0$. But any other initial time could have been chosen as well. $b(0)$ is the photon operator at the initial time and is chosen to coincide with the corresponding operator in the Schrödinger picture. One readily convinces oneself that eq. (8.57) and the corresponding hermitian conjugate satisfy the commutation relations for Bose operators. We now evaluate the coherence functions. We first choose Φ as eigenstate of photon number. We then readily obtain

$$\langle \Phi | b(t) | \Phi \rangle = B(t). \tag{8.58}$$

Of more interest is the correlation function in which two operators are involved, e.g.

$$\langle \Phi | b^+_{\lambda'}(t') b_\lambda(t) | \Phi \rangle. \tag{8.59}$$

Here we now have furnished b^+ and b with mode indices having in mind coherence functions of the form (8.32)–(8.34). Inserting (8.57) and its hermitian conjugate into (8.59) we readily obtain

$$\begin{aligned}(8.59) = {} & B^*_{\lambda'}(t') B_\lambda(t) + B^*_{\lambda'}(t') \langle b_\lambda \rangle \exp[-i\omega_\lambda t] \\ & + B_\lambda(t) \langle b^+_{\lambda'} \rangle \exp[i\omega_{\lambda'} t'] + \langle b^+_{\lambda'} b_\lambda \rangle \exp[i\omega_{\lambda'} t' - i\omega_\lambda t].\end{aligned} \tag{8.60}$$

The final evaluation of this expression depends on the expectation values for the bs in the Schrödinger picture. Using our results we obtained above in this section we readily find the following results. When Φ is a photon eigenstate we obtain

$$(8.59) = B^*_{\lambda'}(t')B_\lambda(t) + n_\lambda \delta_{\lambda\lambda'} \exp[i\omega_\lambda(t' - t)]. \tag{8.61}$$

If Φ is a coherent field

$$(8.59) = (B^*_{\lambda'}(t') + \beta^*_{\lambda'} \exp[i\omega_{\lambda'} t'])(B_\lambda(t) + \beta_\lambda \exp[-i\omega_\lambda t]) \tag{8.62}$$

results. While in eq. (8.61) n_λ is the number of photons of the mode λ present at time $t = 0$, β_λ, β^*_λ in (8.62) are the amplitudes of the coherent states of the corresponding modes. It is interesting to study how the physical properties of the source, i.e. its specific time dependence, is transferred by the field modes to the detector, i.e. how this time dependence of f is reflected by coherence functions. From our results we see that the source term f enters into the Bs, but in addition certain properties of the free fields also appear as is visible from eqs. (8.61) and (8.62).

Let us now study a few explicit examples for $f(t)$. The simplest form of f is that of a harmonic oscillation

$$f(t) = A \exp[-i\omega_0 t], \tag{8.63}$$

in which case we readily obtain

$$B(t) = \frac{-i}{\omega - \omega_0}\{\exp[-i\omega_0 t] - \exp[-i\omega t]\}. \tag{8.64}$$

In the case of resonance, eq. (8.64) reduces to

$$B(t) = t \exp[-i\omega t], \tag{8.65}$$

i.e. the field mode amplitude increases proportional to time. We leave it as an exercise to the reader to calculate similar expressions and the coherence functions for the following cases: (1) $f(t)$ is, in addition to eq. (8.63), exponentially damped and (2) $f(t)$ consists of a sum of exponential functions within a frequency interval around ω and with random phases. What coherence functions result when the phase average is performed?

8.3. Coherence properties of spontaneously emitted light

We wish to show explicitly, how the formalism of the coherence functions (8.28) allows us to directly determine the coherence properties of spontaneously emitted light. To this end we calculate

$$\Gamma(t + \tau, t) = \langle \Phi(0) | \boldsymbol{E}^{(-)}(\boldsymbol{x}, t + \tau) \boldsymbol{E}^{(+)}(\boldsymbol{x}, t) | \Phi(0) \rangle. \tag{8.66}$$

It is most important that the atom should serve as source and not as a counter. Consequently $\boldsymbol{E}^{(-)}$ and $\boldsymbol{E}^{(+)}$ (or $\boldsymbol{A}^{(-)}$ and $\boldsymbol{A}^{(+)}$) must be the full

Heisenberg operators:

$$E^{(-)}(\boldsymbol{x},t) = \exp[iHt/\hbar] \times \left\{ \frac{-i}{\sqrt{V}} \sum_{\lambda} b_{\lambda}^{+} \sqrt{\frac{\hbar\omega_{\lambda}}{2\varepsilon_0}}\, \boldsymbol{e}_{\lambda} \exp[-i\boldsymbol{k}_{\lambda}\boldsymbol{x}] \right\} \exp[-iHt/\hbar], \tag{8.67}$$

where H is the complete Hamiltonian (field + source). Since $\boldsymbol{E}^{(-)}$ and $\boldsymbol{E}^{(+)}$ are decomposed into a linear combination of operators b^{+}, b, we first investigate:

$$\langle \Phi(0) | b_{\lambda}^{+}(t+\tau) b_{\lambda'}(t) | \Phi(0) \rangle, \tag{8.68}$$

where

$$b_{\lambda}^{+}(t) = \exp[iHt/\hbar] b_{\lambda}^{+}(0) \exp[-iHt/\hbar],$$

$$b_{\lambda}(t) = \exp[iHt/\hbar] b_{\lambda}(0) \exp[-iHt/\hbar]. \tag{8.69}$$

$\Phi(0)$ is the initial state (one atom excited, no photon present). We insert (8.69) into (8.68) and observe that

$$\exp[-iHt/\hbar]\Phi(0) = \Phi(t), \tag{8.70}$$

so that

$$(8.68) = \langle \Phi(t+\tau) | b_{\lambda}^{+} \exp[-iH\tau/\hbar] | \Phi(t) \rangle. \tag{8.71}$$

We now use the explicit form, eq. (7.267) of $\Phi(t)$ and find after some elementary algebra:

$$(8.68) = c_{\lambda}^{*}(t+\tau) \exp[i\omega_{\lambda}(t+\tau)] c_{\lambda'}(t) \exp[-i\omega_{\lambda'}t + i\varepsilon_1\tau] \times \langle \Phi_0 | a_1 \exp[-iH\tau/\hbar] a_1^{+} | \Phi_0 \rangle. \tag{8.72}$$

Since $a_1^{+}\Phi_0$ represents a state in which the electron is in its groundstate and no photon is present, H_I can cause only virtual transitions. Also, since they give rise only to renormalization effects, we need not consider those transitions, so that of $\exp[-iH\tau/\hbar]$ only $\exp[-iH_{at}\tau/\hbar]$ is left.

Therefore we finally find

$$(8.68) = c_{\lambda}^{*}(t+\tau) c_{\lambda}(t) \exp[i(\omega_{\lambda} - \omega_{\lambda'})t + i\omega_{\lambda}\tau]. \tag{8.73}$$

When we insert eq. (8.73) into eq. (8.66) we obtain a double sum over λ, λ' of expressions (8.73), still multiplied with coefficients occurring in eq. (8.67). This evaluation is rather tedious and because it sheds no light on the quantum theoretical treatment, which has been completely done above, we merely quote the result:

$$\Gamma(t+\tau,t)=\frac{\bar{\omega}^4}{2\pi\varepsilon_0 c^4}\exp\left[-2\gamma\left(t-\frac{r}{c}\right)\right]\frac{\sin^2\Theta}{r^2}|\vartheta_{21}|^2\exp[i\bar{\omega}\tau-\gamma\tau]. \tag{8.74}$$

In it, the different quantities are defined as follows:
$\bar{\omega}$: transition frequency of the atom,
γ: optical half-width calculated in eq. (7.274),
r: distance: atom-detector,
Θ: angle between atomic dipole moment and direction atom-detector,
ϑ_{21}: optical (electric) dipole moment.
In deriving eq. (8.74) we have assumed that

$$\gamma \ll \bar{\omega}, \qquad r=|\boldsymbol{x}| \gg \frac{c}{\bar{\omega}}=\frac{\lambda}{2\pi}, \qquad r<ct. \tag{8.75, 76, 77}$$

The condition (8.75) is always well fulfilled. Equation (8.76) means that we confine ourselves to the wave region of a Hertzian dipole, while Γ would vanish (on account of causality) if eq. (8.77) were not fulfilled. For the explicit derivation of eq. (8.74) the dipole approximation has been used.

It is most remarkable that the classical treatment of a damped Hertzian oscillator leads to exactly the same expression. When we normalize Γ in order to find the complex degree of coherence (compare section 2.2) we obtain immediately:

$$\gamma(t+\tau,t)=\exp[i\bar{\omega}\tau-\gamma\tau]. \tag{8.78}$$

This clearly shows, that the spontaneously emitted wave track has a coherence time $1/\gamma$. Although $\Gamma(t+\tau,t)$ does not vanish, $\langle \boldsymbol{E}^{(+)}(\boldsymbol{x},t)\rangle = \langle \boldsymbol{e}^{(-)}(\boldsymbol{x},t)\rangle$ do. This can be interpreted as meaning, that the initial phase of $\boldsymbol{e}(\boldsymbol{x},t)$ is unknown. This phase cancels out, however, in eq. (8.66).

8.4. Quantum beats

This section is of interest in several respects. First of all it describes an experiment which allows one to measure the splitting of closely spaced energy levels. Secondly it will allow us to demonstrate basic ideas of quantum theory and thirdly it is a nice example of how to evaluate coherence functions. We first briefly describe the experiment. Let us consider the energy level diagram of fig. 8.1. By a short pulse whose duration is shorter than the inverse of the frequency splitting of the two excited levels 2 and 3, we excite these levels. The excited atom is then in a state which can be described by a coherent superposition of the wave

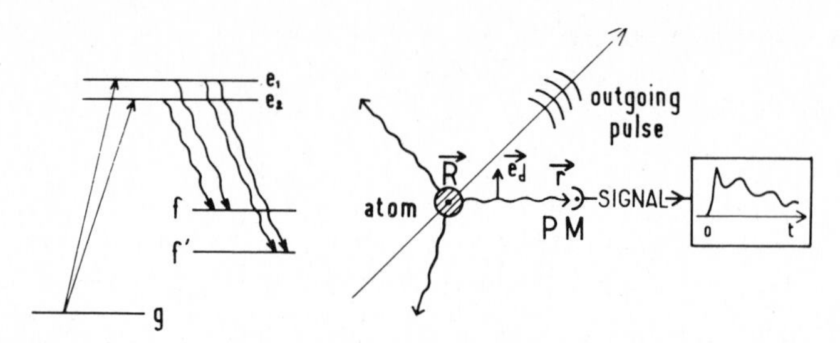

Fig. 8.1. Scheme of quantum beat experiment [after Haroche, in: High Resolution Laser Spectroscopy, ed. K. Shimoda, (Springer 1976).] The left part gives a diagram of the different atomic levels. To compare this diagram with the text identify f and f' with g. The right hand side shows the basic idea of the experiment. The atom is excited by a light pulse in the direction indicated by the outgoing pulse. In a different direction the light emitted from the atom is measured by a photo multiplier and the resulting signal exhibited on a monitor (compare fig. 8.2).

functions of the two excited levels 2 and 3. From there the electron can deexcite to some of the lower states depending on selection rules. According to the different decay channels the excited atom can now emit light at the corresponding frequencies. In the following we assume that we use a detector which is sensitive for photons emitted from levels 2 and 3 to level 1. The essential result of such an experiment is shown in fig. 8.2. In it we plot the intensity of the light absorbed by the detector as a function of time

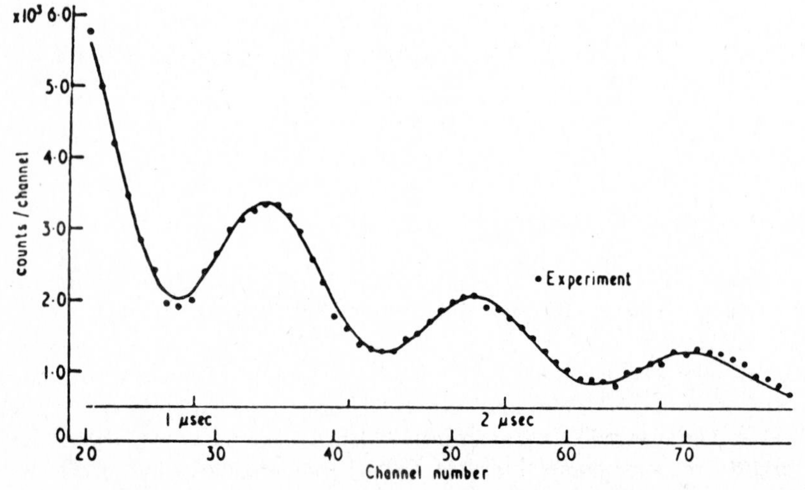

Fig. 8.2. This figure shows the number of counts as a function of time. The oscillatory behaviour of the quantum beat signal is clearly visible. (After Haroche, l.c.)

after the pulse had excited the atom. In contrast to conventional spontaneous emission (compare section 7.6) where simple exponential decay occurs, this decay is now modulated. The modulation frequency is given by $(W_3 - W_2)/\hbar \equiv \omega_{32}$ as we will derive below. Because the modulation frequency can be directly deduced from fig. 8.2 it allows for a measurement of the energy level splitting. This kind of measurement has become an important method to measure the splitting of energy levels.

We now give a quantum theoretical treatment of this effect which on the one hand generalizes the Weisskopf–Wigner theory of spontaneous emission (cf. section 7.9) and on the other hand allows us to show how we can evaluate coherence functions.

In our treatment we shall deal explicitly with the three energy levels 1, 2, 3 of a single atom according to fig. 8.1. We treat the interaction between the atom and the field mode as usual in the rotating wave approximation. Furthermore we start right away in the interaction representation. It is a simple matter to generalize the interaction Hamiltonian of the two level atom whose spontaneous emission we have treated in sections 7.6 and 7.9 to the present case. Since the split energy levels are connected with different wave functions, which may give rise for instance to different directions of polarizations of the emitted light, we distinguish the coupling coefficients g by indices 2, 1; 3, 1 corresponding to the transitions $2 \to 1$ and $3 \to 1$. The interaction Hamiltonian then reads

$$\begin{aligned} \tilde{H}_I = \hbar \sum_{\lambda} \{ & g_{21} a_2^+ a_1 b_\lambda \exp[-i\Delta_{21,\lambda} t] \\ & + g_{21}^* a_1^+ a_2 b_\lambda^+ \exp[i\Delta_{21,\lambda} t] \\ & + g_{31} a_3^+ a_1 b_\lambda \exp[-i\Delta_{31,\lambda} t] \\ & + g_{31}^* a_1^+ a_3 b_\lambda^+ \exp[i\Delta_{31,\lambda} t] \}, \end{aligned} \tag{8.79}$$

where we have used the abbreviation

$$\Delta_{j1,\lambda} \equiv \omega_\lambda - (W_j - W_1)/\hbar. \tag{8.79a}$$

To solve the Schrödinger equation

$$i\hbar \frac{d\tilde{\Phi}}{dt} = \tilde{H}_I \tilde{\Phi} \tag{8.80}$$

in the interaction picture it seems reasonable to make an ansatz which is a straightforward generalization of the Weisskopf–Wigner hypothesis (7.267). We expect $\tilde{\Phi}$ to be a superposition of the excited states 3 and 2 and the level 1 where in addition a photon is present. Thus our present hypothesis reads

$$\tilde{\Phi} = \alpha_3(t) a_3^+ \Phi_0 + \alpha_2(t) a_2^+ \Phi_0 + \sum_\lambda c_\lambda(t) b_\lambda^+ a_1^+ \Phi_0. \tag{8.81}$$

Inserting eq. (8.81) into eq. (8.80) and comparing coefficients of 0 and 1 photon states leaves us with the following set of equations

$$\frac{\mathrm{d}}{\mathrm{d}t}\alpha_3 = -i \sum_\lambda \exp[-i\Delta_{31,\lambda} t] g_{31} c_\lambda(t), \tag{8.82}$$

$$\frac{\mathrm{d}}{\mathrm{d}t}\alpha_2 = -i \sum_\lambda \exp[-i\Delta_{21,\lambda} t] g_{21} c_\lambda(t), \tag{8.83}$$

$$\frac{\mathrm{d}}{\mathrm{d}t} c_\lambda = -i\{ g_{21}^* \alpha_2 \exp[i\Delta_{21,\lambda} t] + g_{31}^* \alpha_3 \exp[i\Delta_{31,\lambda} t]\}. \tag{8.84}$$

We supplement these equations by the initial condition according to which the electron is in a coherent superposition state of levels 3 and 2 at the initial time $t = 0$

$$\tilde{\Phi}(0) = \alpha_3(0) a_3^+ \Phi_0 + \alpha_2(0) a_2^+ \Phi_0. \tag{8.85}$$

Normalization requires

$$|\alpha_3(0)|^2 + |\alpha_2(0)|^2 = 1. \tag{8.86}$$

In the following we will assume that the two levels 2 and 3 are coupled in the same way to the electromagnetic field except that these states may give rise to different directions of polarization. As a consequence we expect that the two levels decay with the same rate constant γ. Accordingly we make the hypothesis

$$\alpha_j(t) = \alpha_j(0) \exp[-\gamma t] \tag{8.87}$$

Inserting eq. (8.87) into (8.84) and integrating these equations yields

$$c_\lambda(t) = -i \sum_{j=2}^{3} \frac{g_{j1}^* \alpha_j(0)}{i\Delta_{j1,\lambda} - \gamma} \left(\exp[(i\Delta_{j1,\lambda} - \gamma)t] - 1\right). \tag{8.88}$$

For our following discussion it will be convenient to split the sum over $j = 2, 3$ into its corresponding parts explicitly

$$c_\lambda(t) = c_{2,\lambda}(t) + c_{3,\lambda}(t). \tag{8.89}$$

We now wish to calculate the field intensity measured in the detector at space point $\boldsymbol{x}$ and time t. For simplicity we assume that the atom is located at the origin of the coordinate system. The expectation value for the field intensity is given by

$$K = \langle \Phi(t) | \boldsymbol{E}^{(-)}(\boldsymbol{x}) \boldsymbol{E}^{(+)}(\boldsymbol{x}) | \Phi(t) \rangle . \tag{8.90}$$

In this description we are in the Schrödinger picture. In it $\boldsymbol{E}^{(\pm)}$ are time-independent field operators while the time dependence is fully taken care of by Φ. It is an interesting exercise for the reader to convince himself that eq. (8.90) is a special case of the coherence function (8.28) when we take equal times and proceed simultaneously from the Heisenberg picture to the Schrödinger picture (see exercise). To evaluate eq. (8.90) we decompose the positive and negative frequency parts of the field strengths into plane waves. Because we have an experiment in mind in which a specific polarization is measured, it suffices to confine the following expansion to modes with the corresponding polarization vector $\boldsymbol{e}$.

$$\boldsymbol{E}^{(+)}(\boldsymbol{x}) = \sum_{\boldsymbol{k}} b_{\boldsymbol{k}} \mathfrak{N}_{\boldsymbol{k}} \boldsymbol{e} \exp[i\boldsymbol{k}\boldsymbol{x}], \tag{8.91}$$

$$\boldsymbol{E}^{(-)}(\boldsymbol{x}) = \sum_{\boldsymbol{k}} b_{\boldsymbol{k}}^{+} \mathfrak{N}_{\boldsymbol{k}} \boldsymbol{e} \exp[-i\boldsymbol{k}\boldsymbol{x}]. \tag{8.92}$$

In it we have replaced the usual mode index λ used everywhere in our book by the wave vector $\boldsymbol{k}$. $b_{\boldsymbol{k}}$, $b_{\boldsymbol{k}}^{+}$ are the photon annihilation and creation operators. $\mathfrak{N}_{\boldsymbol{k}}$ is the normalization factor. Inserting eqs. (8.91) and (8.92) into (8.90) yields

$$K = \sum_{\boldsymbol{k}', \boldsymbol{k}} \mathfrak{N}_{\boldsymbol{k}} \mathfrak{N}_{\boldsymbol{k}'} \langle \Phi(t) | b_{\boldsymbol{k}}^{+} b_{\boldsymbol{k}'} | \Phi(t) \rangle \exp[i(\boldsymbol{k}' - \boldsymbol{k})\boldsymbol{x}]. \tag{8.93}$$

Evidently we are now left with the evaluation of the expectation values of the form

$$\langle \Phi(t) | b_{\boldsymbol{k}}^{+} b_{\boldsymbol{k}'} | \Phi(t) \rangle . \tag{8.94}$$

To this end we have to insert the explicit expression for eq. (8.81) into (8.94). Evaluations of this type have been done at several places in this book (see for instance section 8.2) so that we can immediately write down the final result

$$(8.94) = (c_{2,\boldsymbol{k}}^{*} + c_{3,\boldsymbol{k}}^{*})(c_{2,\boldsymbol{k}'} + c_{3,\boldsymbol{k}'}). \tag{8.95}$$

Note that the coefficients c, c^{*} appearing in eq. (8.95) must all be taken at the same time t.

What should now follow is a somewhat lengthy calculation which involves the performance of summations or integrations over $\boldsymbol{k}$. Since it does not shed any light on the physics we skip the details and just present the final result. In it we make use of the dipole approximation which means that we use the specific form of eq. (7.39a) for the coupling coefficients g. The final result then has the following form

$$K(\boldsymbol{x},t) = (4\pi\varepsilon_0)^{-2}\frac{k_0^4}{|\boldsymbol{x}|^2}\sum_{j,j'}(\boldsymbol{e}\boldsymbol{\vartheta}_{1,j})\alpha_j(0)\alpha_{j'}^*(0)(\boldsymbol{e}\boldsymbol{\vartheta}_{1,j'})$$
$$\times\Theta\left(t-\frac{|\boldsymbol{x}|}{c}\right)\exp\left[(-i\omega_{jj'}-\gamma)\left(t-\frac{|\boldsymbol{x}|}{c}\right)\right], \tag{8.96}$$

where

$$\omega_{jj'} = (W_j - W_{j'})/\hbar, \qquad j = 2,3; \qquad j' = 2,3. \tag{8.97}$$

In it we assume that a special direction of polarization $\boldsymbol{e}$ is measured by the detector. $\boldsymbol{\vartheta}_{1,j}$ are the dipole moment matrix elements for the optical transitions from levels $j = 2$ or 3 to level 1. The field intensity measured at space point $\boldsymbol{x}$ and time t, as given by eq. (8.96), depends on various factors which are worth to be discussed in detail. Starting from the right to the left we observe the following.

The field intensity decays exponentially with time t and the decay constant γ corresponding to the well known spontaneous emission. However, this decay appears modulated by the factor $\exp(-i\omega_{jj'}(t-|\boldsymbol{x}|/c))$. This factor leads really to a modulation as can be seen when we perform the sum over j, j'. We then readily may verify that this exponential function gives rise to a term proportional to $\cos(i\omega_{ij}(t-|\boldsymbol{x}|/c))$. This function causes the modulation of the light signal and occurs with the frequency splitting. Θ is the Heaviside function which vanishes when its argument $t-|\boldsymbol{x}|/c$ is negative and it equals unity for positive $t-|\boldsymbol{x}|/c$. It just reflects the fact that a signal cannot proceed faster than the velocity of light. Of further particular interest is the appearance of $\alpha_j(0)$s, which are the amplitudes of the initial wave functions. Since these amplitudes are complex they still contain certain phases. When we prepare an ensemble of atoms in initial states with random phases we have to average eq. (8.96) over such phases. In such a case the average products $\alpha_i\alpha_j^*$, $i \neq j$ yield 0 and exactly those terms will vanish which give rise to the modulation effect. This shows quite clearly that the modulation effect depends sensitively on the way the initial atom, or more precisely speaking, the initial state of an ensemble of atoms is prepared. For instance, if we excite the atoms incoherently from their groundstate to their excited states by a broad band source such cancellation of contributions in eq. (8.96) will occur. On the other hand a short pulse with a time-limited frequency spread only can cause a coherent superposition of the initial states. We thus see that the outcome of the photon beat experiment depends in a sensitive way on the kind the initial states 2 and 3 are prepared. Space does not allow us to enter the mathematical formulation of that problem.

However, in conclusion of this section we should like to discuss the interpretation of this photon interference experiment which indeed is a beautiful manifestation of typical quantum mechanical processes similar to Young's double slit experiment. Since we can hardly do any better we quote Haroche's description of this effect:

*8.4.1. Physical interpretation of the quantum beat signal**

In the experiment sketched in fig. 8.1, the impinging light pulse is scattered by the atom through two different channels, corresponding each to a given excited state 2, 3. These channels are symbolized by the diagrams of fig. 8.1. When the atom has re-emitted a photon there is no way to distinguish between the two possible channels and to tell whether the photon has been scattered through level 2 (with an amplitude α_2) or through level 3 (with an amplitude α_3). As a general postulate of quantum mechanics, the amplitude corresponding to these two indistinguishable processes must be added and their sum must be squared to yield the expression of the signal. This leads to the interfering term $\alpha_j \alpha_{j'}^*$ in the expression (8.96) of $K(\boldsymbol{x}, t)$. After integration over the photon energy, these interfering terms produce the modulations in the observed signal. As α_2 and α_3 are at resonance modulated, respectively, at frequencies $W_2/\hbar$ and $W_3/\hbar$ their cross product does indeed exhibits a modulation at frequency $(W_3 - W_2)/\hbar$. Quantum beats are thus a typical quantum interference effect quite similar to Young's double slit experiment. In complete analogy with this latter problem, it is quite clear that any attempt to perform an experiment in order to determine through which channel the photon has been scattered will result in a disappearance of the beat pattern. For example, we may try to make use of the polarization selection rules and detect the light emitted with a polarization $\boldsymbol{e}_d$ which can be emitted by only one level 2, for example. In that case, however, the matrix element $(\boldsymbol{e}_d \boldsymbol{\vartheta}_{1,3})$ is equal to zero and there are obviously no beats in eq. (8.96). Another way of determining the emission channel would be to put in front of the detector a narrow band filter centered, for example, around the $(W_2 - W_1)/\hbar$-frequency. As a result of this filtering, the summation over $\boldsymbol{k}$ and $\boldsymbol{k}'$ in eq. (8.93) is now restricted to a small frequency interval excluding the $(W_3 - W_1)/\hbar$-frequency. In that interval the amplitude remains very small, and the beats are again lost.

*The reader is advised to compare the arguments presented here with those of section 1.10.

Exercise on section 8.4

(1) Prove the equivalence between eqs. (8.90) (Schrödinger picture) and (8.28) (Heisenberg picture) where $t = t', x = x'$ and the sum over $i = j = 1, 2, 3$ is taken.
Hint: Use the transformation $\Phi(t) = U\Phi(0)$ with $U = \exp(-iHt/\hbar)$, H: total Hamiltonian.

9. Dissipation and fluctuations in quantum optics

9.1. Damping and fluctuations of classical quantities: Langevin equation and Fokker–Planck equation

In this chapter we will deal with some problems which at first sight seem to have little or nothing to do with quantum optics. Indeed the corresponding ideas and methods were introduced into quantum optics rather late but they have turned out absolutely necessary for the explanation of many phenomena, especially in laser physics and nonlinear optics.

The phenomenon which we have in mind is the Brownian motion. When a particle is immersed in a fluid, the velocity of this particle is slowed down by a force proportional to the velocity of this particle. When one studies the motion of such a particle under a microscope in more detail one realizes that this particle undergoes a zig-zag motion (compare fig. 9.1). This effect was first observed by the biologist Brown. The reason for the zig-zag motion is this. The particle under consideration is steadily pushed by the much smaller particles of the liquid in a random way. Let us describe the whole process from a somewhat more abstract viewpoint. Then we deal with the behaviour of a system (namely the particle) which is coupled to a heatbath or reservior (namely the liquid). The heatbath causes two effects: (1) It decelerates the mean motion of the particle and (2) Simultaneously the heatbath causes statistical fluctuations. As we shall see later atoms or the light field are by no means isolated systems. For instance optically active impurity atoms in solids interact all the time with lattice vibrations. Gas atoms steadily suffer collisions with other atoms which form the heatbath for the individual atom under consideration. The light field in the cavity interacts all the time with the walls forming a heatbath for the light field. From these examples it transpires that the notion of a heatbath plays an important role for systems considered in our book. Its importance will become fully evident when we apply these

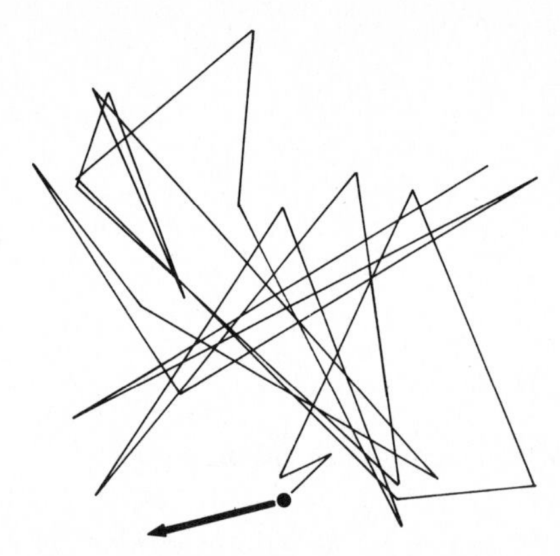

Fig. 9.1. Pathway of a particle undergoing Brownian motion.

methods to laser theory and nonlinear optics. But let us first return to Brownian motion where we consider a one-dimensional example. Let the particle have the mass m. We denote its velocity at time t by $v(t)$. Under the impact of external forces the particle is accelerated according to Newton's law

$$m\frac{\mathrm{d}}{\mathrm{d}t}v = K(t). \tag{9.1}$$

In the present case the force $K(t)$ is composed of two parts. (1) One is the friction force which is proportional to the velocity of the particle but with reversed sign

$$K_{\text{friction}} = -\gamma_0 v. \tag{9.2}$$

(2) The particle is steadily pushed by the particles of the fluid. To develop a model for these pushes we compare this process with a football game (with a soccer game) where the soccer players kick the ball again and again but at each instance for a short time interval only. We idealize these events by assuming an infinitely short time for each individual push. The force exerted on the particle (ball) is then represented by a δ-function. We assume that the individual pushes occur at times t_j. Furthermore, we shall admit that the pushes occur in the right or left direction in a random sequence. According to these ideas, we write the force representing the random pushes in the form

$$F_0(t) = \varphi \sum \delta(t - t_j)(\pm 1)_j. \tag{9.3}$$

An example for F_0 is given in fig. 9.2. φ measures the strength of the pushes. The symbol $(\pm 1)_j$ is meant to indicate the direction of the push at each time t_j. The times t_j form a random sequence whose statistical

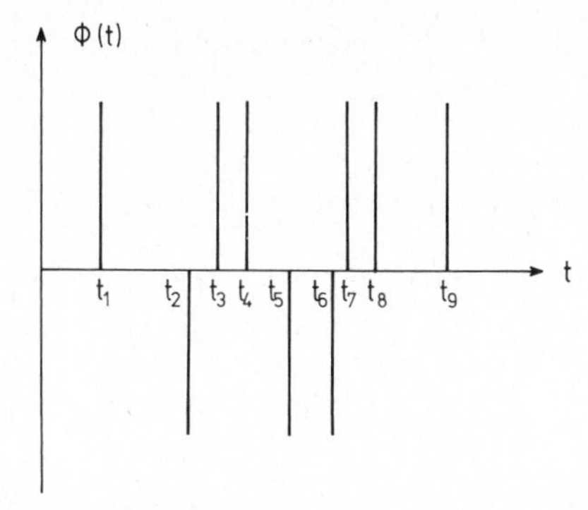

Fig. 9.2. Example of a random sequence of kicks $\Phi(t) \equiv F_0(t)$

properties we shall not discuss here. We assume that the pushes to the left and to the right occur with equal frequencies. Therefore when we average the force $F_0(t)$ over all pushes we obtain

$$\langle F_0 \rangle = 0. \tag{9.4}$$

(For a more detailed presentation of the classical theory of Brownian motion and its general formulations I refer the reader to my book (Haken, Synergetics – An Introduction, 2nd ed., Springer, 1978)). Here we present only some of the basic features of classical Brownian motion. It will turn out that we need a fully quantum mechanical treatment which will be given in the subsequent sections of this book.

In many cases it is important to know the correlation functions of the stochastic force eq. (9.3). Forming the product of these forces at two different times and averaging over all pushes we obtain

$$\langle F_0(t) F_0(t') \rangle = C\delta(t - t'). \tag{9.5}$$

For a proof see, e.g. the above mentioned book. The constant C is proportional to φ^2 and thus a measure for the strength of each push. The force K in eq. (9.1) is composed of the friction force (9.2) and the stochastic force (9.3). This yields the fundamental equation of Brownian motion

$$m\frac{\mathrm{d}v}{\mathrm{d}t} = -\gamma_0 v + F_0(t). \tag{9.6}$$

For simplification of eq. (9.6), we divide this equation by the mass m and introduce new abbreviations

$$\gamma = \gamma_0 / m; \qquad F(t) = F_0(t)/m. \tag{9.7}$$

The equation which we now want to consider thus reads

$$\frac{\mathrm{d}v}{\mathrm{d}t} = -\gamma v + F(t). \tag{9.8}$$

Due to the relation (9.7), the correlation functions (9.4) and (9.5) are replaced by

$$\langle F(t)\rangle = 0$$
$$\langle F(t)F(t')\rangle = Q\delta(t-t'), \tag{9.9}$$

where Q is a measure for the strength of the fluctuations. Equation (9.8) is a differential equation of first order whose formal solution reads

$$v(t) = \int_0^t \exp[-\gamma(t-\tau)]F(\tau)\,d\tau + v(0)\exp[-\gamma t]. \tag{9.10}$$

In it $v(0)$ is the initial velocity prescribed at time $t = 0$. When we wait a sufficiently long time which is large compared to the inverse damping constant γ, the last term on the right-hand side of eq. (9.10) can be neglected. When we then average over all pushes, we obtain by means of eq. (9.9) the relation $\langle v(t)\rangle = 0$. Thus the mean velocity is equal to 0. On the other hand, we observe under a microscope that the particle is steadily pushed around and thus achieves a finite velocity after each push. The vanishing of the mean velocity stems only from the fact that when we perform the average, positive and negative velocities cancel each other. Thus to get a more realistic measure for the average velocity we must form an expression in which the sign of the velocity no longer plays a role. The simplest expression of this kind is $\langle v(t)^2\rangle$. Since it is also interesting to know how long a velocity is preserved, it is advantageous to consider the more general expression

$$\langle v(t)v(t')\rangle. \tag{9.11}$$

Taking $t = t'$ we obtain a measure of the size of the velocity independent of its sign. In addition, for $t \neq t'$ we obtain an expression which tells us how long the velocities remain correlated. The latter can best be seen by means of the following limiting case. Letting the time difference between t and t' become very large the particle at time t will no more be able to "remember" what its velocity was at time t', because it has suffered many random pushes in between. In such a case the velocities at times t and t' have become statistically independent. According to basic rules of probability theory the expression (9.11) then splits into the product

$$\langle v(t)\rangle\langle v(t')\rangle. \tag{9.11a}$$

On the other hand, we have seen above that $\langle v(t)\rangle$ vanishes for $t \to \infty$, so that (9.11) vanishes for great time differences. This result means that at these times there is no longer any correlation. Of course, we expect a continuous transition of the correlation function (9.11) when going from $t = t'$ to $t - t' = \infty$ and we wish to consider this transition more closely.

To this end we form the correlation function (9.11) using the solution (9.10) where we may drop the term $\propto v(0)$ for sufficiently large t, t'

$$\langle v(t)v(t')\rangle = \left\langle \int_0^t \int_0^{t'} \exp[-\gamma(t-\tau)]F(\tau')\,d\tau \right.$$
$$\left. \times \langle \exp[-\gamma(t'-\tau')]F(\tau')\,d\tau'\rangle \right\rangle. \quad (9.12)$$

Since performing the average over the pushes has nothing to do with time integration we may exchange the brackets with the time integration and obain

$$\int_0^t \int_0^{t'} \exp[-\gamma(t+t'-\tau-\tau')]\langle F(\tau)F(\tau')\rangle\,d\tau\,d\tau'. \quad (9.13)$$

In this expression we may use the relation (9.9). Due to the δ-function, the integration of the double integral can be replaced by a single integral. For the case $t > t'$ we obtain especially

$$\langle v(t)v(t')\rangle = Q\int_0^{t'} \exp[-\gamma(t+t')+2\gamma\tau]\,d\tau$$
$$= \frac{Q}{2\gamma}\{\exp[-\gamma(t-t')] - \exp[-\gamma(t+t')]\}. \quad (9.14)$$

We now consider a situation in which the times t and t' are so big that

$$t + t' \gg 1/\gamma, \quad (9.15)$$

but that the time difference $t - t'$ is still of the order

$$|t - t'| \lesssim 1/\gamma. \quad (9.16)$$

In this case, eq. (9.14) acquires the especially simple form (fig. 9.3)

$$\langle v(t)v(t')\rangle = \frac{Q}{2\gamma}\exp[-\gamma(t-t')]. \quad (9.17)$$

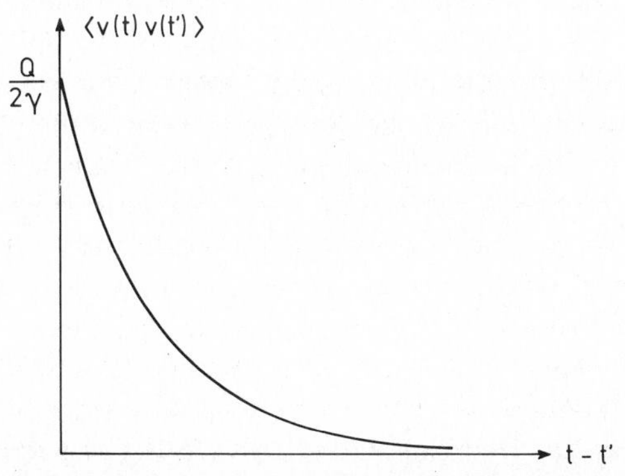

Fig. 9.3. The exponential decay of the correlation function of the velocity of a particle undergoing Brownian motion. The correlation function is plotted versus the time $t - t'$.

We thus have obtained an explicit expression for the correlation function of the velocities which tells us both the size of the average of the squared velocities and the way the correlation of the velocities dies out. Evidently the correlations decay according to an exponential law with decay time $1/\gamma$.

We now want to derive a formula which allows us to derive the factor Q (which is a measure for the size of the pushes) directly. Taking in eq. (9.17) $t = t'$ and multiplying it by $m/2$ we obtain on the left-hand side of eq. (9.17) the mean kinetic energy of the particle. According to fundamental laws of thermodynamics we must assume that the particle is in thermal equilibrium with its surrounding (heatbath). According to thermodynamics the kinetic energy of a single degree of freedom in thermal equilibrium at temperature T is equal to $\frac{1}{2}kT$, where k is Boltzmann's constant and T the absolute temperature. Thus we have quite generally

$$\frac{m}{2}\langle v(t)^2\rangle = \tfrac{1}{2}kT. \tag{9.18}$$

A comparison of eqs. (9.18) and (9.17) yields an explicit expression for Q

$$Q = 2\gamma kT/m. \tag{9.19}$$

As mentioned above Q is a measure for the size of the fluctuations. On the other hand γ is a measure for the damping or, in other words, for the dissipation (due to the damping of the velocity kinetic energy is dissipated). The relation (9.19) is probably the simplest example of a fluctuation-dissipation-theorem. The size of the fluctuations is proportional to the size of the dissipation. The importance of this relation lies in the fact that it allows us to calculate the size of fluctuations by means of the size of dissipation.

Equation (9.8) is called Langevin equation. Such equations and their quantum mechanical generalizations play a fundamental role in laser theory.

Besides the description of Brownian motion by means of Langevin equations, there exists a second approach namely that by the Fokker–Planck equation. The basic idea about the Fokker–Planck equation is this. Let us imagine a large number of identical experiments in which each time a particle of mass m is undergoing a Brownian motion. At a certain time t we can measure the velocities of the particles in each experiment. Taking the results of all the experiments together we shall find a certain number $N(v)\,\mathrm{d}v$ of particles in a given velocity interval $v, v + \mathrm{d}v$. We thus obtain a distribution function $N(v)$ which tells us how frequently the particles have the velocity v. Invoking probability theory we may go over to a probability distribution by normalizing the distribution function $N(v)$ to unity. More precisely speaking we introduce instead of $N(v)$ a new function $f(v)$ which is proportional to $N(v)$ but for which

$$\int_{-\infty}^{\infty} f(v,t)\mathrm{d}v = 1 \tag{9.20}$$

holds. $f(v,t)\,\mathrm{d}v$ has the following meaning (compare fig. 9.4). $f(v,t)\,\mathrm{d}v$ is the probability of finding a particle at time t with a velocity v in the interval v, $v + \mathrm{d}v$. As is shown in statistical physics the probability distribution function f which belongs to Brownian motion, i.e. to the Langevin equation (9.8), obeys the Fokker–Planck equation

$$\frac{\partial}{\partial t} f = \left[\frac{\partial}{\partial v} \gamma v + \frac{Q}{2} \frac{\partial^2}{\partial v^2} \right] f. \tag{9.21}$$

The factor γv is called drift coefficient, the factor Q diffusion coefficient. The stationary solution with $(\partial/\partial t)f = 0$ reads

$$f_{st}(v) = \mathcal{N} \exp[-\gamma v^2/Q]. \tag{9.22}$$

$\mathcal{N}$ is the normalization factor and is given explicitly by

$$\mathcal{N} = \left[\frac{\gamma}{\pi Q} \right]^{1/2}. \tag{9.23}$$

Using the fluctuation–dissipation theorem, eq. (9.19), we can bring eq. (9.22) into the form

$$f_{st}(v) = \mathcal{N} \exp\left[-\frac{m}{2} \frac{v^2}{kT} \right]. \tag{9.24}$$

This is of course the Maxwell–Boltzmann distribution function for the

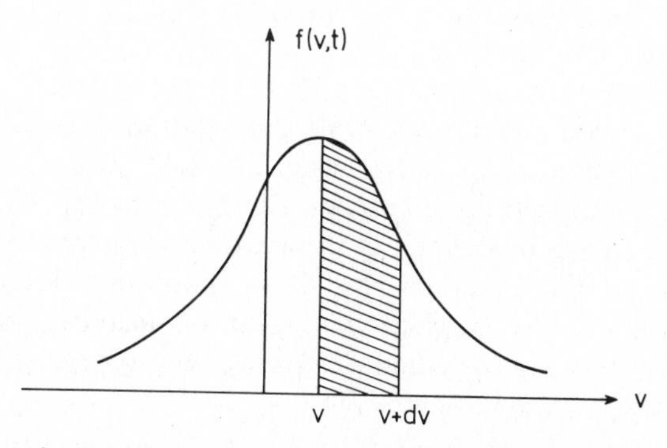

Fig. 9.4. An example of the distribution function obeying the Fokker–Planck equation versus v at a given time t. The shaded area gives us the probability of finding the particle in the velocity range $v \ldots v + \mathrm{d}v$.

velocity of a particle in thermal equilibrium with a heatbath at temperature T.

Later in this book we shall get acquainted with cases in which the system under consideration is coupled to several heatbaths kept at different temperatures. Our considerations can be generalized in several respects, for instance we may consider a Langevin equation which contains a more general force. Simultaneously we replace the velocity by a general variable q. The corresponding Langevin equation with a general but still deterministic force K then reads

$$\frac{\mathrm{d}}{\mathrm{d}t}q = K(q) + F(t). \tag{9.25}$$

$F(t)$ is again the fluctuating force with the properties eq. (9.9). We note that for the derivation of the corresponding Fokker–Planck equation for (9.25) it is necessary that $F(t)$ is, in technical terms, Gaussian. We will not discuss this point here in detail but refer the reader to the literature.

Under these assumptions the Fokker–Planck equation, associated with the Langevin equation (9.25), reads

$$\frac{\partial}{\partial t}f(q,t) = \left[-\frac{\partial}{\partial q}K(q) + \frac{Q}{2}\frac{\partial^2}{\partial q^2} \right] f(q,t). \tag{9.26}$$

Its stationary solution reads

$$f_{st}(q) = \mathfrak{N}\exp\left\{ -\frac{2}{Q}\int^q K(q')\mathrm{d}q' \right\}. \tag{9.27}$$

Q is assumed as a constant independent of q and time. Unfortunately the time-dependent, Fokker–PLanck equation (9.26) cannot be solved explicitly, at least in general, as was possible for a linear force $K \sim q$. Thus computer solutions are required.

We now try to translate our considerations to a damped electric or electromagnetic field. An electric field propagating in a conductor gets damped because it accelerates electrons of the material giving its energy to the electrons. The electrons lose their energy by collisions with lattice vibrations. The energy dissipation of the electric field is proportional to the conductivity σ of the material. This fact can be derived directly from Maxwell's equations. From Maxwell's equations including damping one may derive the telegraph equation describing the propagation of an electromagnetic field in a conductor

$$\left[\frac{\partial^2}{\partial t^2} + \frac{\sigma}{\varepsilon_0}\frac{\partial}{\partial t} - c^2\frac{\partial^2}{\partial x^2} \right] E = 0. \tag{9.28}$$

We have written down the equation in one dimension, the field strength being transverse to the direction of propagation. We now consider a standing wave between two mirrors. For this reason we assume the electric field strength in the form

$$E(x,t) = b(t)\sin kx, \tag{9.29}$$

where the wave number k is chosen in such a way that the wave fits inbetween the 2 mirrors (compare fig. 2.12). By inserting eq. (9.29) in (9.28) and performing the differentiation with respect to x we obtain after dividing the equation by $\sin kx$

$$\left[\frac{\partial^2}{\partial t^2} + 2\kappa\frac{\partial}{\partial t} + \omega_0^2\right] b(t) = 0. \tag{9.30}$$

We have used the abbreviations

$$\frac{\sigma}{\varepsilon_0} = 2\kappa; \qquad kc = \omega_0. \tag{9.30a}$$

The oscillator equation (9.30) can be solved as usual by the ansatz

$$b(t) = a\exp[-i\omega t]. \tag{9.31}$$

By inserting eq. (9.31) in (9.30) yields a relation for the frequency

$$-\omega^2 - 2\kappa\omega i + \omega_0^2 = 0. \tag{9.32}$$

When we assume that the damping constant κ is much smaller than the frequency ω_0 we may write the solution of eq. (9.32) in the form

$$\omega \approx i\kappa \pm \omega_0. \tag{9.33}$$

Thus we obtain the result that the time-dependent amplitude $b(t)$ and thus the electrical field strength (9.29) decays exponentially

$$b(t) = b(0)\exp[-i\omega_0 t - \kappa t]. \tag{9.34}$$

Apart from constant factors which are of no interest the function $b(t)$ is identical with the amplitude $b(t)$ which we used in section 5.1. In order to take into account damping, i.e. the exponential decay of eq. (9.34), we have to supplement the former equation (5.105) by a term $-\kappa b$

$$\frac{\mathrm{d}}{\mathrm{d}t}b(t) = (-i\omega_0 - \kappa)b(t). \tag{9.35}$$

Note, however, that in identifying $b(t)$ of equation (5.105) with that occurring in (9.34) we made a big jump. While eq. (9.34) describes a classical amplitude, b occurring in (9.35) is meant as a quantum mechanical operator! Note further that eq. (9.35) is only a tentative step to incorporate dissipation into quantum mechanics. Namely, we still have to answer the question whether eq. (9.35) yields an operator which fulfills the general laws of quantum mechanics. We will deal with this problem in the subsequent section.

Exercises on section 9.1.

(1) Solve the time-dependent, Fokker–Planck equation (9.21) with the initial condition

$$f(v,0)=\delta(v)$$
$$\delta(v)=\text{Dirac's delta function.}$$

Hint: Try

$$f(v,t)=a(t)\exp\left[-v^2b(t)\right], \qquad (*)$$

where $a(t)$, $b(t)$ are time-dependent functions. In order to determine them, insert $(*)$ into (9.21), divide by $f(v,t)$ and compare coefficients of powers of v. Solve the resulting differential equations for a, b.

(2) Calculate the following average values:

$$\langle v(t)\rangle \equiv \int f(v,t)v\,\mathrm{d}v,$$

$$\langle v^2(t)\rangle \equiv \int f(v,t)v^2\,\mathrm{d}v, \qquad (*)$$

$$\langle v(t)v(t')\rangle \equiv \int f(v,t;v',t')vv'\,\mathrm{d}v\,dv'$$

$$= \int f(v,t|v',t')f(v',t')vv'\,\mathrm{d}v\,dv'.$$

Hints: $f(v,t;v',t')$ is the two-time probability for finding the "particle" around v' at time t' and around v at time t. This probability (– density) may be written as

$$f(v,t;v',t')=f(v,t|v',t')f(v',t'), \qquad t\geqslant t',$$

where $f(v,t|v',t')$, the "conditional probability density", gives the probability for finding the "particle" around v at t provided it was around v' at t'; $f(v',t')$ is the usual single-time probability density. $f(v,t|v',t')$ is a solution to eq. (9.21) with the initial condition $f(v,t'|v',t')=\delta(v-v')$. Show that a formal solution reads

$$f(v,t|v',t')=\exp\left[(t-t')\left(\frac{\partial}{\partial v}\gamma v+\frac{Q}{2}\frac{\partial^2}{\partial v^2}\right)\right]\delta(v-v') \qquad (**)$$

by inserting $(**)$ into eq. (9.21). Then rewrite $(*)$ using this formal expression, expand the exponential into a power series and integrate by parts. The resulting integral may be easily evaluated and gives

$$\langle v(t)v(t')\rangle = \exp[-\gamma(t-t')]\langle v^2(t')\rangle .$$

(3) Derive the telegraph equation (9.28) from Maxwell's equations (5.1), (5.2), putting $\rho = 0$ and $\boldsymbol{j} = \sigma \boldsymbol{E}$.
Hint: Form on both sides of eq. (5.1) curl and insert $(\mathrm{d}/\mathrm{d}t)\mathrm{curl}\,\boldsymbol{H}$ from (5.2). Use the relation $\mathrm{curl}(\mathrm{curl}\,\boldsymbol{E}) = \nabla(\nabla\boldsymbol{E}) - \Delta\boldsymbol{E}$ and eq. (5.5, 6).

9.2. Damping and fluctuations of quantum mechanical variables: Field modes

We now wish to deal with our main problem, namely how we can describe the damping of quantum mechanical quantities. In section 5 we learned how to quantize the light field. At the end of the preceding section we got to know an example of how a classical light field is damped due to its interaction with the medium. It is tempting to translate eq. (9.35) and its corresponding solution (9.34) into quantum mechanics by interpreting b and its Hermitian adjoint b^+ as operators. We repeat the tentative equation for such an operator b

$$\frac{\mathrm{d}}{\mathrm{d}t}b(t) = (-i\omega_0 - \kappa)b(t) \tag{9.36}$$

with the solution

$$b(t) = b(0)\exp[-i\omega_0 t - \kappa t]. \tag{9.37}$$

The operator b^+ has to obey an equation which is conjugate complex to (9.36). Now we recall a fundamental postulate of quantum mechanics. We have seen at several occasions that b and b^+ obey the commutation relation (5.116) for all times. If they did not obey that relation, and for instance the commutation relation equalled 0 for $t \to \infty$ the operators b and b^+ would eventually commute and thus become classical quantities. This, of course, contradicts quantum mechanics. Thus we have to check whether eq. (9.37) and its Hermitian conjugate obey the commutation relation. We obtain

$$(bb^+ - b^+b)_t = (bb^+ - b^+b)_0 \exp[-2\kappa t]. \tag{9.38}$$

The bracket on the right-hand side is the commutation relation valid at initial time $t = 0$. Adopting the Schrödinger picture at that time we know that the bracket equals unity. We thus obtain

$$(9.38) = \exp[-2\kappa t], \tag{9.39}$$

i.e. an expression which vanishes for $t \to \infty$. Thus the commutation relation $bb^+ - b^+b = 1$ is violated and the equation (9.36) cannot be used in quantum mechanics. Quite astonishingly the example of Brownian motion of classical physics which we encountered in section 9.1 offers a solution to this problem. As we have seen the particle keeps its velocity by means of a fluctuating force representing the effect of the heatbath on the particle. In the case of the electromagnetic field we mentioned that its damping is caused by the coupling of the field to another system, for instance to the electrons in a conductor. This leads to the question whether these electrons can exert fluctuating forces on the electric field in much the same way as the liquid caused fluctuating forces acting on the particle. This is indeed so and we shall come back to a detailed treatment of this problem in the next section. Here we anticipate the final result. It will turn out that the effect of a heatbath at temperature T on an oscillator, especially a mode of the electromagnetic field, obeys the equation

$$\frac{\mathrm{d}}{\mathrm{d}t} b(t) = (-i\omega_0 - \kappa) b(t) + F(t). \tag{9.40}$$

In this equation, the b's are operators. As we will see below also the fluctuating forces F are operators having the following properties

$$\begin{aligned} &\langle F(t)\rangle = 0, \qquad \langle F^+(t)\rangle = 0, \\ &\langle F^+(t)F^+(t')\rangle = \langle F(t)F(t')\rangle = 0, \\ &\langle F^+(t)F(t')\rangle = 2\kappa\bar{n}\delta(t-t'), \\ &\langle F(t)F^+(t')\rangle = 2\kappa(\bar{n}+1)\delta(t-t'). \end{aligned} \tag{9.41}$$

The bracket means a quantum-statistical average over the fluctuations, or in other words, over the variables of the heatbath. It is evident from the last line expressions of (9.41) that the correlation functions of F and F^+ depend on the sequence of F and F^+. This reflects the fact that the F's are operators. Insofar, there is an important difference between classical Langevin forces and the quantum mechanical Langevin forces treated here. As we shall see, in applications we don't need to know how the average is defined in detail, although we will give an explicit example below. All we need to know are the properties of eq. (9.41), and perhaps the Gaussian property which we will explain later.

$\bar{n}$ is the mean number of photons present at temperature T. We now want to show that the fluctuating forces having the properties (9.41) give rise to solutions of eq. (9.40) which obey the quantum mechanical commutation relations. The solution of (9.40) has exactly the same form as that of equation (9.8), namely the form (9.10). Inserting these solutions into the commutation relation $bb^+ - b^+b$ and averaging over the heatbath variables we obtain after some calculation the result

$$\langle bb^+ - b^+b\rangle = 1. \tag{9.42}$$

In contrast to eq. (9.39), the new operators b and b^+ obey the commutation relation, at least when averaged over the heat bath. It can be shown that this is all we need to require. When we average b^+ or b over the heatbath variables we obtain on account of (9.41).

$$\langle b^+(t)\rangle = \langle b^+(0)\rangle \exp[-\kappa t + i\omega_0 t] \tag{9.43}$$

$$\langle b(t)\rangle = \langle b(0)\rangle \exp[-\kappa t - i\omega_0 t]. \tag{9.44}$$

Thus the averaged operators behave exactly as the classical quantities. In complete analogy to (9.12) we can construct the correlation functions of b, b^+

$$\langle b^+(t)b(t')\rangle = \bar{n}\exp[-\kappa|t-t'|]\exp[i\omega_0(t-t')]. \tag{9.45}$$

Again $\bar{n}$ is the number of photons present in thermal equilibrium. Clearly the correlation between b^+ and b at times t, t' decays with the time constant κ.

Exercises on section 9.2.

(1) Prove eq. (9.42).
Hint: Find the solution of (9.40) in a form analogous to (9.10). Insert it into (9.42) and use (9.41).

(2) Derive eq. (9.45)
Hint: Use the solution $b(t)$ constructed in exercise (1) and the relations (9.41).

(3) The mode operators b, b^+ are coupled to a heatbath and a driving force $f(t)$, $f^*(t)$, respectively: $f(t) = f_0 \exp(i\omega' t)$. Calculate

$$\langle b(t)\rangle, \qquad \langle b^+(t)b(t')\rangle.$$

Hint: Start from the equation

$$\frac{\mathrm{d}}{\mathrm{d}t} = (-i\omega - \kappa)b + F(t) + f(t)$$

and its Hermitian conjugate. $F(t)$ is the fluctuating force. Solve these equations and use (9.41).

9.3. Quantum mechanical Langevin equations. The origin of quantum mechanical fluctuating forces*

In the preceding two sections we got acquainted with the idea of fluctuating forces and we learned about their properties. In this section we want to show by means of an exactly solvable example how these fluctuating forces

*This section is somewhat more difficult to read and can be skipped.

arise explicitly. In the sense of thermodynamics we assume that any heatbath (reservoir) with some general properties which we will specify below, produces fluctuating forces with the same typical properties. For this reason it will be sufficient to choose a heatbath which can be treated in an explicit manner. Such an example is provided by a heatbath composed of an infinite set of harmonic oscillators. We couple the harmonic oscillator under consideration which represents the particle of Brownian motion (provided it is coupled harmonically to the origin), or the field mode, to this heatbath. The Hamiltonian of the total quantum system thus consists of three parts, namely that of the considered oscillator ("system")

$$H_0 = \hbar\omega_0 b^+ b, \tag{9.46}$$

the Hamiltonian of the reservoir oscillators with creation and annihilation operators B_ω^+, B_ω

$$H_B = \sum_\omega \hbar\omega B_\omega^+ B_\omega \tag{9.47}$$

and the interaction. We assume that the interaction is linear in the operators B_ω and b^+ and their hermitian conjugates. We assume further that the creation of a quantum of the harmonic oscillator of the system is connected with the annihilation of a quantum of the reservoir and vice versa. Thus the interaction Hamiltonian acquires the form

$$H_I = \hbar\sum_\omega \{g_\omega b^+ B_\omega + g_\omega^* B_\omega^+ b\}. \tag{9.48}$$

The Heisenberg equations (compare section 5.6) for b^+ and B_ω^+ read

$$\frac{\mathrm{d}b^+}{\mathrm{d}t} = i\omega_0 b^+ + i\sum_\omega g_\omega^* B_\omega^+, \tag{9.49}$$

$$\frac{\mathrm{d}B_\omega^+}{\mathrm{d}t} = i\omega B_\omega^+ + ig_\omega b^+. \tag{9.50}$$

Equation (9.50) immediately allows for the solution

$$B_\omega^+(t) = i\int_0^t b^+(\tau) g_\omega \exp[i\omega(t-\tau)]\,\mathrm{d}\tau + B_\omega^+(0)\exp[i\omega t], \tag{9.51}$$

where $B_\omega^+(0)$ is the operator at time $t_0 = 0$. By inserting eq. (9.51) into (9.49) we arrive at

$$\frac{\mathrm{d}b^+}{\mathrm{d}t} = i\omega_0 b^+ - \int_0^t b^+(\tau)\sum_\omega |g_\omega|^2 \exp[i\omega(t-\tau)]\,\mathrm{d}\tau$$
$$+ i\sum_\omega g_\omega^* B_\omega^+(0)\exp[i\omega t]. \tag{9.52}$$

We want to show that the integral occurring in eq. (9.52) can be easily evaluated, at least in a good approximation. Since $b^+(\tau)$ appears under it, we must first discuss the dependence of b^+ on τ. When we assume a weak interaction between b^+, b and the bath, we expect that the last two expressions in eq. (9.52) do not change the time dependence of b, b^+ too dramatically. Consequently we expect b^+ to have the form

$$b^+(t) = e^{i\omega_0 t}\tilde{b}^+(t), \tag{9.53}$$

where $\tilde{b}^+(t)$ changes much more slowly in time than its exponential factor. Furthermore, we replace the first sum over ω in eq. (9.52) by an integral:

$$\sum_\omega |g_\omega|^2 \to \int_0^\infty |\tilde{g}_\omega|^2 \exp(i\omega(t-\tau) - i\omega_0(t-\tau))\,\mathrm{d}\omega \tag{9.54}$$

where $|\tilde{g}_\omega|^2$ differs from $|g_\omega|^2$ by a factor which stems from the numeration of ω's.

A characteristic feature of heatbaths is that they provide very many (practically infinitely many) degrees of freedom, which must be coupled to the system under consideration, in our case to a harmonic oscillator. This implies that the coupling constants g_ω are of comparable size. We consider the idealized case in which $|\tilde{g}_\omega|^2$ is independent of ω. Furthermore, we shift the integration variable $\omega \to \omega'$ so that

$$\omega - \omega_0 = \omega'$$

and the lower limit of the integral lies at $\omega' = -\omega_0$. Since ω_0 is large we replace the lower limit by $\omega' = -\infty$. The thus resulting integral is proportional to Dirac's δ-function

$$\int_{-\infty}^{\infty} |\tilde{g}|^2 \exp[i\omega(t-\tau)]\,\mathrm{d}\omega = 2\kappa\delta(t-t'), \tag{9.55}$$

where $\kappa = \pi|\tilde{g}|^2$. For an evaluation of the integrals it must be noted that the δ-function has the property

$$\int_0^t \delta(t-\tau)\,\mathrm{d}\tau = \tfrac{1}{2}. \tag{9.56}$$

By means of these results the integral in eq. (9.52) can be replaced by $-\kappa b^+(t)$

$$\frac{\mathrm{d}b^+}{\mathrm{d}t} = i\omega_0 b^+ - \kappa b^+ + \underbrace{i\sum_\omega g_\omega^* B_\omega^+(0)\exp[i\omega t]}_{F^+(t)} \tag{9.57}$$

where the last term is evidently the fluctuating force. We determine the properties of this fluctuating force and consider the average of $F^+(t)$ over the heatbath variables

$$\langle F^+(t)\rangle_B = i\sum_\omega g^*_\omega \exp[i\omega t]\langle B^+_\omega(0)\rangle_{B,\omega}. \tag{9.58}$$

The index B at $\langle\ldots\rangle_B$ means "bath average", the index ω at $\langle\ldots\rangle_{B,\omega}$ means "bath average over the bath variables B^+_ω, B_ω alone." The average $\langle B^+_\omega(0)\rangle_{B,\omega}$ can be explicitly evaluated. We assume that at initial time $t = 0$ the heatbath and thus each of its oscillators is in thermal equilibrium. According to quantum-statistics this average is defined by

$$\langle B^+_\omega(0)\rangle_{B,\omega} = Z_\omega^{-1}\sum_{n=0}^{\infty} \exp[-n\hbar\omega/(kT)]\langle\Phi_n|B^+_\omega(0)|\Phi_n\rangle. \tag{9.59}$$

(Readers not familiar with such quantum statistical averages will find a motivation of this definition in section 9.5.) Z_ω is the partition function defined by

$$Z_\omega = \sum_{n=0}^{\infty} \exp[-n\hbar\omega/(kT)]. \tag{9.60}$$

k is Boltzmann's constant and T the absolute temperature. $\langle\Phi_n|B^+_\omega(0)|\Phi_n\rangle$ is the quantum mechanical expectation value of the operator B^+_ω in the Schrödinger picture, taken with respect to the oscillator wave function Φ_n. Because we know that

$$\langle\Phi_n|B^+_\omega(0)|\Phi_n\rangle = 0, \tag{9.61}$$

we obtain

$$\langle F^+(t)\rangle_{B,\omega} = 0. \tag{9.62}$$

In a similar way we find

$$\langle F(t)\rangle_{B,\omega} = 0. \tag{9.63}$$

For later use it will be convenient to write averages of the form (9.59) more concisely. Let Ω be an arbitrary operator (composed of B_ω and B^+_ω for a fixed ω), then we define

$$\langle\Omega\rangle_{B,\omega} = Z_\omega^{-1}\, Tr_\omega(\Omega \exp[-H_{B,\omega}/|(kT)]), \tag{9.64}$$

where $H_{B,\omega} = \hbar\omega B^+_\omega B_\omega$. "$Tr$" means trace. The trace of an operator $\tilde{\Omega}$ is defined by

$$Tr\,\tilde{\Omega} = \sum_{n=0}^{\infty} \langle\Phi_n|\tilde{\Omega}|\Phi_n\rangle, \tag{9.64a}$$

where in the present case the Φ_n's are, as before, the eigenfunctions of the harmonic oscillator with frequency ω. The equivalence of eq. (9.64),

$\Omega = B_\omega^+$, with (9.59) will be shown in the exercise. In a similar way, we can evaluate

$$\langle F^+(t)F(t')\rangle_B = \sum_\omega |g_\omega|^2 \exp[i\omega(t-t')]$$
$$\times Tr\left(B_\omega^+ B_\omega \exp\left[-\frac{H_B}{kT}\right]\right) Z_\omega^{-1} = \sum_\omega |g_\omega|^2 \exp[i\omega(t-t')]\bar{n}_\omega(T) \tag{9.65}$$

where H_B is given by (9.47) and the trace refers to the multi-oscillator eigenfunctions belonging to H_B. For some intermediate steps leading to (9.65) cf. exercise 2 below. $\bar{n}_\omega(T)$ is the mean number of quanta of an oscillator with frequency ω and at temperature T:

$$\bar{n}_\omega(T) = Z_\omega^{-1} Tr(B_\omega^+ B_\omega \exp[-H_B/(kT)]). \tag{9.66}$$

In a similar way we obtain

$$\langle F(t)F^+(t')\rangle_B = \sum_\omega |g_\omega|^2 \exp[-i\omega(t-t')](\bar{n}_\omega(T)+1). \tag{9.67}$$

It follows from eqs. (9.65) and (9.67) that the bath average over the commutator has the form

$$\langle [F(t), F^+(t')]\rangle_B = 2\kappa\delta(t-t'). \tag{9.68}$$

One may even show that (9.68) holds without the bath average, i.e.

$$[F(t), F^+(t')] = 2\kappa\delta(t-t') \tag{9.68a}$$

(cf. exercise 3). Equations (9.65) and (9.67) can be directly expressed by the damping constant κ [cf. (9.55)] only for the temperature $T = 0$. For higher temperatures it must be noted that in practical calculations

$$\langle F^+(t)F(t')\rangle_B = \sum_\omega |g_\omega|^2 \exp[i\omega(t-t')]\langle B_\omega^+ B_\omega\rangle_{B,\omega} \tag{9.69}$$

and

$$\langle F(t)F^+(t')\rangle_B = \sum_\omega |g_\omega|^2 \exp[-i\omega(t-t')]\langle B_\omega B_\omega^+\rangle_{B,\omega} \tag{9.70}$$

always appear under an integral which contains a factor $\exp[i\omega_0 t]$, where ω_0 is the frequency of the harmonic oscillator (light field). If this fact is taken into account, eq. (9.69) may be expressed in the form [compare the steps leading to eq. (9.55)]

$$\langle F^+(t)F(t')\rangle_B = \sum_\omega |g_\omega|^2 \exp[i\omega(t-t')]\bar{n}_\omega(T) = 2\kappa\bar{n}_{\omega_0}(T)\delta(t-t'), \tag{9.71}$$

and similarly

$$\langle F(t)F^+(t')\rangle = 2\kappa\big(\bar{n}_{\omega_0}(T) + 1\big)\delta(t - t'). \tag{9.72}$$

The δ-function in eqs. (9.71) and (9.72) expresses the fact that the heatbath has a very short memory. As may be seen from eqs. (9.69) and (9.70), it is essential for the derivation that the heatbath frequencies have a spread which covers a whole range around the oscillator frequency ω_0. It is interesting to derive equations for the average values $\langle b^+\rangle_B$, $\langle b\rangle_B$, $\langle b^+b\rangle_B$, and $\langle[b, b^+]\rangle_B$ where the average over the heatbath variables is taken. By taking the average over equation (9.57) we obtain by use of (9.62)

$$\frac{\mathrm{d}}{\mathrm{d}t}\langle b^+(t)\rangle_B = (i\omega_0 - \kappa)\langle b^+(t)\rangle_B. \tag{9.73}$$

Quite similarly it follows that

$$\frac{\mathrm{d}}{\mathrm{d}t}\langle b(t)\rangle_B = (-i\omega_0 - \kappa)\langle b(t)\rangle_B. \tag{9.74}$$

To derive an equation for $\langle b^+b\rangle_B$ we solve eq. (9.57) explicitly

$$b^+(t) = b^+(0)\exp\big[(i\omega_0 - \kappa)t\big] + \int_0^t \exp\big[(i\omega_0 - \kappa)(t - \tau)\big]F^+(\tau)\,\mathrm{d}\tau \tag{9.75}$$

and form

$$\begin{aligned}\frac{\mathrm{d}}{\mathrm{d}t}\langle b^+b\rangle_B = \left\langle \frac{\mathrm{d}b^+}{\mathrm{d}t}b + b^+\frac{\mathrm{d}b}{\mathrm{d}t}\right\rangle_B &= -2\kappa\langle b^+b\rangle_B \\ &+ \int_0^t \langle F^+(t)F(\tau)\rangle_B \exp\big[-(i\omega_0 + \kappa)(t - \tau)\big]\,\mathrm{d}\tau \\ &+ \int_0^t \langle F^+(\tau)F(t)\rangle_B \exp\big[(i\omega_0 - \kappa)(t - \tau)\big]\,\mathrm{d}\tau \\ &+ b^+(0)\exp\big[(i\omega_0 - \kappa)t\big]\langle F(t)\rangle_B \\ &+ \langle F^+(t)\rangle_B b(0)\exp\big[(-i\omega_0 - \kappa)t\big].\end{aligned} \tag{9.76}$$

The integrals may be evaluated using eqs. (9.71) and (9.72), and the last two terms in (9.76) vanish on account of (9.62) and (9.63) so that we find

$$\frac{\mathrm{d}}{\mathrm{d}t}\langle b^+b\rangle_B = -2\kappa\langle b^+b\rangle_B + 2\kappa\bar{n}(T). \tag{9.77}$$

$\bar{n}(T)$ ($\equiv \bar{n}_{\omega_0}(T)$) is the mean number of quanta of the system oscillator with frequency ω_0 at temperature T. In a similar way we show how the fluctuating forces restore the commutation relation. In analogy to eq.

(9.76) we can derive

$$\frac{\mathrm{d}}{\mathrm{d}t}\langle[b,b^+]\rangle_B = -2\kappa\langle[b,b^+]\rangle_B + 2\kappa. \tag{9.78}$$

Its general solution reads

$$\langle[b,b^+]\rangle_{B,t} = C\exp[-2\kappa t] + 1. \tag{9.79}$$

When we insert the initial condition

$$\langle[b,b^+]\rangle_{B,0} = 1 \tag{9.80}$$

we find

$$C = 0 \tag{9.81}$$

which means that the commutation relation is preserved for all times.

So far we considered only a single particle (a harmonic oscillator coupled to a reservoir). The whole procedure can be extended to a set of particles (harmonic oscillators), which are in contact with statistically independent reservoirs at temperature T. Again we start from a Hamiltonian composed of the Hamiltonians of these particles, of their individual heatbaths and of the interactions between particles and heatbaths. The heatbath variables can be eliminated and we end up with the following set of quantum mechanical Langevin equations

$$\frac{\mathrm{d}b_\lambda^+}{\mathrm{d}t} = (i\omega_\lambda - \kappa_\lambda)b_\lambda^+ + F_\lambda^+. \tag{9.82}$$

The fluctuating forces have the following properties

$$\langle F_\lambda(t)\rangle = \langle F_\lambda^+(t)\rangle = 0, \tag{9.83}$$

$$\langle F_\lambda^+(t)F_{\lambda'}^+(t')\rangle = \langle F_\lambda(t)F_{\lambda'}(t')\rangle = 0, \tag{9.84}$$

$$\langle F_\lambda^+(t)F_{\lambda'}(t')\rangle = \bar{n}_\lambda(T)2\kappa_\lambda\delta(t-t')\delta_{\lambda\lambda'}, \tag{9.85}$$

$$\langle F_\lambda(t)F_{\lambda'}^+(t')\rangle = (\bar{n}_\lambda(T)+1)2\kappa_\lambda\delta(t-t')\delta_{\lambda\lambda'}. \tag{9.86}$$

The occurrence of the Kronecker symbol $\delta_{\lambda\lambda'}$ stems from the fact that the reservoirs are assumed to be statistically independent. We conclude this section with a general remark on the statistical properties of F_λ, F_λ^+. As can be seen from eq. (9.57), F_λ^+ (and F_λ) is composed of very many individual contributions which are statistically independent. If F_λ were a classical and real quantity, q, one may apply the central limit theorem which states that q is "Gaussian distributed". This means that the probability that the classical random variable q has the value q is given by

$$P(q) = \mathcal{N}\exp(-aq^2).$$

One may show that such a concept can be generalized to our operators F_λ,

F_λ^+. In an appropriately generalized sense, the F_λ's are then "Gaussian distributed" (for more details see the second volume).

Exercises on section 9.3.

(1) Prove the equivalence of eq. (9.64), (with $\Omega = B_\omega^+$) with (9.59).
Hints: Use the definition of the trace and show that

$$e^{-H_{B,\omega}/(kT)}\Phi_n = e^{-\hbar\omega n/(kT)}\Phi_n.$$

(2) Derive eq. (9.65)
Hint: Insert F^+ according to eq. (9.57) and its Hermitian conjugate into

$$\langle \dots \rangle_B = \prod_\omega Z_\omega^{-1} Tr\left(e^{-H_B/(kT)}(F^+F)\right)$$

$$= \prod_\omega Z_\omega^{-1} \sum_{\{n\}} \langle \Phi_{\{n\}} | e^{-H_B/(kT)} F^+ F | \Phi_{\{n\}} \rangle .$$

Show that

$$Tr\left(e^{-H_B/(kT)} B_{\omega_1}^+(0) B_{\omega_2}(0)\right) \prod_\omega Z_\omega^{-1} = 0$$

for $\omega_1 \neq \omega_2$ and

$$= Z_{\omega_1}^{-1} \sum_{n=0}^{\infty} n e^{-\hbar\omega_1 n/(kT)} \equiv (9.66)$$

$$\equiv \bar{n}_{\omega_1}(T) \text{ for } \omega_1 = \omega_2 .$$

Hint: Split H_B into a sum and $\Phi_{\{n\}}$ into a product over ω.
(3) Prove eq. (9.68a)
Hint: Insert $F(t)$ and $F^+(t)$ and use the commutation relations of B_ω, B_ω^+. Use eq. (9.55).
(4) Calculate $\langle b_\lambda^+(t) b_{\lambda'}(t') \rangle$, $\lambda \neq \lambda'$, where b_λ^+ obeys eq. (9.82).

9.4. Langevin equations for atoms and general quantum systems

9.4.1 Example of a two-level atom

In the previous sections of this chapter we have been mainly concerned with the question of how to incorporate damping and fluctuations into the treatment of quantum mechanical systems described by Bose operators b,

b^+. This allows us to deal in a proper way with the damping and fluctuations of field modes. On the other hand also atoms in solids or in gases interact with their surrounding which serves as a heatbath. Thus we expect that also atoms or, more precisely speaking, the electrons of atoms show damping and fluctuations. In the previous sections we learned about two methods to incorporate damping and fluctuations into quantum mechanics. We may formulate these two methods as follows:

(a) We start with the Heisenberg equations of motion of the field mode (i.e. the harmonic oscillator) alone. We then incorporate damping and fluctuation terms into these equations. The damping terms are chosen in such a way that the expectation values of the operators obey the analogous classical equations with phenomenological damping terms. The fluctuating forces must be determined in such a way that the quantum mechanical consistency (commutation relations!) is preserved.

(b) We start from the Heisenberg equations of motion of the total system containing the field mode under consideration, the heatbath and the interaction between these two subsystems. We then eliminate the variables of the heatbath giving rise to damping and fluctuations of the subsystems we are interested in (namely the field mode).

In the literature both methods have been applied also to atoms and to even more general quantum systems. Since the method analogous to (a) is more elegant and simpler, we will present it here.

To explain the basic idea let us consider a two-level atom whose electron may occupy the quantum states $j = 1$ or $j = 2$. Using again creation and annihilation operators for electrons in the corresponding states j, i.e. a_j^+, a_j, we can define the mean occupation number of level j by the expectation value

$$n_j = \langle \Phi | a_j^+ a_j | \Phi \rangle ; \qquad j = 1, 2. \tag{9.87}$$

Now let us consider the coupling of this atom to its surrounding, for instance to lattice vibrations or to an (incoherent) electromagnetic field. Then in general these interactions will cause transitions between the two levels so that the occupation numbers change. Consider for instance a level $j = 2$. Then its occupation number increases in the course of time due to transitions from the lower level to the upper level and it decreases due to the inverse process. The transition rates will be proportional in each case to the occupation number of the initial level. We therefore find

$$\frac{\mathrm{d}n_2}{\mathrm{d}t} = W_{21} n_1 - W_{12} n_2. \tag{9.88}$$

Since the electron must be in either one of the two levels (even in a superposition state where the electron "jumps" in a statistical fashion between the two levels, it must be found in any of the two levels when the occupation number is measured) we must have $n_1 + n_2 = 1$. Thus we can immediately write down also the equation for level 1

$$\frac{dn_1}{dt} = -W_{21}n_1 + W_{12}n_1. \tag{9.89}$$

As we have seen in section 4.6, the transition of an electron may be often connected with an electric dipole moment. Leaving aside all constant factors, we can take as the operators for the positive or negative frequency part of the dipole moment

$$a_2^+ a_1 \tag{9.90}$$

and

$$a_1^+ a_2, \tag{9.91}$$

respectively.

Let us consider the corresponding expectation values which we abbreviate by α and α^*

$$\langle\Phi|a_1^+ a_2|\Phi\rangle = \alpha, \qquad \langle\Phi|a_2^+ a_1|\Phi\rangle = \alpha^*. \tag{9.92, 93}$$

When the atom is not coupled to heatbaths, α oscillates at a frequency $-\omega_{21}$. When we couple the atom to a reservoir, for instance to the continuum of field modes (compare sections 7.9, 8.3), the dipole moment is damped and decays exponentially. Thus we are led to the conclusion that the expectation value of the dipole moment obeys an equation of the form

$$\frac{d\alpha}{dt} = -i\omega_{21}\alpha - \gamma_{12}\alpha. \tag{9.94}$$

Now let us reconsider equations (9.88), (9.89), (9.94) and its conjugate complex equation. Having in mind the definitions (9.87), (9.92) and (9.93) we can put these atomic equations in analogy with the equations (9.73) and (9.74) describing the damping of the expectation value of a field mode operator b^+ or b. Now let us do exactly the same as we have done with respect to the Bose operators b and b^+ before. There we translated the equations for expectation values into those for operators b, b^+. We saw that we could conserve quantum mechanical consistency only by suitable fluctuating forces. Doing exactly the same in our case we immediately find the following equations

$$\frac{d}{dt}a_2^+ a_2 = W_{21}a_1^+ a_1 - W_{12}a_2^+ a_2 + \Gamma_{22}(t), \tag{9.95}$$

$$\frac{d}{dt}a_1^+ a_1 = -W_{21}a_1^+ a_1 + W_{12}a_2^+ a_2 + \Gamma_{11}(t), \tag{9.96}$$

$$\frac{d}{dt}a_2^+ a_1 = (i\omega_{21} - \gamma_{21})a_2^+ a_1 + \Gamma_{21}(t), \tag{9.97}$$

$$\frac{d}{dt}a_1^+ a_2 = (-i\omega_{21} - \gamma_{12})a_1^+ a_2 + \Gamma_{12}(t). \tag{9.98}$$

In order to go back from eqs. (9.95)–(9.98) to (9.88), (9.89) and (9.94) we must require

$$\langle \Gamma_{ik}(t) \rangle = 0. \tag{9.99}$$

The meaning of the brackets needs some discussion. In section 9.3 we have derived the corresponding fluctuating forces for Bose operators b, b^+. There we have seen that the fluctuating forces depend on the variables of the heatbath. This leads us to the conclusion that also in our present case the brackets mean averaging over the variables of the heatbath and in addition taking the expectation value with respect to Φ of the atomic operators.

Now let us ask the question how we can determine the correlation functions between the fluctuating forces Γ. First of all we shall assume again that the heatbaths contain a broad spectrum of frequencies so that we may assume that the Γ's are δ-correlated.

$$\langle \Gamma_{ik}(t)\Gamma_{lm}(t') \rangle = G_{ik,lm}\delta(t - t'). \tag{9.100}$$

This can actually be justified by an explicit derivation of such fluctuating forces in a way analogous to section 9.3. Here, however, we introduce this as a hypothesis in the manner of a model.

The question how to formulate quantum mechanical consistency is not quite so simple. At a first sight one might think that one should require the conservation of the Fermi commutation relations (6.15). However, it should be noted that in eqs. (9.95)–(9.98) only products of a creation and an annihilation operator occur. This leads us to the idea to seek appropriate quantum mechanical relations for such products

$$a_i^+ a_k, \qquad a_l^+ a_m. \tag{9.101}$$

Such a product has the following property. When we apply $a_i^+ a_k$ to a state of the form

$$\Phi = a_j^+ \Phi_0 \tag{9.102}$$

(Φ_0: vacuum state) it transforms the state j into a new state i provided $j = k$ and it gives 0 otherwise. This property reoccurs when we take the product of expressions of the form (9.101). We then readily establish

$$a_i^+ a_k a_l^+ a_m = a_i^+ a_m \delta_{kl}, \tag{9.103}$$

using the Fermi commutation relations and the fact that we deal exclusively with states containing only one electron. It turns out that the relation (9.103) is exactly the requirement of quantum mechanical consistency we need. We will assume in the following that this relation (9.103) holds at least when it is averaged over the heatbath variables. We shall show below that this requirement allows us to calculate the coefficients $G_{ik,lm}$ uniquely and explicitly. Since the corresponding derivation is somewhat lengthy we first write down the final result for a two-level atom. For the special case of equations (9.95)–(9.98) we obtain the following relations

$$G_{11,11} = W_{12} n_2 + W_{21} n_1, \tag{9.104}$$

$$G_{11,22} = -W_{21} n_1 - W_{12} n_2 = G_{22,11}, \tag{9.105}$$

$$G_{22,22} = W_{21} n_1 + W_{12} n_2, \tag{9.106}$$

$$G_{12,12} = G_{21,21} = 0, \tag{9.107}$$

$$G_{12,21} = W_{12} n_2 - W_{21} n_1 + (\gamma_{12} + \gamma_{21}) n_1, \tag{9.108}$$

$$G_{21,21} = W_{21} n_1 - W_{12} n_2 + (\gamma_{12} + \gamma_{21}) n_2. \tag{9.109}$$

These relations must be understood as follows. We have to insert into them the solutions of the averaged equations (9.88), (9.89) and (9.94) so that, in general, n_1 and n_2 are time-dependent functions. In many cases of practical interest we deal with steady states. In such a case

$$\frac{dn_1}{dt} = \frac{dn_2}{dt} = 0. \tag{9.110}$$

As a consequence, by use of (9.88), (9.89) the relations (9.108) and (9.109) can be simplified

$$G_{12,21} = (\gamma_{12} + \gamma_{21}) n_1 \equiv 2\gamma n_1, \tag{9.111}$$

$$G_{21,12} = (\gamma_{12} + \gamma_{21}) n_2 \equiv 2\gamma n_2, \tag{9.112}$$

where we have introduced the abbreviation

$$\gamma_{12} = \gamma_{21} = \gamma. \tag{9.112a}$$

Similarly we could also simplify eqs. (9.104)–(9.106) by expressing, for instance, n_1 by n_2 or vice versa. We will illustrate the application of this formalism by some examples treated in the exercises. Readers not so much interested in mathematical details or in the general case can stop reading this section here, but they are advised to have a look at the exercises. On the other hand, readers interested in the general result but not in its derivation may proceed as follows: go on until eq. (9.118). The final result for $G_{ik,lm}$ (9.100) is represented in eq. (9.141).

9.4.2. *General case of a quantum system*

Let us now turn to the explicit determination of the coefficients $G_{ik,lm}$ in (9.100). We may do this not only for a single electron occupying states $k = 1, 2, \ldots, N$ in an atom but for an arbitrary quantum system. The reason for this is as follows. First we may introduce as an abbreviation

$$a_i^+ a_k = P_{ik}. \tag{9.113}$$

However, the operator P_{ik} has a very general meaning in quantum mechanics. When it is applied to a quantum state k it projects this state k onto a state i. One may show that one can formulate the quantum mechanical equations quite generally by means of such projection operators. We shall not give here a derivation of the corresponding equations for projection operators because the reader may always take the example of a single electron state as described by the relation (9.102) and the former equations (9.95)–(9.98). Projection operators obey the relations

$$P_{ik} P_{lm} = P_{im} \delta_{kl}. \tag{9.114}$$

[compare (9.103)]. We require that these relations remain valid when the quantum system is coupled to heat reservoirs provided we average over the reservoirs

$$\langle P_{ik} P_{lm} \rangle = \delta_{kl} \langle P_{im} \rangle \tag{9.115}$$

and take the expectation value with respect to an arbitrary quantum state of the system itself. Generalizing the equations (9.88), (9.89) and (9.94) to the present problem we start from equations of the form

$$\frac{\mathrm{d}}{\mathrm{d}t} \langle P_{i_1, i_2} \rangle = \sum_{j_1, j_2} \langle M_{i_1 i_2, j_1 j_2} P_{j_1 j_2} \rangle. \tag{9.116}$$

The corresponding quantum mechanical equations containing fluctuating forces are

$$\frac{\mathrm{d}}{\mathrm{d}t} P_{i_1 i_2} = \sum_{j_1 j_2} M_{i_1 i_2, j_1 j_2} P_{j_1 j_2} + \Gamma_{i_1 i_2}. \tag{9.117}$$

We subject the fluctuating forces Γ to the conditions (9.99) and (9.100).

Furthermore we take into account that the quantum system must be in one of the states $i = 1 \ldots N$ (compare the discussion leading to (9.89)). One may show that this implies

$$\sum_{i=1}^{N} P_{ii} = 1. \tag{9.118}$$

For the following it is convenient to introduce new abbreviations. We consider $P_{i_1 i_2}$ as components of a vector A

$$\left(P_{i_1 i_2}\right) = A. \tag{9.119}$$

Similarly we consider the M's as the components of a matrix

$$\left(M_{i_1 i_2, j_1 j_2}\right) = M. \tag{9.120}$$

Especially we have for $t = 0$

$$A(0) = \left(P_{i_1 i_2}(0)\right). \tag{9.121}$$

This allows us to represent equation (9.116) as

$$\frac{\mathrm{d}}{\mathrm{d}t}\langle A\rangle = \langle MA\rangle. \tag{9.122}$$

In it, M contains the transition probabilities (especially the losses, the nondiagonal phase-destroying terms produced by the heatbaths and "coherent" driving fields). Since these driving fields, if treated quantum mechanically, may still depend in an implicit way on the heatbath, the average over the heatbath has to comprise both M and A. We now go one step back and consider the motion of the unaveraged operators A by adding fluctuating forces, Γ.

$$\frac{\mathrm{d}}{\mathrm{d}t}A = MA + \Gamma. \tag{9.123}$$

The driving forces can have any form and in particular can still depend on $a_i^+ a_k$, for instance, in the following form

$$\Gamma(t) = L(t)A + N(t), \tag{9.124}$$

where L and N do not depend on A. In order to come back from (9.123) to (9.122) we must assume

$$\langle \Gamma\rangle = 0. \tag{9.125}$$

The formal solution of eq. (9.123) can be written as an integral representation

$$A = \int^t K(t,\tau)\Gamma(\tau)\,\mathrm{d}\tau + A_h \tag{9.126}$$

where A_h is a solution of the homogeneous part of eq. (9.123). We postulate as usual that the kernel K has the property

$$K(t,t) = E\ (= \text{unity matrix}). \tag{9.127}$$

The initial condition requires

$$A(0) = \left(P_{k_1k_2}(0)\right), \tag{9.128}$$

where the $P_{k_1k_2}$'s are operators in the Schrödinger picture. We now construct

$$\langle \tilde{A}BA \rangle = \left\langle \left(\int^t \tilde{\Gamma}(\tau)\tilde{K}(t,\tau)\,\mathrm{d}\tau + \tilde{A}_h \right) B \left(\int^t K(t,\tau)\Gamma(\tau)\,\mathrm{d}\tau + A_h \right) \right\rangle, \tag{9.129}$$

where $\tilde{}$ denotes the adjoint matrix. B is assumed as a constant matrix

$$B = (B_{k,l}) \tag{9.130}$$

having only one nonvanishing element $(i,j) = (i_1i_2, j_1j_2)$:

$$B_{k_1k_2,l_1l_2} = \delta_{k_1i_1}\delta_{k_2i_2}\delta_{l_1j_1}\delta_{l_2j_2}. \tag{9.131}$$

Using the property (9.115) we find for the left-hand side of eq. (9.129) by differentiation

$$\delta_{i_2j_1}\frac{\mathrm{d}}{\mathrm{d}t}\langle P_{i_1j_2}\rangle = \delta_{i_2j_1}\sum_{m,n}\langle M_{i_1j_2,mn}P_{mn}\rangle. \tag{9.132}$$

By differentiation of the right-hand side of eq. (9.129) and by using the properties of (9.123) we obtain the following terms

$$\langle \tilde{\Gamma}(t)BA \rangle_{ij}, \qquad \langle \tilde{A}B\Gamma(\tau) \rangle_{ij}, \tag{9.133, 134}$$

$$\langle \tilde{A}\tilde{M}BA \rangle_{ij}, \qquad \langle \tilde{A}BMA \rangle_{ij}. \tag{9.135,136}$$

For the evaluation of expressions (9.133) and (9.134) we use the Markovian property, i.e.

$$\langle \Gamma_{i_1i_2}(t)\Gamma_{j_1j_2}(t') \rangle = G_{i_1i_2,j_1j_2}\delta(t-t') \tag{9.137}$$

and

$$\langle \Gamma_{i_1i_2}(t) \rangle = 0. \tag{9.138}$$

We now have to perform the average over the heatbaths where we know that the fluctuations have only a very short memory. We assume that the response of the particle (or spin) is slow enough for K and A_h to be taken out of the average. (Note that M contains at most the heatbath coordinates indirectly over the quantum mechanical ("coherent") fields, but in all the loss terms and phase memory destroying parts they do not appear.) (9.133) then gives the contribution

$$\tfrac{1}{2}G_{i,j} \tag{9.139}$$

and the same quantity follows from (9.134) in a similar manner. In order to evaluate (9.135) we contract the product of the A's by means of the rule (9.115), where we make use of the commutativity of the operators P with the operators of the driving fields within M, which holds in the Heisenberg picture. We thus obtain

$$\left\langle \sum_{m,n} P_{mn} \tilde{M}_{mn,i_1i_2} P_{j_1j_2} \right\rangle = \sum_m \langle P_{mj_2} M_{i_1i_2,mj_1} \rangle \tag{9.140}$$

from eq. (9.135) and a similar one from eq. (9.136). We now compare the results of both sides of (9.129), thus obtaining the following equation for the G's which occur in eqs. (9.137) and (9.140).

$$G_{i_1i_2,j_1j_2} = \sum_{m,n} \langle\{\delta_{i_2j_1} M_{i_1j_2,mn} - \delta_{n_1j_2} M_{i_1i_2,mj_1}$$

$$- \delta_{m,i_1} M_{j_1j_2,i_2n}\}\rangle\langle P_{mn}\rangle . \tag{9.141}$$

This equation represents the required result. It allows us to calculate all correlation functions of the type (9.137) if the coefficients M are given and the solutions $\langle P_{ik}\rangle$ of the averaged equations are known. We now show that in (9.141) all terms containing external fields cancel each other, so that only transition rates and damping constants need to be used for the M's. We came across such terms in eqs. (9.95)–(9.98). In these equations the transition rates W_{ik} and damping constants γ_{ik} stemmed from the coupling of the system to heatbaths. However, we may also imagine that the system is coupled in addition to external fields, for instance the classical light field, or to other quantum systems. In such a case additional terms occur on the right-hand side of eqs. (9.95)–(9.98) which stem from that coupling. The explicit form of the corresponding terms can be found quite easily. To this end we have only to consider the left-hand sides of eqs. (9.95)–(9.98). The left-hand sides are just the time derivatives of operators. However, we know that such time derivatives can be calculated by means of the Heisenberg equations. Thus the effect of external driving forces or the coupling to other quantum systems (but not heatbaths!) can be incorporated into (9.95)–(9.98) by adding the corresponding commutators of $a_i^+ a_k$ with the interaction Hamiltonian to the right-hand sides of (9.95)–(9.98). In the present context we have, of course, more general quantum systems in mind but still the coherent part stems from a commutator with a

Hamiltonian. We want to show that such terms do not affect the relations (9.141) or, in other words, that the coherent part of M drops out from (9.141). Now let us formulate this problem more explicitly. The coherent part of M stems from the commutation of $P_{i_1 i_2}$ with a Hamiltonian which is of the form

$$H = \sum_{mn} c_{mn} P_{mn}. \tag{9.142}$$

The evaluation of the commutator

$$\frac{i}{\hbar}\left[H, P_{i_1 i_2}\right] = \frac{i}{\hbar}\sum_{m,n} c_{mn}\left(\delta_{n i_1} P_{m i_2} - \delta_{i_2 m} P_{i_1 n}\right) \tag{9.143}$$

yields immediately

$$\sum_{j_1 j_2} \underbrace{\frac{i}{\hbar}\left(c_{j_1 i_1}\delta_{j_2 i_2} - c_{i_2 j_2}\delta_{j_1 i_1}\right)}_{M^{\text{coh}}_{i_1 i_2, j_1 j_2}} P_{j_1 j_2}. \tag{9.144}$$

A comparison of this expression with the right-hand side of eq. (9.116) show that

$$M^{\text{coh}}_{i_1 i_2, j_1 j_2} = \frac{i}{\hbar}\left(c_{j_1 i_1}\delta_{j_2 i_2} - c_{i_2 j_2}\delta_{j_1 i_1}\right). \tag{9.145}$$

When this explicit form is inserted into the right-hand side of (9.141) these terms cancel each other. In the case of the laser, the incoherent part of the atomic equations often reads

$$\begin{aligned}
\frac{\mathrm{d}}{\mathrm{d}t} P_{jj} &= \sum_k W_{jk} P_{kk} - \sum_k W_{kj} P_{jj} + \Gamma_{jj}(t),\\
\frac{\mathrm{d}}{\mathrm{d}t} P_{jk} &= -\gamma_{jk} P_{jk} + \Gamma_{jk}(t), \qquad j \neq k.
\end{aligned} \tag{9.146}$$

By specializing formula (9.141) we obtain

$$\begin{aligned}
G_{ii,jj} = \delta_{ij}&\left\{\sum_k \left(W_{ki}\langle P_{kk}\rangle + W_{ik}\langle P_{ii}\rangle\right)\right\}\\
&- W_{ji}\langle P_{ii}\rangle - W_{ij}\langle P_{jj}\rangle,
\end{aligned} \tag{9.147}$$

$$G_{ij,ij} = 0, \qquad i \neq j, \tag{9.148}$$

$$G_{ij,ji} = \sum_k \left\{W_{ik}\langle P_{kk}\rangle - W_{ki}\langle P_{ii}\rangle\right\} + (\gamma_{ij} + \gamma_{ji})\langle P_{ii}\rangle,\ i \neq j. \tag{9.149}$$

For many applications, e.g. laser theory, a knowledge of the second moments of the fluctuating forces, i.e. (9.149) is completely sufficient, because the fields interact with many independent atoms. Thus the results, e.g. on the linewidth, depend only on a sum over very many independently fluctuating forces, which possesses a Gaussian distribution (the fluctuating forces of a single atom are not Gaussian, however).

So far we have shown that eq. (9.141) follows from condition (9.115). It is rather simple to show that the opposite also holds.

Exercises on section 9.4

(1) From eqs. (9.95) and (9.96) derive an equation for $s = a_2^+ a_2 - a_1^+ a_1$. Solve that equation as well as those for $a_2^+ a_1$, $a_1^+ a_2$ and calculate

$$\langle a_2^+ a_1 \rangle, \qquad \langle s \rangle, \qquad \langle (a_2^+ a_1)_t (a_1^+ a_2)_{t'} \rangle,$$
$$\langle (a_2^+ a_1)_t (a_2^+ a_1)_{t'} \rangle, \qquad \langle s_t s_{t'} \rangle.$$

The indices t, t' are the arguments of a_j^+, a_k in the Heisenberg picture.
Hint: Solve the equations in the same way as ordinary differential equations treating $a_j^+ a_k$ as classical quantities. (This is justified because the equations are linear in $a_j^+ a_k$.) When forming correlation functions, be careful in keeping the sequence of operators.

(2) In section 4.9, exercise (1), we encountered the Bloch equations used in spin resonance. These equations contained both the applied magnetic field and incoherent damping terms. As we know from section 4.8, an analogy exists between a two-level atom and spin $\frac{1}{2}$. This leads us to the following exercise. Put in the Bloch equation $\boldsymbol{B} = (0, 0, B_z)$. Compare the Bloch equations of section 4.9, exercise (1) with eqs. (9.88), (9.89) and (9.94) and its conjugate complex.
Hint: Try the following analogies:

$$\alpha (9.92) \leftrightarrow \langle s_x \rangle - i \langle s_y \rangle,$$

$$\alpha^* (9.93) \leftrightarrow \langle s_x \rangle + i \langle s_y \rangle,$$

$$\tfrac{1}{2}(n_2 - n_1)(9.87) \leftrightarrow \langle s_z \rangle.$$

(3) Consider the same situation as in exercise (2). Derive quantum mechanical Langevin equations for the spin operators

$$s_+ = s_x + i s_y, \qquad s_- = s_x - i s_y, \qquad s_z.$$

Hint: Use the correspondence

$$s_+ \leftrightarrow a_2^+ a_1, \qquad s_- \leftrightarrow a_1^+ a_2, \qquad s_z = \tfrac{1}{2}(a_2^+ a_2 - a_1^+ a_1)$$

and transform the eqs. (9.95)–(9.99) correspondingly. In this way the correlation functions of the corresponding Γ's can be determined also.

(4) A two-level atom is coupled to heatbaths and a coherent classical electric field, $\boldsymbol{E} = \boldsymbol{E}_0 \cos \omega t$. Derive the quantum mechanical Langevin equations for $a_j^+ a_k$, $j, k = 1, 2$ and the correlation functions (9.100). Assume that the dipole moments $\boldsymbol{\vartheta}_{11} = \boldsymbol{\vartheta}_{22} = 0$, but $\boldsymbol{\vartheta}_{12} \neq 0$.

Hint: Write

$$\frac{\mathrm{d}}{\mathrm{d}t}(a_j^+ a_k) = \frac{\mathrm{d}}{\mathrm{d}t}(a_j^+ a_k)_{\text{incoherent}} + \frac{\mathrm{d}}{\mathrm{d}t}(a_j^+ a_k)_{\text{coherent}}.$$

In it $(\mathrm{d}/\mathrm{d}t)(a_j^+ a_k)_{\text{incoherent}}$ is defined by the r.h.s. of equations (9.95)–(9.98), whereas $(\mathrm{d}/\mathrm{d}t)(a_j^+ a_k)_{\text{coherent}}$ is defined by $(i/\hbar)[H, a_j^+ a_k]$, where

$$H = \sum_j \hbar \varepsilon_j a_j^+ a_j + \hbar g a_2^+ a_1 + \hbar g^* a_1^+ a_2.$$

What is the explicit form of g? Show that $g \propto \cos \omega t$.

Hints: Use the results (9.141) and the discussion thereafter. To determine g, use sections 7.3, 7.5. (To check your results, consult the following exercise.)

(5) A two-level atom is coupled to heatbaths and a coherent, resonant classical electric field $\boldsymbol{E} = \boldsymbol{E}_0 \cos \omega t$. Its quantum-mechanical Langevin equations are given by

$$\frac{\mathrm{d}}{\mathrm{d}t}(a_1^+ a_2) = -i(\varepsilon_2 - \varepsilon_1) a_1^+ a_2 + ig(a_2^+ a_2 - a_1^+ a_1) - \gamma_{12} a_1^+ a_2 + \Gamma_{12}(t),$$

$$\frac{\mathrm{d}}{\mathrm{d}t}(a_2^+ a_1) = i(\varepsilon_2 - \varepsilon_1) a_2^+ a_1 - ig^*(a_2^+ a_2 - a_1^+ a_1) - \gamma_{21} a_2^+ a_1 + \Gamma_{21}(t),$$

$$\frac{\mathrm{d}}{\mathrm{d}t}(a_2^+ a_2) = ig^* a_1^+ a_2 - ig a_2^+ a_1 + W_{21} a_1^+ a_1 - W_{12} a_2^+ a_2 + \Gamma_{22}(t),$$

$$\frac{\mathrm{d}}{\mathrm{d}t}(a_1^+ a_1) = ig a_2^+ a_1 - ig^* a_1^+ a_2 - W_{21} a_1^+ a_1 + W_{12} a_2^+ a_2 + \Gamma_{11}(t),$$

$$g = \boldsymbol{E}_0 \boldsymbol{\vartheta}_{12} \cos \omega t, \qquad \boldsymbol{\vartheta}_{12} = \int \varphi_1^*(\boldsymbol{x}) e\boldsymbol{x} \varphi_2(\boldsymbol{x}) \mathrm{d}^3 \boldsymbol{x},$$

$$\varphi_j(\boldsymbol{x}) \text{ atomic wave function.}$$

Solve the equations for $a_j^+ a_j$, $a_j^+ a_k$; $j, k = 1, 2$ in the rotating wave approximation ($\varepsilon_2 - \varepsilon_1 \equiv \omega_{21} = \omega$) and calculate

$$\langle a_1^+ a_2 \rangle, \qquad \langle\langle (a_1^+ a_2) \rangle_t, \langle (a_1^+ a_2)_{t'} \rangle,$$
$$\langle (a_1^+ a_2)_t (a_2^+ a_1)_{t'} \rangle.$$

Hint: Take the Fourier transform of the equations.

9.5. The density matrix

In section 9.1 we saw that we may mathematically treat Brownian motion either by means of the Langevin equation (9.8) or in entirely equivalent fashion by the Fokker–Planck equation (9.21). Then in section 9.2 we recognized that the classical Langevin equation (9.8) possesses an analogue in quantum mechanics. Thus our results can be summarized by means of the following table 1:

Table 1.

	Langevin	Fokker–Planck
Classical	$\frac{\mathrm{d}q}{\mathrm{d}t} = -\kappa q + F(t)$	$\frac{\partial f}{\partial t} = \left[\frac{\partial}{\partial q}\kappa q + \frac{Q}{2}\frac{\partial^2}{\partial q^2}\right] t$
Quantum mechanical	$\frac{\mathrm{d}b}{\mathrm{d}t} = -\kappa b + F(t)$	?

The empty box in the lower right corner immediately leads to the question whether there exists in quantum mechanics an equation which corresponds to the Fokker–Planck equation. As we saw above the solution of the Fokker–Planck equation has certain advantages. Knowing its solutions we easily can calculate average values of q and of its powers at any time t by means of mere integrations

$$\langle q^n \rangle = \int q^n f(q,t)\,\mathrm{d}q. \tag{9.150}$$

There exists indeed a quantity in quantum mechanics which directly corresponds to the distribution function f and which allows us to calculate average values or, in the sense of quantum mechanics, expectation values. The quantity corresponding to f is called density matrix and is often denoted by ρ. As we know, in quantum mechanics we have to attribute operators Ω to the observables, say coordinate or momentum. Let Ω correspond to q^n of eq. (9.150). The quantum mechanical analogue of (9.150) then reads

$$\langle \Omega \rangle = Tr(\Omega\rho) \tag{9.151}$$

where "Tr" means trace which we will define below [(9.158), cf. also (9.64a)]. We first want to motivate the definition (9.151), then define ρ, and eventually derive an equation for ρ. Equation (9.151) can be considered as a generalization of the quantum mechanical average or expectation value well known to us:

$$\overline{\Omega} = \int \varphi_n^*(\boldsymbol{x})\Omega\varphi_n(\boldsymbol{x})\,\mathrm{d}^3\boldsymbol{x}. \tag{9.152}$$

In the following we shall use Dirac's bra and ket notation:

$$\bar{\Omega} = \langle n|\Omega|n\rangle . \tag{9.153}$$

This notation means a considerable generalization compared to eq. (9.152) because we can use quite general wave functions such as those constructed by means of creation operators. But still, the $|n\rangle$'s are the usual wave functions of quantum mechanics and describe in this sense pure states. Now we wish to incorporate thermodynamics or, more precisely speaking, statistics into the formulation of average values. To this end consider, for instance, an ensemble of oscillators in thermal equilibrium with a heatbath at temperature T. In this case the quantum mechanical oscillators are not all in a definite pure quantum state. Rather we can only say that a state n with energy W_n is occupied with a certain probability p_n. When we want to calculate the average value of an observable of such an ensemble, we have to weight the individual expectation values (9.153) by p_n. This leads us to the definition of the quantum statistical average

$$\langle\Omega\rangle = \sum_n p_n \langle n|\Omega|n\rangle . \tag{9.154}$$

As usual we assume

$$\sum_n p_n = 1 . \tag{9.155}$$

We adopt the definition (9.154) for general quantum systems and not only for oscillators. In classical physics, the p_n's are well known for a system in thermal equilibrium with a heatbath at temperature T. According to Boltzmann we have

$$p_n = Z^{-1} \exp(-W_n/(kT)), \tag{9.156}$$

where the partition function Z is given by

$$Z = \sum_n \exp(-W_n/(kT))$$

and takes care of the normalization of p_n [cf. eq. (9.155)]. We show that eq. (9.154) can be considered as a special case of (9.151) with the density matrix ρ given by

$$\rho = \sum_n p_n |n\rangle\langle n| . \tag{9.157}$$

The reader should note the sequence of $|n\rangle$ and $\langle n|$, i.e. ket–bra! We insert eq. (9.157) into (9.151) and use the definition of the trace of an arbitrary operator $\tilde{\Omega}$

$$Tr\,\tilde{\Omega} = \sum_{n'} \langle n'|\tilde{\Omega}|n'\rangle . \tag{9.158}$$

We obtain (with $\tilde{\Omega} = \Omega\rho$)

$$\begin{aligned}\langle\Omega\rangle &= \sum_{n'} \langle n'|\Omega \sum_{n} p_n |n\rangle\langle n|n'\rangle \\ &= \sum_{n'}\sum_{n} p_n \langle n'|\Omega|n\rangle\langle n|n'\rangle .\end{aligned} \tag{9.159}$$

Owing to the orthogonality of wave functions,

$$\langle n|n'\rangle = \delta_{nn'},$$

eq. (9.159) reduces to

$$\langle\Omega\rangle = \sum_{n} p_n \langle n|\Omega|n\rangle ,$$

i.e. eq. (9.154). We can generalize our considerations. The density matrix (9.157) is still insofar too special, as it contains only terms of the form $|n\rangle\langle n|$ and not of the form $|n\rangle\langle m|$, i.e. with two quantum numbers n *and* m. Since in practical cases such combinations may also occur, we define the density matrix by

$$\rho = \sum_{mn} \rho_{mn} |n\rangle\langle m| \tag{9.160}$$

where the ρ_{mn}'s are constants, i.e. classical numbers. Furthermore, we have seen in section 3.2 that we may form expectation values not only for time-independent but also for time dependent wave functions. Consequently we now consider the case that the wave function $|n\rangle$ is time dependent and obeys the Schrödinger equation

$$H|n\rangle = i\hbar \frac{\mathrm{d}}{\mathrm{d}t}|n\rangle . \tag{9.161}$$

Similarly, $\langle m|$ obeys the "adjoint" Schrödinger equation

$$\langle m|H = -i\hbar \frac{\mathrm{d}}{\mathrm{d}t}\langle m| . \tag{9.162}$$

This form may look to some reader unfamiliar, but we may introduce the equation (9.162) just as equation defining $\langle m|$! Since $|n\rangle$ and $\langle m|$ are time dependent, ρ is time dependent too. We are now in the position to derive an evolution equation for ρ. To this end we differentiate ρ with respect to time

$$\frac{\mathrm{d}\rho}{\mathrm{d}t} = \sum_{mn} \rho_{mn} \left(\frac{\mathrm{d}|n\rangle}{\mathrm{d}t}\langle m| + |n\rangle \frac{\mathrm{d}\langle m|}{\mathrm{d}t} \right). \tag{9.163}$$

By use of eqs. (9.161) and (9.162) we obtain

$$\frac{d\rho}{dt} = -\frac{i}{\hbar}\sum_{mn}\rho_{mn}(H|n\rangle\langle m| - |n\rangle\langle m|H) \tag{9.164}$$

or, extracting H to the left and right of the sum, respectively

$$\frac{d\rho}{dt} = -\frac{i}{\hbar}\left\{H\sum_{mn}\rho_{mn}|n\rangle\langle m| - \sum_{mn}\rho_{mn}|n\rangle\langle m|H\right\}. \tag{9.165}$$

Using on the r.h.s. of eq. (9.165) the original definition of ρ, (9.160), we obtain the fundamental density matrix equation

$$\frac{d\rho}{dt} = -\frac{i}{\hbar}(H\rho - \rho H)$$

or, in short

$$\frac{d\rho}{dt} = -\frac{i}{\hbar}[H,\rho]. \tag{9.166}$$

So far we have established a general frame. We now want to apply this formalism to treat the dynamics of systems coupled to reservoirs. This will allow us to fill the empty box in table 1. Since the mathematical treatment will become somewhat involved we will first give a short description of its spirit and an explicit example. The basic idea is this (cf. also section 9.5.2): We start from the density matrix ρ_{tot} and its equation (9.166) for the total system composed of the "proper" system (for instance the field oscillator), the heatbaths and their mutual interaction. Then we derive an equation for the reduced density matrix ρ of the "proper" system alone by eliminating the bath variables. Such an equation describes damping and fluctuations of the "proper" system. We will present the general reduction scheme below.

To elucidate the general features of such reduced density matrix equations and their solutions we first write down the result for the damped harmonic oscillator whose quantum mechanical Langevin equation we had considered in section 9.2 and derived in section 9.3. This density matrix equation reads

$$\begin{aligned}\frac{d\rho}{dt} = {} & -i\omega[b^+b,\rho] + \delta\{[b^+\rho,b] + [b^+,\rho b]\} \\ & + \xi\{[b\rho,b^+] + [b,\rho b^+]\},\end{aligned} \tag{9.167}$$

where the bracket $[A,B]$ is defined as usual by

$$[A,B] = AB - BA.$$

For later purposes we write (9.167) more explicitly in the form

$$\begin{aligned}\frac{d\rho}{dt} = {} & -i\omega(b^+b\rho - \rho b^+b) + \delta\{2b^+\rho b - bb^+\rho - \rho bb^+\} \\ & + \xi\{2b\rho b^+ - b^+b\rho - \rho b^+b\}.\end{aligned} \tag{9.168}$$

To get an insight into the meaning of the constants δ and ξ we derive by means of (9.167) or (9.168) an equation for the average values of b^+ and b^+b, respectively. To this end we multiply eq. (9.168) by b^+ from the left and take the average value which we define according to (9.151). Since traces have the cyclic property

$$Tr(\Omega_1\Omega_2) = Tr(\Omega_2\Omega_1) \tag{9.169}$$

one easily finds for the average light field amplitude the equation

$$\frac{d\langle b^+\rangle}{dt} = \{i\omega - (\xi - \delta)\}\langle b^+\rangle. \tag{9.170}$$

In an analogous manner we find equations for the photon number

$$\frac{d\langle b^+b\rangle}{dt} = 2\delta - 2(\xi - \delta)\langle b^+b\rangle, \tag{9.171}$$

and for the commutator

$$\frac{d}{dt}\langle(bb^+ - b^+b)\rangle = 2(\xi - \delta) + 2(\delta - \xi)\langle(bb^+ - b^+b)\rangle. \tag{9.171a}$$

Eq. (9.170) allows for the solution

$$\langle b^+\rangle = \langle b^+\rangle_0 \exp([i\omega - (\xi - \delta)]t), \tag{9.172}$$

whereas eq. (9.171) is solved by

$$\langle b^+b\rangle = \langle b^+b\rangle_0 \exp[-2(\xi - \delta)t] + \frac{\delta}{\xi - \delta}(1 - \exp[-2(\xi - \delta)t]). \tag{9.173}$$

Under the initial condition $\langle(bb^+ - b^+b)\rangle = 1$ at $t = 0$, the solution of (9.171a) reads

$$\langle(bb^+ - b^+b)\rangle = 1, \qquad \text{for all } t, \tag{9.173a}$$

i.e. the commutation relation is preserved for all times, at least in the sense of an average. We can compare the solution (9.172) with (9.43) of section 9.2. This allows us to identify $\xi - \delta$ with the former decay constant κ. On the other hand letting $t \to \infty$ in eq. (9.173) we find the stationary solution when the heatbath is at a temperature T. The average photon number resulting from (9.173) must be that in thermal equilibrium: $\bar{n}$. This allows us to make the identifications

$$\xi - \delta = \kappa, \qquad \frac{\delta}{\xi - \delta} = \left[\exp\left[\frac{\hbar\omega}{kT}\right] - 1\right]^{-1} = \bar{n}. \tag{9.174, 175}$$

9.5.1. An example of the solution of the density matrix equation

We will not try to solve the general time-dependent density matrix equation (9.167). However, it is of some interest to consider at least a typical example. Here it is convenient to use Dirac's bra and ket notation. We make the following ansatz for the density matrix

$$\rho(t) = \sum_{n=0}^{\infty} |n\rangle \rho_n(t) \langle n| \tag{9.176}$$

where $|n\rangle$ and $\langle n|$ are time independent. (Incidentally, we learn from this example that we may shift the time dependence of the $|n\rangle$'s to one of the "expansion" coefficients, $\rho_n(t)$. This is quite analogous to the transition from the Schrödinger picture to the Heisenberg picture. Eq. (9.176) which covers, as we shall see below, at least the stationary state. It covers, however, also a certain class of time-dependent states but we shall not discuss this question here. When we insert eq. (9.176) into the density matrix equation (9.167) or (9.168) it is useful to have the following relations in mind

$$b^+|n\rangle = \sqrt{n+1}\,|n+1\rangle, \qquad b|n\rangle = \sqrt{n}\,|n-1\rangle, \qquad (9.177, 178)$$

$$\langle n|b = \sqrt{n+1}\,\langle n+1|, \qquad \langle n|b^+ = \sqrt{n}\,\langle n-1|. \qquad (9.179, 180)$$

The relations (9.177) and (9.178) are well known to us (compare section 3.3). The relations (9.179) and (9.180) are just the Hermitian conjugates of the relations (9.177) and (9.178). By using (9.177) and (9.179) we can easily evaluate

$$b^+\rho b. \tag{9.181}$$

By inserting an individual term of (9.176) into (9.181) we thus obtain

$$b^+|n\rangle\langle n|b = \sqrt{n+1}\,|n+1\rangle\langle n+1|\sqrt{n+1}\,. \tag{9.182}$$

In an analogous way we can evaluate

$$b^+b\rho. \tag{9.183}$$

A typical term then reads

$$\begin{aligned} b^+b|n\rangle\langle n| &= b^+\sqrt{n}\,|n-1\rangle\langle n| \\ &= \sqrt{n}\,\sqrt{n}\,|n\rangle\langle n|. \end{aligned} \tag{9.184}$$

Proceeding with all the other terms in a similar way we may immediately evaluate (9.168) with (9.176) which yields

$$\begin{aligned} \sum_{n=0}^{\infty} \frac{d\rho_n}{dt}|n\rangle\langle n| &= 2\delta \sum_{n=0}^{\infty} \rho_n(n+1)\{|n+1\rangle\langle n+1| - |n\rangle\langle n|\} \\ &\quad + 2\xi \sum_{n=0}^{\infty} \rho_n n\{|n-1\rangle\langle n-1| - |n\rangle\langle n|\}. \end{aligned} \tag{9.185}$$

In this equation, the ρ_n's are still unknown functions of time to be determined. Now we must remember that the bra and kets, or more precisely speaking,

$$|n\rangle\langle n| \tag{9.186}$$

are linearly independent. That means any equation of the form of (9.185) can be solved only when the coefficients of each term (9.186) vanish. To derive such equations for the coefficients we must transform terms of the form

$$|n+1\rangle\langle n+1| \tag{9.186a}$$

into those of the form (9.186). Take as an example

$$\sum_{n=0}^{\infty} \rho_n (n+1)|n+1\rangle\langle n+1|, \tag{9.187}$$

then make the replacement

$$n+1=n' \tag{9.188}$$

which transforms (9.187) into

$$\sum_{n'=1}^{\infty} \rho_{n'-1} n' |n'\rangle\langle n'|. \tag{9.189}$$

Since the term with $n'=0$ vanishes we may let the sum start from $n'=0$. Dropping eventually the prime we then find instead of eq. (9.187)

$$\sum_{n=0}^{\infty} \rho_{n-1} n |n\rangle\langle n|. \tag{9.190}$$

Making the same procedure with terms containing $n+1$ we eventually find

$$\sum_{n=0}^{\infty} \frac{d\rho_n}{dt}|n\rangle\langle n| = 2\sum_{n=0}^{\infty} |n\rangle\langle n| \{\delta n \rho_{n-1} - \delta(n+1)\rho_n + \xi(n+1)\rho_{n+1} - \xi n \rho_n\}. \tag{9.191}$$

Due to the above-mentioned linear independence of the expressions (9.186) we are then left with a set of equations

$$\frac{d\rho_n}{dt} = 2\{\delta n \rho_{n-1} + \xi(n+1)\rho_{n+1} - [\delta(n+1) + \xi n]\rho_n\}. \tag{9.192}$$

We shall be concerned with its stationary solution only where we put

$$\frac{d\rho_n}{dt} = 0 \tag{9.193}$$

so that (after a rearrangement of terms) we have to solve the equations

$$n[\delta\rho_{n-1} - \xi\rho_n] - (n+1)[\delta\rho_n - \xi\rho_{n+1}] = 0, \qquad n=0,1,2,\ldots \tag{9.194}$$

For $n = 0$ we find

$$\delta\rho_0 - \xi\rho_1 = 0 \tag{9.195}$$

or

$$\rho_1 = (\delta/\xi)\rho_0.$$

For $n = 1$ we obtain

$$[\delta\rho_0 - \xi\rho_1] - 2[\delta\rho_1 - \xi\rho_2] = 0$$

which due to eq. (9.195) reduces to $\delta\rho_1 - \xi\rho_2 = 0$ or $\rho_2 = (\delta/\xi)\rho_1$. This procedure can be continued leading to

$$\delta\rho_n - \xi\rho_{n+1} = 0$$

or

$$\rho_{n+1} = (\delta/\xi)\rho_n. \tag{9.196}$$

This is a recursion formula. Expressing ρ_n by ρ_{n-1}, ρ_{n-1} by ρ_{n-2} etc. we readily find

$$\rho_n = \rho_0(\delta/\xi)^n. \tag{9.197}$$

By putting

$$\delta/\xi = e^{-\alpha}, \tag{9.198}$$

eq. (9.197) acquires the form

$$\rho_n = \rho_0 e^{-\alpha n}. \tag{9.199}$$

As we have seen above, cf. eqs. (9.174) and (9.175), δ and ξ can be expressed by directly observable quantities, namely the damping constant κ and the thermal photon number $\bar{n}$, or, equivalently, by $\hbar\omega/(kT)$. This allows us to determine α. After some algebra we obtain

$$\alpha = \hbar\omega/(kT). \tag{9.200}$$

However, with eq. (9.200), the expression (9.199) is just the Boltzmann distribution we introduced in section 2.3 (cf. eqs. (2.59) and (2.58)). Even ρ_0 coincides with Z^{-1} as can be seen as follows. Since the density matrix is the quantum mechanical analogue of a distribution function the normalization condition of such a distribution fu ction also applies, namely in the form

$$Tr\,\rho = 1. \tag{9.201}$$

Inserting (9.176) with (9.199) into (9.201) brings us to the relation

$$\rho_0 = \left(\sum_{n=0}^{\infty} \exp[-\alpha n]\right)^{-1} \tag{9.202}$$

or, after performing the sum:

$$\rho_0 = 1 - \exp[-\alpha]. \tag{9.203}$$

Inserting the result (9.202)–(9.203) into (9.176), we find the explicit form of the density matrix

$$\rho = (1 - \exp[-\alpha]) \sum_{n=0}^{\infty} \exp[-\alpha n]|n\rangle\langle n|. \tag{9.204}$$

We hope that the reader now has a certain feeling of how to deal with a typical density matrix equation and what some typical properties of a density matrix are. We now turn to the final step of our treatment, namely we wish to show how one may derive a density matrix equation for an arbitrary quantum system.

9.5.2. *General derivation*

We consider the interaction of a free field (e.g. the light mode or the electron of an atom) with a heatbath or a set of heatbaths. The density matrix of the total system obeys the equation

$$\frac{\mathrm{d}}{\mathrm{d}t}\rho_{\mathrm{tot}} = -\frac{i}{\hbar}[H, \rho_{\mathrm{tot}}], \tag{9.205}$$

where

$$H = H_0 + H_{F-B} \tag{9.206}$$

and

$$H_0 = H_F + H_B. \tag{9.207}$$

H_F is the Hamiltonian of the free field and reads explicitly

$$H_F = \begin{cases} \hbar\omega b^+ b & \text{light-mode} \\ \sum_i \hbar\varepsilon_i a_i^+ a_i & \text{atom} \\ \sum_i \hbar\varepsilon_i P_{ii} & \text{arbitrary quantum system described by projection operators.} \end{cases} \tag{9.208}$$

The Hamiltonian of the heatbath H_B may consist of a sum of several Hamiltonians corresponding to different heatbaths

$$H_B = \sum_j H_B^{(j)}. \tag{9.209}$$

H_{F-B} represents the interaction between the two systems. For our analysis it is convenient to proceed to the interaction representation

$$\tilde{\rho}_{tot} = \exp[iH_0t/\hbar]\rho_{tot}\exp[-iH_0t/\hbar], \tag{9.210}$$

$$\tilde{H}_{F-B} = \exp[iH_0t/\hbar]H_{F-B}\exp[-iH_0t/\hbar]. \tag{9.211}$$

The interaction Hamiltonian has the form

$$\tilde{H}_{F-B}(t) = \hbar\sum_k V_k(t)B_k(t)$$

$$= \begin{cases} \underbrace{b\exp[-i\omega t]}_{V_1(t)}\ \underbrace{B^+(t)}_{\hbar B_1} + \underbrace{b^+\exp[i\omega t]}_{V_2(t)}\ \underbrace{B(t)}_{\hbar B_2} & (9.212) \\ \sum_{ij}\underbrace{P_{ij}\exp[i\omega_{ij}t]}_{V_k(t)}\ \underbrace{B_{ij}(t)}_{B_k}. & (9.213) \end{cases}$$

$$\tilde{P}_{jk} \equiv \exp[iH_0t/\hbar]P_{jk}\exp[-iH_0t/\hbar]$$

$$= P_{jk}\exp[i\omega_{jk}t], \tag{9.214}$$

where

$$\omega_{jk} = \varepsilon_j - \varepsilon_k. \tag{9.215}$$

The explicit form of the V's suggests that we write

$$V_k(t) = V_k\exp[i\Delta\omega_k t]. \tag{9.216}$$

We treat the interaction by perturbation theory up to second order which yields

$$\tilde{\rho}_{tot} = \rho_0 + \left(\frac{-i}{\hbar}\right)\int_0^t d\tau[\tilde{H}_{F-B}(\tau),\rho_0] + \left(\frac{-i}{\hbar}\right)^2\int_0^t d\tau_2\int_0^{\tau_2} d\tau_1\{\ldots\}, \tag{9.217}$$

where $\rho_0 = \tilde{\rho}_{tot}(0)$ and

$$\{\ldots\} = \{\tilde{H}_{F-B}(\tau_2)\tilde{H}_{F-B}(\tau_1)\rho_0 - \tilde{H}_{F-B}(\tau_2)\rho_0\tilde{H}_{F-B}(\tau_1)$$
$$- \tilde{H}_{F-B}(\tau_1)\rho_0\tilde{H}_{F-B}(\tau_2) + \rho_0\tilde{H}_{F-B}(\tau_1)\tilde{H}_{F-B}(\tau_2)\}. \tag{9.218}$$

Since we are ultimately interested only in the variables of the field we eliminate the bath variables from (9.217) by taking the trace over the heatbaths

$$\tilde{\rho} = Tr_B\tilde{\rho}_{tot}. \tag{9.219}$$

Therefore, $\tilde{\rho}$ depends only on the field operators. It is essential that at time $t = 0$ the total density matrix ρ_{tot} factorizes in that of the free field, $\rho(0)$

and that of the heatbath, ρ_B:

$$\rho_0 \equiv \tilde{\rho}_{\text{tot}}(0) = \rho(0)\rho_B. \tag{9.220}$$

ρ_B may itself factorize into the density matrices of the individual baths

$$\rho_B = \rho_{B,1}\rho_{B,2}\cdots. \tag{9.221}$$

We allow that these baths are kept at different temperatures T_j, so that

$$\rho_{B,j} = Z_j^{-1}\exp\left[-H_B^{(j)}/(kT_j)\right], \tag{9.222}$$

where

$$Z_j = Tr_{B_j}\left(\exp\left[-H_B^{(j)}/(kT_j)\right]\right). \tag{9.223}$$

For simplification of the further analysis, we may assume

$$Tr_B(B_k\rho) = 0 \tag{9.224}$$

because the corresponding terms in (9.217) would lead to mere energy shifts, but no damping effects. After taking the trace over eq. (9.217) we find

$$\tilde{\rho}(t) = \rho(0) - \int_0^t d\tau_2 \int_0^{\tau_2} d\tau_1 \{\dots\},$$

$$\begin{aligned}\{\dots\} = \sum_{kk'} \{ & V_k(\tau_2)V_{k'}(\tau_1)\rho(0)Tr_B(B_k(\tau_2)B_{k'}(\tau_1)\rho_B(0)) \\ & - V_k(\tau_2)\rho(0)V_{k'}(\tau_1)Tr_B(B_k(\tau_2)\rho_B(0)B_{k'}(\tau_1)) \\ & - V_k(\tau_1)\rho(0)V_{k'}(\tau_2)Tr_B(B_k(\tau_1)\rho_B(0)B_{k'}(\tau_2)) \\ & + \rho(0)V_k(\tau_1)V_{k'}(\tau_2)Tr_B(\rho_B(0)B_k(\tau_1)B_{k'}(\tau_2))\}.\end{aligned} \tag{9.225}$$

The second term on the right hand side of eq. (9.217) has now been dropped because of eq. (9.224). In order to further simplify eq. (9.225), we discuss the expressions

$$\int_0^t d\tau_2 \exp[i\Delta\omega_k\tau_2] \int_0^{\tau_2} d\tau_1 \exp[i\Delta\omega_{k'}\tau_1] K_{kk'}(\tau_2 - \tau_1), \tag{9.226}$$

where K is an abbreviation for

$$K_{kk'}(\tau_2, \tau_1) = K_{kk'}(\tau_2 - \tau_1) = Tr_B(B_k(\tau_2)B_{k'}(\tau_1)\rho_B(0)). \tag{9.227}$$

The first part of eq. (9.227) expresses the fact that the correlation function K depends only on the time difference, provided, that the interaction is stationary.* A simple coordinate transformation

*The proof runs as follows: Stationary means that in the Schrödinger picture B_k, $B_{k'}$ and H_B are time independent. We distinguish between two cases: (a) B_k and $B_{k'}$ belong to different heatbaths. The trace (9.227) then factorizes into $K_k K_{k'}$ where e.g. $K_k = Tr_{B,k}\exp[iH_B^{(k)}\tau_2/\hbar]B(0)\exp[-iH_B^{(k)}\tau_2/\hbar]\exp[-H_B^{(k)}/(kT_k)]Z_k^{-1}$ where (9.222) was used. Because the trace is cyclic: $Tr(AB) = Tr(BA)$ we immediately find $K_k =$

$$\tau_2 - \tau_1 = \tau' \tag{9.228}$$

yields for (9.226)

$$\int_0^t \mathrm{d}\tau_2 \exp[i(\Delta\omega_k + \Delta\omega_{k'})\tau_2] \int_0^{\tau_2} \exp[-i\Delta\omega_{k'}\tau] K_{kk'}(\tau)\,\mathrm{d}\tau. \tag{9.229}$$

We assume that the heatbath has a short memory so that (9.227) is only nonvanishing for

$$|\tau_2 - \tau_1| < \tau_0. \tag{9.230}$$

In the following we consider times t which are large compared to the correlation time τ_0. Then we may simply replace the upper limit τ_2 by infinity, so that we finally find for (9.226)

$$\int_0^t \mathrm{d}\tau_2 \exp[i(\Delta\omega_k + \Delta\omega_{k'})\tau_2] \int_0^{\infty} \exp[-i\Delta\omega_{k'}\tau] K_{kk'}(\tau)\,\mathrm{d}\tau. \tag{9.231}$$

If

$$\Delta\omega_k + \Delta\omega_{k'} = 0 \tag{9.232}$$

the expression (9.231) becomes

$$t\int_0^{\infty} \exp[-i\Delta\omega_{k'}\tau] K_{kk'}(\tau)\,\mathrm{d}\tau. \tag{9.233}$$

In the following we may simply assume that (9.232) holds because otherwise there appear rapidly oscillating terms which cancel the corresponding expressions. We now make the second essential assumption, namely, that the interaction with the heatbath is so small that during the time $\tau > \tau_0$ the density matrix has changed very little. This allows us to make the replacement

$$\frac{\tilde{\rho}(t) - \tilde{\rho}(0)}{t} = \frac{\mathrm{d}\tilde{\rho}(t)}{\mathrm{d}t}. \tag{9.234}$$

If we further use the cyclic property of traces

$$Tr_B(B_k(\tau_1)\rho_B(0)B_{k'}(\tau_2)) = Tr_B(B_{k'}(\tau_2)B_k(\tau_1)\rho_B(0)) \tag{9.235}$$

$Tr_{B,k}(B_k(0)\exp[-H_B^{(k)}/(kT_k)]) = 0$ due to (9.224). (b) B_k and $B_{k'}$ belong to the same heatbath. Using (9.222) we write (9.227) in the form

$$Z_k^{-1}Tr_B(\exp[iH_B^{(k)}\tau_2/\hbar]B_k(0)\exp[-iH_B^{(k)}\tau_2/\hbar]$$
$$\times \exp[iH_B^{(k)}\tau_1/\hbar]B_{k'}(0)\exp[-iH_B^{(k)}\tau_1/\hbar]\exp[-H_B^{(k)}/(kT_k)]).$$

Since the trace is cyclic (9.227) transforms into

$$Z_k^{-1}Tr_B(\exp[iH_B^{(k)}(\tau_2-\tau_1)/\hbar]B_k(0)\exp[-iH_B^{(k)}(\tau_2-\tau_1)/\hbar]$$
$$\times B_{k'}(0)\exp[-H_B^{(k)}/(kT_k)])$$

which depends only on $\tau_2 - \tau_1$.

we may rewrite (9.225) in the form

$$\frac{d\tilde{\rho}(t)}{dt} = -\sum_{kk'} \Big\{ V_k V_{k'} \rho(0) \exp[i(\Delta\omega_k + \Delta\omega_{k'})t]$$

$$\times \int_0^\infty \exp[-i\Delta\omega_k \tau] Tr_B(B_k(\tau) B_{k'}(0) \rho_B(0))\, d\tau$$

$$- V_k \rho(0) V_{k'} \exp[i(\Delta\omega_k + \Delta\omega_{k'})t]$$

$$\times \int_0^\infty \exp[-i\Delta\omega_k \tau] Tr_B(B_{k'}(\tau) B_k(0) \rho_B(0))\, d\tau$$

$$+ \rho(0) V_k V_{k'} \exp[i(\Delta\omega_k + \Delta\omega_{k'})t]$$

$$\times \int_0^\infty \exp[-i\Delta\omega_k \tau] Tr_B(B_k(0) B_{k'}(\tau) \rho_B(0))\, d\tau$$

$$- V_k \rho(0) V_{k'} \exp[i(\Delta\omega_k + \Delta\omega_{k'})t]$$

$$\times \int_0^\infty \exp[-i\Delta\omega_k \tau] Tr_B(B_{k'}(0) B_k(\tau) \rho_B(0))\, d\tau \Big\}. \tag{9.236}$$

We now assume that the iteration step from time $t = 0$ to the time t may be repeated at consecutive times so that we may replace the initial time $t = 0$ by the arbitrary time. The physical meaning of this assumption, which is quite essential for the whole procedure, can be justified by the following consideration: Let us assume that the heatbaths are themselves coupled to still larger heatbaths and so on. Then these "other" heatbaths will again and again bring back the original heatbaths to a truly thermodynamic, i.e. *random* state with the originally introduced temperatures T_j. The reader should be warned that there is a whole literature on statistical mechanics dealing with this and related problems. The idea of heatbath hierarchies, to my knowledge, is not employed there, but it seems essential that the heatbaths are themselves open systems. Otherwise the following argument cannot be disproved: Let us consider the total system: field + heatbaths as a closed one. Then according to fundamental laws of thermodynamics the total system eventually tends to a state with a unique temperature, T, in contrast to our assumption that the heatbaths retain their individual temperatures T_j. Since the integral expressions of the first and second sum agree provided k is exchanged with k', and the same is true for the third and fourth term, eq. (9.236) takes the very simple form

$$\frac{d\tilde{\rho}}{dt} = \sum_{kk'} \{ [V_{k'}\tilde{\rho}(t), V_k] A_{kk'} + [V_{k'}, \tilde{\rho}(t) V_k] A'_{kk'} \}, \tag{9.237}$$

where the coefficients A and A' are defined by

$$A_{kk'} = \exp[i(\Delta\omega_k + \Delta\omega_{k'})t] \times \int_0^\infty \exp[-i\Delta\omega_{k'}\tau] Tr_B(B_k(\tau)B_{k'}(0)\rho_B(0))\,d\tau \tag{9.238}$$

$$A'_{kk'} = \exp[i(\Delta\omega_k + \Delta\omega_{k'})t] \times \int_0^\infty \exp[-i\Delta\omega_k\tau] Tr_B(B_k(0)B_{k'}(\tau)\rho_B(0))\,d\tau. \tag{9.239}$$

A and A' are in a close internal connection: We consider the particular case that

$$\Delta\omega_k + \Delta\omega_{k'} = 0; \qquad B_k^+ = B_{k'} \tag{9.240}$$

holds. This is suggested by the detailed consideration of the interaction Hamiltonian (9.212), (9.213) since it must be Hermitian. In the atomic case for instance, we always find the following combination

$$a_i^+a_j\exp[i\omega_{ij}t]B_{ij} + a_j^+a_i\exp[-i\omega_{ij}t]B_{ij}^+. \tag{9.241}$$

It follows that k and k' are connected by

$$(i,j) = k \qquad \text{and} \qquad (j,i) = k'. \tag{9.242}$$

Under this assumption we obtain

$$A'_{kk'} = \int_0^\infty \exp[-i\Delta\omega_k\tau] Tr_B(\rho_B(0)B_{k'}^+(0)B_{k'}(\tau))\,d\tau$$
$$= \left(\int_0^\infty \exp[-i\Delta\omega_{k'}\tau] Tr_B(B_{k'}^+(\tau)B_{k'}(0)\rho_B(0))\,d\tau\right)^* = A_{kk'}^*. \tag{9.243}$$

It is therefore sufficient to discuss the properties of the A's only. We evaluate the trace over the heatbath in the energy representation of the heatbath alone. (We denote the energy of state n by $\hbar\Omega_n$.) This yields

$$A_{kk'} = \int_0^\infty \exp[-i\Delta\omega_{k'}\tau] \sum_{n,m} \langle n|B_k|m\rangle\langle m|B_k^+|n\rangle \times \exp[i(\Omega_n - \Omega_m)\tau - \hbar\Omega_n/(kT)]\,d\tau Z^{-1}$$
$$= \sum_{nm} |\langle n|B_k|m\rangle|^2 \exp[-\hbar\Omega_n/(kT)]\left\{\pi\delta(\Omega_n - \Omega_m - \Delta\omega_{k'}) + \frac{iP}{\Omega_m - \Omega_n - \Delta\omega_{k'}}\right\}Z^{-1} \tag{9.244}^*$$

where Z is the normalization of the heatbath trace. In the same way we

*P means principal value [cf. eq. (7.273)].

obtain

$$A_{k',k} = \sum_{nm} |\langle n|B_k|m\rangle|^2 \exp[-\hbar\Omega_m/(kT)] \left\{ \pi\delta(\Omega_m - \Omega_n + \Delta\omega_{k'}) + \frac{iP}{\Omega_m - \Omega_n - \Delta\omega_k} \right\} Z^{-1} \tag{9.245}$$

where use was made of eq. (9.222). A comparison of eqs. (9.245) and (9.244) leads to the important relation

$$Re\, A_{kk'} = Re\, A_{k'k} \exp[\hbar\omega_{k'}/(kT)] \tag{9.246}$$

which we discuss both for the light field and the atom: (1) light field: with eq. (9.212) we have e.g. $k = 1$, $k' = 2$, $\Delta\omega_{k'} = \omega$

$$Re\, A_{21} = Re\, A_{12} \exp[\hbar\omega/(kT)]. \tag{9.247}$$

(2) atom: with (9.213), (9.241) and (9.242) we have

$$Re\, A_{\underbrace{ij}_{k'},\underbrace{ji}_{k}} = Re\, A_{\underbrace{ji}_{k},\underbrace{ij}_{k'}} \exp[\hbar\omega_{ij}/(kT)]. \tag{9.248}$$

Specialization of eq. (9.237). (a) Light mode. We make the following identifications

$$\begin{aligned} V_1 &= b, & V_2 &= b^+, \\ \Delta\omega_1 &= -\omega, & \Delta\omega_2 &= \omega, \\ B_1 &= B^+, & B_2 &= B. \end{aligned} \tag{9.249}$$

Equation (9.237) then takes the form

$$\frac{d\tilde{\rho}}{dt} = [b\tilde{\rho}, b^+]A_{21} + [b^+\tilde{\rho}, b]A_{12} + [b, \tilde{\rho}b^+]A_{21}^* + [b^+, \tilde{\rho}b]A_{12}^*. \tag{9.250}$$

Since the imaginary parts of A give rise to mere frequency shifts which can be absorbed into the frequency of the actual oscillator we keep only the real parts and put

$$A_{21} = \xi, \qquad A_{12} = \delta = \xi \exp[-\hbar\omega/(kT)]. \tag{9.251, 252}$$

The final equation for the density matrix of the light field alone thus takes the form

$$\frac{d\tilde{\rho}}{dt} = \delta([b^+\tilde{\rho}, b] + [b^+, \tilde{\rho}b]) + \xi([b\tilde{\rho}, b^+] + [b, \tilde{\rho}b^+]). \tag{9.253}$$

Finally, we may transfrom $\tilde{\rho}$ (interaction representation) into ρ (Schrödinger representation), cf. eq. (9.210), and obtain eq. (9.167) above.

(b) Atom. Here we make the following identifications

$$V_k \to a_i^+ a_j, \qquad \Delta\omega_k \to \omega_i - \omega_j, \tag{9.254}$$

$$B_k = B_{ij}, B_{ji} = B_{ij}^+.$$

The density matrix equation takes the simple form

$$\frac{d\tilde{\rho}}{dt} = \sum_{ij} \{ [a_i^+ a_j \tilde{\rho}, a_j^+ a_i] A_{ji,ij} + [a_i^+ a_j, \tilde{\rho} a_j^+ a_i] A_{ji,ij}^* \}. \tag{9.255}$$

We remind the reader that $\tilde{\rho}$ is the density matrix of the "free field" in the interaction representation. According to (9.214) $a_i^+ a_j$ are operators in the Schrödinger picture. As a final step of our analysis we transform the whole equation (9.255) back into the Schrödinger picture. To this end we make the substitution

$$\tilde{\rho} = U^{-1} \rho U,$$

where U is given by

$$U = \exp[-iH_F t/\hbar],$$

and obtain

$$(i/\hbar)(H_F \tilde{\rho} - \tilde{\rho} H_F) + U^{-1} \frac{d\rho}{dt} U$$

$$= \sum_{i_j} \{ [a_i^+ a_j U^{-1} \tilde{\rho}^u, a_j^+ a_i] A_{ji,ij} + [a_i^+ a_j, U^{-1} \tilde{\rho} U a_j^+ a_i] A_{ji,ij}^* \}. \tag{9.255a}$$

We multiply this equation from the left by U and from the right by U^{-1}. For the first bracket on the left-hand side of (9.255a) we obtain

$$(i/\hbar)[H_F i\rho].$$

The second term on the left-hand side transforms into d_ρ/dt. One may readily convince oneself that in the commutators on the right-hand side of (9.255a) now expressions of the form

$$U a_i^+ a_j U^{-1}$$

must be evaluated. As we know [cf. eq. (9.214)] these latter expressions are transformed into

$$a_i^+ a_j \exp[i\omega_{ij} t]$$

where ω_{ij} is given by eq. (9.215). Collecting all the transformed terms we may rewrite eq. (9.255a) in the form

$$\frac{d\rho}{dt} = -(i/\hbar)[H_F,\rho] + \sum_{ij}\{[a_i^+ a_j\rho i a_j^+ a_i]A_{ji,ij} + [a_i^+ a_j, \rho a_j^+ a_i]A^*_{ji,ij}\}. \tag{9.255b}$$

The time evolution of ρ is now explicitly determined by two different terms. The commutator between H_F and ρ describes the coherent evolution of the system whereas the other terms under the sum describe the incoherent processes.

We derive the equation for the average of the operator $a_m^+ a_n$ by multiplying both sides of eq. (9.255b) by $a_m^+ a_n$ and taking the trace:

$$\frac{d\langle a_m^+ a_n\rangle}{dt} = Tr\Big\{\sum_{ij}\big[(a_m^+ a_n a_i^+ a_j\bar\rho a_j^+ a_i - a_m^+ a_n a_j^+ a_i a_i^+ a_j\bar\rho)A_{ji,ij} + (a_m^+ a_n a_i^+ a_j\bar\rho a_j^+ a_i - a_m^+ a_n\bar\rho a_j^+ a_i a_i^+ a_j)A^*_{ji,ij}\big] \times (i/\hbar)a_m^+ a_n[H_F,\rho]\Big\}. \tag{9.256}$$

For the further evaluation of the trace we use its cyclic property and arrange the operators in such a way that the density matrix $\tilde\rho$ stands on the right-hand side. We then use the theorem

$$a_{i_1}^+ a_{j_1} a_{i_2}^+ a_{j_2}\dots a_{i_n}^+ a_{j_n} = a_{i_1}^+ a_{j_n}\delta_{j_1 i_2}\delta_{j_2 i_3}\dots\delta_{j_{n-1} i_n} \tag{9.257}$$

which holds if only one electron is present. The relation (9.257) can be proved as follows: consider within the product the terms which bear the indices λ and $\lambda+1$. Exchanging a_{j_λ} with $a_{i_{\lambda+1}}^+$ and using the commutation relations we find

$$a_{i_1}^+ a_{j_1}\dots a_{j_\lambda} a_{i_{\lambda+1}}^+ a_{j_{\lambda+1}}\dots a_{i_n}^+ a_{j_n} = a_{i_1}^+ a_{j_1}\dots(\delta_{j_\lambda i_{\lambda+1}} - a_{i_{\lambda+1}}^+ a_{j_\lambda})a_{j_{\lambda+1}}\dots a_{i_n}^+ a_{j_n}. \tag{9.258}$$

The term which still contains $a_{i_{\lambda+1}}^+ a_{j_\lambda}$ contains two subsequent annihilation operators. These operators are applied to a state which contains only one electron. The corresponding wave function Φ vanishes. If (9.258) is performed for all indices, (9.257) results. Using (9.257) we transform (9.256) into

$$\frac{d}{dt}\langle a_m^+ a_n\rangle = \langle a_m^+ a_n\rangle(i\omega_{mn} + i\Delta\omega_{mn} - \gamma_{mn}), \quad \text{for } m \neq n, \tag{9.259}$$

$$\frac{d}{dt}\langle a_m^+ a_m\rangle = \sum_j\{\langle a_j^+ a_j\rangle W_{jm} - \langle a_m^+ a_m\rangle W_{mj}\}, \tag{9.260}$$

where we have used the abbreviations

$$W_{jm} = A_{jm,mj} + A^*_{jm,mj}, \tag{9.261}$$

$$\gamma_{mn} = \sum_i Re(A_{mi,im} + A^*_{ni,in}) = \tfrac{1}{2}\sum_i (W_{ni} + W_{in}), \tag{9.262}$$

$$\Delta\omega_{mn} = -\sum_i Im(A_{ni,in} + A^*_{mi,im}). \tag{9.263}$$

W_{jm} is evidently the transition rate from the state j to the state m, as may be seen by considering eq. (9.260). γ_{mn} is the phase halfwidth from the second part of eq. (9.259). We see that the halfwidth is determined by half the sum of the transition rates out of the two states n and m which are considered. Equation (9.262) contains also an additional term which stems from phase fluctuations alone which are not accompanied by real transitions, i.e. W_{nn}, eq. (9.261), contains frequency shifts which we shall neglect, however, in the following. We now show that (9.259), (9.260) can also be interpreted as an equation for the density matrix itself if the latter one is written in the occupation number representation. We put

$$\rho(t) = \sum_{m,n} \rho_{mn}(t) a_m^+ a_n \tag{9.264}$$

and consider

$$Tr(a_m^+ a_n \rho(t)) = \sum_l \langle \Phi_0 a_l a_m^+ a_n \sum_{m'n'} \rho_{m'n'}(t) a_{m'}^+ a_{n'} a_l^+ \Phi_0 \rangle, \tag{9.265}$$

where Φ_0 is the vacuum state and $a_l^+ \Phi_0$ a one-electron state. Using (9.257) we find immediately

$$Tr(a_m^+ a_n \rho(t)) = \rho_{nm}(t), \tag{9.266}$$

so that we can identify

$$\langle a_m^+ a_n \rangle = \rho_{nm}(t). \tag{9.267}$$

Exercises on section 9.5.

(1) A quantum mechanical harmonic oscillator (e.g. a field mode) described by operators b^+, b is coupled to a reservoir and a coherent source. What is its density matrix equation?

Hint: Write

$$\frac{d\rho}{dt} = \left(\frac{d\rho}{dt}\right)_{\text{incoherent}} + \left(\frac{d\rho}{dt}\right)_{\text{coherent}}$$

$(d\rho/dt)_{\text{incoherent}}$ is given by the r.h.s. of (9.156) and

$$\left(\frac{d\rho}{dt}\right)_{\text{coherent}} = -\frac{i}{\hbar}[H, \rho],$$

where

$$H = \hbar\omega_0 b^+ b + \gamma_0 \exp(i\omega t) b + \gamma_0 \exp(-i\omega t) b^+ .$$

(2) Calculate $\langle b \rangle$, $\langle b^+ b \rangle$ using $\rho =$ (9.204).
Hint: Remember that $\langle \Omega \rangle = Tr(\Omega\rho)$
(3) A two-level atom is coupled to reservoirs and a coherent driving field. Derive its density matrix equation.
Hint: Same as for exercise (1).

Mathematical appendix. Dirac's δ-function and related functions

In this book we encountered, at several occasions, Dirac's δ-function and functions related to it. Here we give a list of these functions, their definitions and their most important properties. It should be noted that these functions are quantities which are meaningful only when they occur under an integral. When these quantities occur in equations without integral it is understood that eventually an integration is performed. The δ-function can be defined in various ways. The one occurring in this book is

$$\delta(\omega) = \frac{1}{2\pi} \lim_{t\to\infty} \int_{-t}^{t} e^{-i\omega\tau} \mathrm{d}\tau = \frac{1}{\pi} \lim_{t\to\infty} \frac{\sin \omega t}{\omega}. \tag{A.1}$$

The limit $t \to \infty$ of the function $\sin \omega t/\omega$ does not exist in itself. But when we multiply (A.1) by a continuous function $f(\omega)$ and integrate over ω over an interval which contains $\omega = 0$ we obtain the following relation

$$\int_{-\omega_1}^{\omega_2} \delta(\omega) f(\omega)\, \mathrm{d}\omega = \frac{1}{\pi} \lim_{t\to\infty} \int_{-\omega_1}^{\omega_2} \frac{\sin \omega t}{\omega} f(\omega)\, \mathrm{d}\omega \tag{A.2}$$

which by means of the transformation $\omega t = \omega'$ is transformed into

$$\frac{1}{\pi} \lim_{t\to\infty} \int_{-\omega_1 t}^{\omega_2 t} \frac{\sin \omega'}{\omega'} f(\omega'/t)\, \mathrm{d}\omega'. \tag{A.3}$$

The effect of the limit $t \to \infty$ can be visualized as follows. $f(\omega'/t)$ tends to $f(0)$ while the limits of the integral tend to $-\infty$ and $+\infty$, respectively. Therefore (A.3) reduces to

$$f(0) \frac{1}{\pi} \int_{-\infty}^{+\infty} \frac{\sin \omega'}{\omega'} \mathrm{d}\omega' \tag{A.4}$$

which can also be proven mathematically rigorously. The integral over ω' exists and is equal to π. Therefore equating the left-hand side of (A.2) to

(A.4) we obtain the fundamental property of the δ-function

$$\int_{-\omega_1}^{\omega_2} \delta(\omega) f(\omega)\,\mathrm{d}\omega = f(0). \tag{A.5}$$

It holds if f is continuous and the integration comprises the point $\omega = 0$. As can be seen directly from the definition (A.1), δ is an even function

$$\delta(-\omega) = \delta(\omega). \tag{A.6}$$

Due to this fact, one may readily deduce from (A.5) the relation

$$\int_0^{\omega_2} \delta(\omega)\,\mathrm{d}\omega = \tfrac{1}{2}. \tag{A.7}$$

The δ-function can also be defined in three dimensions by

$$\delta(\mathbf{x}) = \frac{1}{(2\pi)^3} \int \exp(i\mathbf{kx})\,\mathrm{d}^3\mathbf{k}. \tag{A.8}$$

Provided $f(x)$ is a continuous function and the volume of integration comprises the origin $\mathbf{x} = 0$ we obtain in analogy to (A5)

$$\int \delta(\mathbf{x}) f(\mathbf{x})\,\mathrm{d}^3\mathbf{x} = f(0). \tag{A.9}$$

A function closely related to the δ-function is the δ_+-function in which the integration over τ, which occurred in (A.1), runs only over positive values. Thus the δ_+-function is defined by

$$\delta_+(\omega) = \frac{1}{2\pi} \lim_{t\to\infty} \int_0^t e^{-i\omega\tau}\,\mathrm{d}\tau = \frac{1}{2\pi} \lim_{t\to\infty} \frac{1 - e^{-i\omega t}}{i\omega}. \tag{A.10}$$

When we split the exponential function into cosine and sine we obtain

$$\delta_+(\omega) = \frac{1}{2\pi i} \lim_{t\to\infty} \frac{1 - \cos\omega t}{\omega} + \frac{1}{2\pi} \lim_{t\to\infty} \frac{\sin\omega t}{\omega}. \tag{A.11}$$

While the second part on the right hand side is immediately recognized as the δ-function, the first represents a new function which we abbreviate by P/ω.

$$\lim_{t\to\infty} \frac{1 - \cos\omega t}{\omega} = \frac{P}{\omega}. \tag{A.12}$$

To get an insight into its properties we multiply (A.12) by a continuous function $f(\omega)$ and integrate over an interval $-\omega_1 \ldots \omega_2$

$$\int_{-\omega_1}^{\omega_2} f(\omega) \frac{P}{\omega}\,\mathrm{d}\omega. \tag{A.13}$$

Let us consider first the effect of $\cos\omega t$. When t is very large, $\cos\omega t$

becomes a very rapidly oscillating function so that when integrated over ω the positive and negative contributions cancel each other (compare also fig. 4.3 on page 138 where a quite similar discussion is made). Even if $\cos \omega t$ is multiplied by a continuous function $f(\omega)$ this statement remains valid. On the other hand $(1 - \cos \omega t)/\omega$ vanishes for any finite t when ω goes to 0. As a consequence of these considerations we recognize that the effect of P/ω is as follows. Outside of $\omega = 0$, P/ω can be replaced by $1/\omega$. On the other hand the point $\omega = 0$ is cut out. Therefore we may write

$$P\int_{-\omega_1}^{\omega_2} f(\omega)\frac{1}{\omega}\,d\omega = \lim_{\epsilon \to +0}\left(\int_{-\omega_1}^{-\epsilon}\frac{f(\omega)}{\omega}\,d\omega + \int_{\epsilon}^{\omega_2}\frac{f(\omega)}{\omega}\,d\omega\right). \tag{A.14}$$

P has, therefore, the property of denoting the principal value of the integral. Knowing the meaning of P/ω and the δ-function we may finally write (A.11) in short as

$$\delta_+(\omega) \equiv \frac{1}{2\pi i}\lim_{t\to\infty}\frac{1-e^{-i\omega t}}{\omega} = \frac{1}{2\pi i}\frac{P}{\omega} + \tfrac{1}{2}\delta(\omega). \tag{A.15}$$

In addition to δ_+ one may define δ_- which is related to δ_+ by

$$\delta_-(\omega) = \delta_+^*(\omega). \tag{A.16}$$

The functions $\delta(\omega)$, δ_+, δ_- and $P/(2\pi i\omega)$ can also be defined in the complex ω-plane. Each time these functions equal $1/(2\pi i\omega)$ but the paths of integration are different.

Let $f(\omega)$ be a function which contains no singularity at $\omega = 0$. When we perform an integration on a path which is a small circle around $\omega = 0$ we obtain according to Cauchy's theorem

$$\frac{1}{2\pi i}\oint\frac{f(\omega)}{\omega}\,d\omega = f(0). \tag{A.17}$$

Comparing (A.17) with (A.5) we recognize that we may define the δ-function as $1/(2\pi i\omega)$ with the additional prescription to perform the integration over a small circle around $\omega = 0$. P is the prescription to leave out the interval $-\epsilon$ to $+\epsilon$ when we integrate along the real axis.

To define δ_+ we proceed as follows. The clockwise integration over $1/(2\pi i\omega)$ along a semicircle in the lower half of the complex plane yields $\frac{1}{2}$, i.e. just half of the value we would obtain by use of the δ-function. Let us now consider an integration over $1/(2\pi i\omega)$ along the real axis from $-\infty$ till $+\infty$ where we leave out the point $\omega = 0$ using a small semicircle in the lower half plane around $\omega = 0$. The result of this operation is the same as that of integrating along the real axis over

$$\frac{1}{2\pi i}\frac{P}{\omega} + \tfrac{1}{2}\delta(\omega). \tag{A.18}$$

(A.18) is identical with the definition of the δ_+-function. On the other hand we may replace the path of integration containing the circle by a path along the whole real axis when we shift ω by a small amount $-is$. This leads us to the final result

$$\delta_+ = \frac{1}{2\pi i} \lim_{s \to 0} \frac{1}{\omega - is} = \frac{1}{2\pi i} \frac{P}{\omega} + \frac{1}{2}\delta(\omega). \tag{A.19}$$

Similarly we obtain

$$\delta_- = -\frac{1}{2\pi i} \lim_{s \to 0} \frac{1}{\omega + is} = -\frac{1}{2\pi i} \frac{P}{\omega} + \frac{1}{2}\delta(\omega). \tag{A.20}$$

References and further reading

The topics treated in this volume comprise a vast field with an enormous number of publications. Since this book is meant as a textbook for students, I am giving here those references which are of the greatest interest to them: namely, first textbooks on related subjects so that students can learn more about details, in particular on atomic physics and quantum theory. Second, I have listed references to the original work of the pioneers of this field. Finally, I included those references which I directly made use of in my book or which contain related material.

There are a number of scientific journals dealing with modern optics. Quite excellent review articles may be found in particular in the following series dedicated particularly to modern optics:

Wolf, E. Ed: Progress in Optics. North-Holland, Amsterdam
Springer Series in Optical Sciences. Springer, Berlin

Textbooks on QUANTUM MECHANICS:
Davydov, A. S.: Quantum Mechanics. Addison Wesley, 1965
Dirac, P. A. M.: The Principles of Quantum Mechanics. 4th ed. Clarendon, Oxford 1958
Gasiorowicz, S.: Quantum Physics. Wiley, 1974
Landau, L. D. and *Lifshitz, E. M.*: Quantum Mechanics. Nonrelativistic Theory. Addison Wesley, 1965
Merzbacher, E.: Quantum Mechanics. 2nd ed. Wiley, 1970
Messiah, A.: Quantum Mechanics. Vol I, II. Wiley, 1968
Sakurai, J. J.: Advanced Quantum Mechanics. Addison Wesley, 1967
Schiff, L. I.: Quantum Mechanics. 3rd ed. McGraw Hill, 1968

Textbooks on SPIN RESONANCE:
Abragam, A.: The Principles of Nuclear Magnetism. Clarendon, Oxford 1961

Abragam, A. and *Bleaney, B.*: Electron Paramagnetic Resonance of Transition Ions. Clarendon, Oxford 1970
Poole, jr., C. P. and *Farach, H. A.*: The Theory of Magnetic Resonance. Wiley, 1972
Slichter, C. P.: Principles of Magnetic Resonance. Springer, New York, 1978

Textbooks on QUANTIZATION OF LIGHT FIELD:
Bogoliubov, N. N. and *Shirkov, D. V.*: An Introduction to the Theory of Quantized Fields. 3rd ed. Wiley, New York 1980
Fain, V. M. and *Khanin, Ya. I.*: Quantum Electronics. Vol I, II. Pergamon, Oxford 1969
Feynman, R. P.: Quantum Electronics. Benjamin, New York 1962
Haken, H.: Quantum Field Theory of Solids. North-Holland, Amsterdam 1976
Heitler, W. H.: The Quantum Theory of Radiation. 3rd ed. Clarendon, Oxford 1954
Jauch, J. M. and *Rohrlich, F.*: The Theory of Photons and Electrons. 2nd ed. Springer 1976
Klauder, J. R. and *Sudarshan, E. C. G.*: Fundamentals of Quantum Optics. Benjamin, New York 1968
Louisell W. H.: Radiation and Noise in Quantum Electronics. McGraw Hill, 1964
Louisell W. H.: Quantum Statistical Properties of Radiation. Wiley, New York 1973
Schweber S. S.: An Introduction to Relativistic Quantum Field Theory. Harper & Row, New York 1961

For Chapter 1:
§1.5. The early quantum theory of matter and light
Bohr, N.: Phil. Mag. 26, 476, 857 (1913)
Einstein, A.: Ann. d. Phys. 17, 132 (1905); 20, 199 (1906)
Einstein, A.: Phys. Z. 18, 121 (1917)
Planck, M.: Verh. d. deut. phys. Gesellsch. 2, 237 (1900); Ann. d. Phys. 4, 553 (1901)

cf. textbooks: *QUANTUM MECHANICS*

§1.6. and 1.7. Quantum mechanics
DeBroglie, L.: Nature 112, 540 (1923)
Dirac, P. A. M.: Proc. Roy. Soc. A117, 610 (1928)
Heisenberg, W.: Z. Phys. 33, 879 (1925)
Schrödinger, E.: Ann. d. Phys. 79, 361, 489, 734 (1926)

(a) Perturbation theory:
(i) Time-independent
Schrödinger, E.: Ann. d. Phys. 80, 437 (1926)
(ii) Time-dependent
Dirac, P. A. M.: Proc. Roy. Soc. A112, 661 (1926); A114, 243 (1927)

cf. textbooks: *QUANTUM MECHANICS*

§1.8. Quantum electrodynamics
Fermi, E.: Rev. Mod. Phys. 4, 87 (1932)
Heisenberg, W. and *Pauli, W.*: Z. Phys. 56, 1 (1926); 59, 168 (1930)

cf. textbooks: *QUANTUM MECHANICS; QUANTIZATION OF LIGHT FIELD*

§1.9. Probabilistic interpretation of quantum theory
Bell, J. S.: Physics 1, 195 (1964)
Bohr, N.: Nature 121, 580 (1928)
Born, M.: Z. Phys. 37, 863 (1926); Nature 119, 354 (1927)
Clauser, J F. and *Shimoni, A.*: Rep. Progr. i. Phys. 41, 1881 (1978)
Einstein, A., and *Podolsky, B.* and *Rosen, N.*: Phys. Rev. 47, 777 (1935)
d'Espagnat, B.: Sci. Am. Nov. 79, 128 (1979)
Heisenberg, W.: Z. Phys. 43, 172 (1927)

cf. textbooks: *QUANTUM MECHANICS*

§1.11. Coherence
Beran, M. and *Parrent, G. B.*: Theory of Partial Coherence. Englewood Cliffs, New York 1964
Born, M. and *Wolf, E.*: Principles of Optics. Pergamon, Oxford 1964
Francon, M.: Optical Interferometry. Academic Press, New York 1966
Glauber, R. J.: In: Quantum Optics. Eds. *DeWitt, C.*, *Blandin, A.*, and *Cohen-Tannoudji, C.*; Gordon & Breach, New York 1965
Glauber, R. J.: In: Proc. of the Int. School of Physics "Enrico Fermi". Varenna 1967. Academic Press, New York
Hanbury Brown, R. and *Twiss, R. Q.*: Nature 117, 27 (1956)
Klauder, J. R. and *Sudarshan, E. C. G.*: Fundamentals of Quantum Optics. Benjamin, New York 1968
Mandel, L. and *Wolf, E.*: Rev. Mod. Phys. 37, 231 (1965)
Mandel, L. and *Wolf, E.*: Selected Papers on Coherence and Fluctuations of Light. Vols I, II. Dover Publ., New York 1970

§1.15. Lamb shift
Lamb, W. E. and *Retherford, R. C.*: Phys. Rev. 72, 241 (1947); 79, 549 (1950)

§1.17. Nonlinear Optics
Bloembergen, N.: Nonlinear optics. Benjamim, New York 1965

For Chapter 2:
§2.1. Waves
Young, T.: *Phil.* Trans. Roy. Soc. 12, 387 (1802)

§2.2. Coherence
Same as for §1.11

§2.3. Planck's radiation law
Kubo, R.: Statistical Mechanics. North-Holland, Amsterdam 1965
Kubo, R.: Thermodynamics. North-Holland, Amsterdam 1968
Planck, M.: Verh. d. deut. phys. Gesellsch. 2, 237 (1900); Ann. d. Phys. 4, 553 (1901)
Tien, C. J. and *Lienhard, J. H.*: Statistical Thermodynamics. Holt, Rinehart and Winston, New York 1971

§2.4. and 2.5. Photons
Same as for §1.5

For Chapter 3
§3.1. Wave equation for matter: Schrödinger equation
Davisson, C. and *Germer, L. H.*: Phys. Rev. 30, 705 (1927)

cf. textbooks: *QUANTUM MECHANICS*

§3.2. Measurements and expectation values

cf. textbooks: *QUANTUM MECHANICS*

§3.3. Harmonic oscillator
Born, M., Heisenberg, W. and Jordan, P.: Z. Phys. 35, 557 (1925)
Ludwig, G.: Z. Phys. 130, 468 (1951)

cf. textbooks: *QUANTUM MECHANICS*

§3.4. Hydrogen atom
Bethe, *H.* and *Salpeter*, *E. E.*: Quantum Mechanics of one and two Electron Atoms. Springer, Berlin; Academic, New York 1957

cf. textbooks: *QUANTUM MECHANICS*

§3.5. Atoms
Barrow, *G. M.*: Introduction to Molecular Spectroscopy. McGraw Hill, New York 1962
Haken, *H.* and *Wolf*, *H. C.*: Atom- and Quantenphysik. Springer, Berlin 1980
Herzberg, *G.*: Infrared and Raman Spectra of Polyatomic Molecules. D. van Nostrand, Princeton 1964
Kuhn, *H. G.*: Atomic Spectra. Longmans, London 1964
Landau, *L. D.*, *Akhiezer*, *A. I.* and *Lifshitz*, *E. M.*: General Physics, Mechanics and Molecular Physics. Pergamon, Oxford 1967
Sommerfeld, *A.*: Atombau und Spektrallinien. Vol I, II. Vieweg, Braunschweig 1960
White, *H. E.*: Introduction to Atomic Spectra. McGraw Hill, New York 1934

§3.6. Solids
Haken, *H.*: Quantum Field Theory of Solids. North-Holland, Amsterdam 1976
Haug, *A.*: Theoretical Solid State Physics. Vol I, II. Pergamon, Oxford 1972
Jones, *W.* and *March*, *N. H.*: Theoretical Solid State Physics. Vol I, II. Wiley, New York 1973
Kittel, *C.*: Quantum Theory of Solids. Wiley, New York 1964
Kittel, *C.*: Introduction to Solid State Physics. Wiley, New York 1966
Peierls, *R. E.*: Quantum Theory of Solids. Oxford at the Clarendon Press, 1955
Pines D.: Elementary Excitations in Solids. Benjamin, New York 1963
Ziman, *J. M.*: Principles of the Theory of Solids. Cambridge Univ. Press, 1965
Ziman, *J. M.*: Electrons and Phonons. Oxford at the Clarendon Press, 1967

§3.7. Nuclei
Bohr, *A.* and *Mottelson*, *B. R.*: Nuclear Structure. Benjamin, New York 1969

§3.8. Spin
McWeeny, R.: Spins in Chemistry. Academic Press, New York 1970
Pauli, W.: Z. Phys. 43, 601 (1927)
Uhlenbeck, G. E. and *Goudsmith, S.*: Naturwiss. 13, 953 (1925); Nature 117, 264 (1926)

cf. textbooks: *SPIN RESONANCE*

For Chapter 4:
§4.1.-4.3. Two-level atom in external fields

cf. textbooks: *QUANTUM MECHANICS; QUANTIZATION OF LIGHT FIELD*

§4.4. Two-photon absorption
Biraben, F., Cagnac, B. and *Gruenberg, G.*: Phys. Rev. Lett. 32, 643 (1974)
Goeppert-Mayer, M.: Ann. Phys. 9, 273 (1931)
Hänsch, T. W., Harvey, K. C., Meisel, G. and *Schawlow, A. L.*: Opt. Comm. 11, 50 (1974)
Hopfield, J. J., Worlock, J. M. and *Park, K.*: Phys. Rev. Lett. 11, 414 (1963)
Hopfield, J. J. and *Worlock, J. M.*: Phys. Rev. A137, 1455 (1965)
Kaiser, W. and *Garrett, C. G. B.*: Phys. Rev. Lett. 7, 229 (1961)
Levenson, M. D. and Bloembergen, N.: Phys. Rev. Lett. 32, 645 (1974)
Mahr, H.: Two Photon Absorption Spectroscopy. p. 285. In: Quantum Electronics, Vol. 1A. Eds. Rabin, H. and Tang, C. L., Academic Press, New York 1975

§4.5. Non-resonant perturbations

cf. textbooks: *QUANTUM MECHANICS*

§4.6.-4.9. Analogy two-level atom and spin $\frac{1}{2}$
Feynman, R. P., Vernon, F. L. and *Hellwarth, R. W.*: J. Appl. Phys. 28, 49 (1957)
Haken, H.: Laser Theory. Encycl. of Phys. XXV/2c. Springer, Berlin 1970

(a) Rotating wave approximation:
Bloch, F. and *Siegert, A. J.*: Phys. Rev. 57, 522 (1940)
Rabi, J. J., Ramey, N. F. and *Schwinger, J.*: Rev. Mod. Phys. 26, 107 (1954)

(b) Photon-echo experiment
Kurnit, N. A., Abella, I. D. and *Hartmann, R. S.*: Phys. Rev. Lett 13, 567 (1964); Phys. Rev. 141, 391 (1966)

(c) Free introduction decay, nutation
Brewer, R. G.: In: Coherence in Spectroscopy and Modern Physics, Eds. *Arecchi, F. T., Bonifacio, R.* and *Scully, M. O.* Plenum, New York 1978

cf. textbooks: *SPIN RESONANCE*

For Chapter 5
§5.1.-5.4. Quantization of light field
Fermi, E.: Rev. Mod. Phys. 4, 87 (1932)
Jackson, J. D.: Classical Electrodynamics. Wiley, New York 1967

cf. textbooks: *QUANTUM MECHANICS; QUANTIZATION OF LIGHT FIELD*

§5.5. Coherent states
Glauber, R. J.: Phys. Rev. Lett. 10, 84 (1963); Phys. Rev. 130, 2529 (1963); 131, 2766 (1963)
Schrödinger, E.: Ann. Phys. 87, 570 (1928)
Schwinger, J.: Phys. Rev. 91, 728 (1953)

cf. textbooks: *QUANTIZATION OF LIGHT FIELD; QUANTUM MECHANICS*

5.6. Heisenberg picture

cf. textbooks: *QUANTUM MECHANICS*

5.7. Driven harmonic oscillator
Ludwig, G.: Z. Phys. 130, 468 (1951)

cf. textbooks: *QUANTUM MECHANICS*

5.8. Quantization of light field: Multimode case
Same as for § 5.1.-5.4

5.9. Uncertainty relations, limits of measurability
Carruthers, *P.* and *Nieto*, *M. M.*: Rev. Mod. Phys. 40, 411 (1968)
Susskind, *L.* and *Glowgower*, *J.*: Physics 1, 49 (1964)

cf. textbooks: *QUANTIZATION OF LIGHT FIELD; QUANTUM MECHANICS*

For Chapter 6:
§6.2. Quantization of electron wave field
Jordan, *P.* and *Wigner*, *E.*: Z. Phys. 47, 631 (1928)
Pauli, *W.*: Z. Phys. 31, 765 (1925)

cf. textbooks: *QUANTIZATION OF LIGHT FIELD*

For Chapter 7:
§7.1.-7.4. Interaction light-matter
Jackson, *D.*: Classical Electrodynamics, Wiley, New York 1967

cf. textbooks: *QUANTIZATION OF LIGHT FIELD*

§7.5. Dipole approximation
Fiutak, *J.*: Can. J. Phys. 41, 12 (1963)
Goeppert-Mayer, *M.*: Ann. Phys. 9, 273 (1931)

cf. textbooks: *QUANTIZATION OF LIGHT FIELD*

§7.6. Spontaneous and stimulated emission, absorption
Weisskopf, *V.* and *Wigner*, *E.*: Z. Phys. 63, 54 (1930); 65, 18 (1930)

cf. textbooks: *QUANTIZATION OF LIGHT FIELD*

§7.7. Feynman diagrams
Power, *E. A.*: Introductory Quantum Electrodynamics. Longmans, Green & Co, London 1964
Todorov, *T. I.*: Analytic Properties of Feynman Diagrams in Quantum Field Theory. Pergamon, Oxford 1971

§7.8. Lamb shift
Bethe, *H. A.*: Phys. Rev. 72, 339 (1947)
Bethe, *H.* and *Salpeter*, *E. E.*: Quantum Mechanics of one and two electron atoms, Springer, Berlin; Academic, New York 1957

Hänsch, *T. W.*, *Lee*, *S. A.*, *Wallenstein*, *R.* and *Wieman*, *C.*: Phys. Rev. Lett. 34, 307 (1975)
Lamb, *W. E.* and *Retherford*, *R. C.*: Phys. Rev. 72, 241 (1947); 79, 549 (1950)
Lee, *S. A.*, *Wallenstein*, *R.* and *Hänsch*, *T. W.*: Phys. Rev. Lett. 35, 1262 (1975); Phys. Rev. Lett. 34, 307 (1975)

cf. textbooks: *QUANTIZATION OF LIGHT FIELD*

§7.9. Damping and linewidth
Same as for §7.6

§7.10. Absorption and emission of single mode

cf. textbooks: *QUANTIZATION OF LIGHT FIELD*

§7.11. Dynamic Stark effect
(a) Our approach of section 7.11 is related to that of Agarwal, but uses a coherent state representation of the driving field.
Agarwal, *G. S.*: Quantum Statistical Theories of Spontaneous Emission and their Relation to other Approaches, Springer Tracts in Modern Physics Vol. 70, Springer, Berlin 1974

(b) For different approaches and experiments see:
Bonifacio, *R.* and *Lugiato*, *L. A.*: Opt. Comm. 19, 172 (1976); Phys. Rev. A18, 1129 (1978)
Carmichael, *H. J.* and *Walls*, *D. F.*: *J. Phys.* B9, 1199 (1976)
Chang, *C. S.* and *Stehle*, *P.*: Phys. Rev. A4, 641 (1971)
Hartig, *W.*, *Rasmurren*, *W.*, *Schieder*, *W. R.* and *Walther*, *H.*: Z. Phys. A278, 205 (1976)
Hassan, *S. S.* and *Bullough*, *R. K.*: *J. Phys.* B8, L147 (1975)
Mollow, *B. R.*: Phys. Rev. 188, 1969 (1969); A12, 1919 (1975); J. Phys. A8, L11 (1975)
Schuda, *F.*, *Stroud*, *jr.*, *C. R.* and *Hercher*, *M.*: *J. Phys.* B7, L198 (1974)
Stroud, *jr.*, *C. R.*: In: Proceedings of the 3rd Rochester Conference on Quantum Electronics. Eds. *Mandel*, *L.* and *Wolf*, *E.* Plenum, New York 1973
Wu, *F. Y.*, *Grove*, *R. E.* and *Ezekiel*, *S.*: Phys. Rev. Lett. 35, 1426 (1975)

For Chapter 8:
§8.1.-8.3. Quantum mechanical coherence functions
Same as for §1.11

§8.4. Quantum beats
Haroche, *S*.: in: High Resolution Laser Spectroscopy. Ed. *Shimoda*, *K*. Springer, 1976

For Chapter 9:
§9.1. Langevin equation, Fokker-Planck equation
Bharucha Reid, *A*. *T*.: Theory of Markov Processes. McGraw Hill, Toronto 1960
Chandrasekkar, *S*.: Rev. Mod. Phys. 15, 1 (1973)
Doob, *J*. *L*.: Stochastic Processes. Wiley, New York 1963
Haken, *H*.: Synergetics. An Introduction. 2nd ed. Springer, Berlin 1978
Stratonovich, *R*. *L*.: Topics in the Theory of Random Noise, Vols I, II. Gordon & Breach, New York 1967
Uhlenbeck, *G*. *E*. and *Ornstein*, *L*. *S*.: Phys. Rev. 36, 823 (1930)
Wax, *N*.: Noise and Stochastic Processes, Dover, New York 1954

§9.2.-9.4. Damping and fluctuations in quantum mechanics
Haken, *H*. and *Weidlich*, *W*.: Z. Phys. 189, 1 (1966)
Haken, *H*.: Laser Theory. Encycl. of Phys. XXV/2c. Springer, Berlin 1970
Senitzky, *J*. *R*.: Phys. Rev. 119, 670 (1960); 124, 642 (1961)

cf. textbooks: *QUANTIZATION OF LIGHT FIELD*

§9.5 Density matrix
Argyres, *P*. *N*. and *Kelley*, *P*. *L*.: Phys. Rev. 134A, 98 (1964)
Haake, *F*.: Statistical Treatment of Open Systems by Generalized Master Equations. Springer Tracts in Modern Physics 66. Springer, Berlin 1973.
Haken, *H*.: Laser Theory. Encycl. of Phys. XXV/2c. Springer, Berlin 1970
Mori, *H*.: Progr. Theor. Phys. 33, 423 (1965); 34, 394 (1965)
Nakajima, *S*.: Progr. Theor. Phys. 20, 948 (1958)
Wangsness, *R*. *K*. and *Bloch*, *F*.: Phys. Rev. 89, 728 (1953)
Weidlich, *W*. and *Haake*, *F*.: Z. Phys. 185, 30 (1965)
Zwanzig, *R*.: J. Chem. Phys. 33, 1338 (1960); Lect. Theor. Phys. (Boulder) 3, 106 (1960)

Subject index

absorption 6, 125, 222, 229
absorption coefficient 229
angular momentum operators 78, 101
annihilation operator for bosons 83
~~~ fermions 197, 308
~~~ field mode 162
atomic dipole moment 306
average, statistical 287

Baker–Hausdorff theorem 168
band structures, schematic 106
bare mass 245
beat frequency 279
Bloch's equations 117, 145, 148, 156, 314
~ theorem 104
Bloch wave function 105, 109
Bohr's postulate 10, 60
Boltzmann's constant 290, 300
~ distribution 54, 323
boundary conditions, for field in cavity 186
bra 73, 77, 318
Brownian motion 187, 285, 287, 289, 296

cavity modes 16, 159, 179, 181, 184, 187, 271
central limit theorem 303
classical approach 201
coherence 24
coherence function 28, 44, 49
~ functions, quantum mechanical 265, 271
coherence, spatial 47
~ temporal 47
coherent light 17
~ states 31, 170, 172, 173, 188, 190, 255
~ states of radiation field 273
commutation relation 79, 182, 295
~ relation for bosons 84
~ relation, Heisenberg picture 176
~ relations between fermions and bosons 210
~~ for creation and annihilation operators 166
~~~ fermions 198
completeness relation 250
complete set 127
complex degree of coherence for spontaneous emission 277
~~~ mutual coherence 4, 46
Compton effect 57
conditional probability density 294
conductivity 292
continuity equation 203
correlation function 47, 288
~ function of fluctuations 287
~ functions, higher order 270
~ functions of field operators 274
~~~ quant. mech. fluct. forces 301
Coulomb gauge 163, 207
coupling coefficient electron-light field 209
creation and annihilation operators 165
~ operator for bosons 3, 83
~~~ fermions 197, 308
~~~ field mode 162

damped harmonic oscillator 7, 202
damping in quantum systems 29
~ of field modes 295
Davisson–Germer experiment 64
De Broglie 13, 64
decay time 290
δ-function 299
density matrix 316
~ matrix equation 319
~~~

~~ equation for damped harmonic oscillator 319
~ of states 52, 53
~ operator 316
dephasing effects 155
diffraction 3, 40
diffusion coefficient 291
dipole approximation 218, 269, 281
~ matrix element 209, 220
~~ elements, for hydrogen states 100
~ moment 306
Dirac 12, 72
Dirac's δ-function 204, 299
~ notation 72, 317
dispersion, nonlinear 36
dispersion, classical theory 135
displaced harmonic oscillator 213
dissipation in general quantum systems 309
drift coefficient 291
driven atomic oscillator 135
Drude–Lorentz-model of matter 4
dynamic Stark effect 256

echo signal 154
Einstein 10, 58
Einstein's coefficients 12, 59, 122, 126, 226
electromagnetic waves 37
electron wave field, quantized 195
emission 6
energy density of electromagnetic field 42, 161, 180
~~~ electron wave field 197
~ levels of atoms 10
~~~ harmonic oscillator 54
~ levels, quantized 13
ensemble 317
ESR (electron spin resonance) 141
exciton 108
exclusion principle 197
expectation value 71, 164, 199

Fermi commutation relations 198
Feynman graphs 230
fluctuating forces for fermions 307
~~~ bosons 296, 303
~ forces, quantum mechanical 299
fluctuation–dissipation theorem 290, 291
fluctuations 285, 286
fluctuations in quantum systems 29, 309
~ of field modes 295
Fokker–Planck equation 285, 290, 292, 294
free induction decay 148, 154
frequency mixing 133, 136
~ shift, in dynamic Stark effect 261
Fresnel 3
friction force 286

Gaussian distribution 303, 314
geometrical optics 1
Glowgower, Susskind 191

Hamilton equations 160, 205
~ function 67
Hamiltonian 160
Hamiltonian, electron in external field 205
~ of free fields 209
Hamiltonian, second quantization 197
harmonic oscillator 17, 53, 80, 87, 160
~ oscillator, displaced 172
~~, eigenfunctions 5, 85, 87
~~, energy levels 85
Hartree–Fock procedure 103
Hartree procedure 103
heatbath 30, 285, 290, 292, 296, 317
heatbath average 300
~ composed of oscillators 298, 299
Heaviside function 282
Heisenberg 12, 18
Heisenberg equations 176, 177
~ equations for Brownian particle 298
~~~ driven field modes 274
~~~ field operators 272
~ picture 174, 188, 268
helium atom, level scheme 102
Hermitian operator 74
~ polynomial 87
Hertz 4
Hertzian dipole 277
Huygens 1
hydrogen atom 93
~ atom, eigenfunctions 94, 100
~~, energy levels 94, 99

incoherent light 124
induction equation 157
insulator 106
intensity 42
interaction electron-light field in second quantization 208
~~~

~ field-matter in second quantization 204
~ light field-matter 201
~ representation 212
interference 3, 21, 25, 40

ket 73, 77, 318
Kronecker symbol 124

Lamb shift 33, 216, 242, 249
Langevin equation 285, 287, 292
~ equation for spin operators 314
~~ in quantum mechanics 297, 303
~ equations for atoms 305
~~~ two-level systems 315
Laplace operator 66
~ transform 262
~ transformation 260
laser 34, 313
level splitting 277
lifetime 253
lifetime of atom 61
linewidth, homogeneous 251, 261
Lorentzian line 253, 263
Lorentz force 204, 206

Markovian property 311
mass renormalization 237, 245
matrix elements 73
~ elements, dipole moments 122
Maxwell 3
Maxwell's equations 37, 157, 179
Maxwell–Boltzmann distribution 291
measurements in quantum systems 19, 71, 75, 191
metal 106
momentum operator 72
multimode expansion of electromagnetic field 181
mutual coherence, complex degree of 46
~ coherence function 44

negative frequency part 38, 182, 265
Newton 1
Newton's law 286
NMR (nuclear magnetic resonance) 141
non-resonant perturbations 133
nonlinear optics 34
nonresonant processes 215
normalization condition 170
~ factor of cavity mode 181
nucleus, energy scheme 111

occupation number 305
operators 72
optical matrix elements 242
~ nutation 148
orthonormality condition 198
oscillating dipole moments 92, 155
~ dipoles 5, 34, 36, 149
oscillator, electronic 4
oscillator model of matter 15

partition function 54, 300, 317
Pauli's principle 197
periodic boundary conditions 69, 104
perturbation Hamiltonian 127, 136
~ theory 123, 126, 230, 233
~ theory, general result 130
phase average 124, 126
photoelectric effect 11, 56
photon 11, 17, 56, 164
photon echo 149, 152
~ energy 12
~ momentum 12
~ number eigenfunction 190
~~ representation 272
π-pulse 144, 149
$\pi/2$-pulse 144, 149
Planck 8
Planck's constant 9, 10, 53, 63
~ formula 9, 50, 55, 58, 61
plane waves 185
plasma frequency 203
~ oscillations 203
Poisson distribution 171
polarization density 204
positive frequency part 38, 182, 265
potential of point charge 247
principal value 253
probability distribution 290, 294, 316
projection operator 309
pure state 317

quantization of electromagnetic field, plane waves 184
~~ light field 157
~~~ field, multimode case 179
quantum beats 277, 283

~ beats, observed signal 278, 282
~ electrodynamics 16
~ mechanical coherence functions 271
~~ consistency 296, 305, 307
~~ fluctuations 303
~ mechanics 12
~ noise 2
~ statistical average 317

Rabi frequency 138
rate equation, for photons 61
~ equations for occupation numbers 305
Rayleigh–Jeans' law 60
reduced density matrix 319, 324
~~ matrix, equations of motion 324
reflection 1
refraction 3
relaxation times T_1, T_2 156
renormalization 33
resonant processes 215
rotating wave approximation 138, 144, 216, 267, 315
Rutherford 8

scalar potential 12, 182, 204
Schrödinger 12
Schrödinger equation 63, 66, 127
~ equation, formal solution 232
~ equation for single field mode 164, 169
~ equation, interaction representation 213
~~, two-level atom and single mode 211
~ equation, two-level atom in multimode field 216
~ picture 174
selection rules 100
self-energy 234, 236, 241, 242, 244
semiconductor 107
single electron state 198
solids 104
Sommerfeld 12
spectrum, dynamic Stark effect 263
spherical wave 44
spin 112, 147
spin echo 149, 150, 152
~ flipping 144
~ functions 113, 141
~ Hamiltonian 114
~ matrices 104, 113, 114
~ operators 314
~ precession 143, 147
~ resonance 141
spontaneous emission 11, 16, 17, 28, 31, 59, 222, 251
~ emission, coherence properties 275
standing wave 293
stationary solution, Brownian motion 291
~ solution of Fokker-Planck eq. 292
statistical operator 316
stimulated emission 11, 15, 59, 125, 222, 226, 231
superposition principle 38

telegraph equation 292
temperature 290, 317
thermal equilibrium 60, 290, 297, 300, 317
~ photon number 301, 320
three-level atom in multimode field 241
time average 265
trace 300, 316
trace, cyclic property 320
transition probability 155, 224, 310
~ probability for spontaneous emission 253
~ rate 305
two-level atom and single field mode 214
~ atom, compared to spin 1/2 145
~ atom coupled to heatbath 305
~~ in classical electric field 214
~~~ multimode field 216
~ atom, resonant coherent external field 137
two-photon absorption 130, 132
two-time probability density 294
two-level atom, in external field 119
~~~~ incoherent external field 123

uncertainty relations 18, 32, 75, 189

vacuum state, bosons 84, 86, 184
~~, for fermions 197
vector potential 163, 182, 185, 204, 218, 268
velocity of light 5
vertex 230
virtual processes 33, 215, 233
~ transitions 276
visibility of fringes, definition 44

wave-particle-dualism 1, 18, 21
wave function 13
~ function, normalization condition 69

~~, probabilistic interpretation 7, 20, 68
~ number 293
~ packet 14, 90, 146, 149, 212
~ packet of spin functions 115
~ track 48
Weisskopf–Wigner 280
Weisskopf-Wigner theory 251
Wiens's law 50

Young's experiment 3, 21, 41, 47, 283

zero point energy 182